BioMEMS and Biomedical Nanotechnology

Volume III
Therapeutic Micro/Nanotechnology

BioMEMS and Biomedical Nanotechnology

Mauro Ferrari, Ph.D., Editor-in-Chief
Professor, Brown Institute of Molecular Medicine Chairman
Department of Biomedical Engineering
University of Texas Health Science Center, Houston, TX

Professor of Experimental Therapeutics
University of Texas M.D. Anderson Cancer Center, Houston, TX

Professor of Bioengineering
Rice University, Houston, TX

Professor of Biochemistry and Molecular Biology
University of Texas Medical Branch, Galveston, TX

President, the Texas Alliance for NanoHealth
Houston, TX

Volume III
Therapeutic Micro/Nanotechnology

Edited by

Tejal Desai
Dept. of Bioengineering and Physiology
University of California, San Francisco, CA

Sangeeta Bhatia
Harvard—MIT Division of Health Sciences & Technology,
Electrical Engineering & Computer Science,
Massachusetts Institute of Technology, MA

 Springer

Tejal Desai
University of California, San Francisco
San Francisco, California

Sangeeta Bhatia
University of California, San Diego
San Diego, California

Mauro Ferrari
Ohio State University
Columbus, Ohio

Library of Congress Cataloging-in-Publication Data

Volume III
ISBN-10: 0-387-25565-6   e-ISBN 10: 0-387-25844-2   Printed on acid-free paper.
ISBN-13: 978-0387-25565-1   e-ISBN-13: 978-0387-25844-7
Set
ISBN-10: 0-387-25661-3   e-ISBN:10: 0-387-25749-7
ISBN-13: 978-0387-25561-3   e-ISBN:13: 978-0387-25749-5

9 8 7 6 5 4 3 2 1   SPIN 11407171

springer.com

Dedicated to Richard Smalley (1943–2005), in Memoriam

To Rick,

father founder of nanotechnology
prime inspiration for its applications to medicine
gracious mentor to its researchers
our light—forever in the trenches with us

(Rick Smalley received the 1996 Chemistry Nobel Prize
for the co-discovery of carbon-60 buckeyballs)

Contents

Tejal A. Desai, Sadhana Sharma, Robbie J. Walczak, Anthony Boiarski,
Michael Cohen, John Shapiro, Teri West, Kristie Melnik, Carlo Cosentino,
Piyush M. Sinha, and Mauro Ferrari

List of Contributors

VOLUME III

Dirk R. Albrecht, Health Sciences and Technology, Massachusetts Institute of Technology, Cambridge, Massachusetts USA

Ravi V. Bellamkonda, WHC Dept. of Biomedical Engineering, Georgia Institute of Technology/Emory University, Atlanta, Georgia USA

Edward C. Benzel, Department of Neurosurgery, The Cleveland Clinic Foundation, Cleveland, Ohio USA

Kiran Bhadriraju, Dept. of Biomedical Engineering, Johns Hopkins University School of Medicine, Baltimore, Maryland USA

Sangeeta N. Bhatia, Harvard—MIT Division of Health Sciences & Technology, Electrical Engineering & Computer Science, Massachusetts Institute of Technology, Cambridge, Massachusetts USA

Tony Boiarski, IMEDD Inc., Columbus, Ohio USA

Anthony Bolarski, IMEDD Inc., Columbus, Ohio USA

Warren C.W. Chan, Institute of Biomaterials and Biomedical Engineering, Toronto, Canada

Alice A. Chen, Health Sciences and Technology, Massachusetts Institute of Technology, Cambridge, Massachusetts USA

Christopher S. Chen, Dept. of Biomedical Engineering, Johns Hopkins University School of Medicine, Baltimore, Maryland USA

Michael Cohen, IMEDD Inc., Columbus, Ohio USA

Carlo Cosentino, Dept. of Experimental & Clinical Medicine, University of Catanzaro, Catanzaro, Italy

Frédérique Cunin, UMR CNRS/ENSCM, Montpellier cedex, France

Tejal A. Desai, Dept. of Bioengineering, and Physiology, University of California, San Francisco, California USA

Rebekah A. Drezek, Dept. of Bioengineering, Rice University, Houston, Texas USA

Lisa A. Ferrara, Spine Research Laboratory, The Cleveland Clinic Foundation, Cleveland, Ohio USA

Mauro Ferrari, Ph.D., Professor, Brown Institute of Molecular Medicine Chairman, Department of Biomedical Engineering, University of Texas Health Science Center, Houston, TX; Professor of Experimental Therapeutics, University of Texas M.D. Anderson Cancer Center, Houston, TX; Professor of Bioengineering, Rice University, Houston, TX; Professor of Biochemistry and Molecular Biology, University of Texas Medical Branch, Galveston, TX; President, the Texas Alliance for NanoHealth, Houston, TX

Hans Fischer, Department of Materials Science & Engineering, Institute of Biomaterials and Biomedical Engineering, Toronto, Canada

Aaron J. Fleischman, Dept. of Biomedical Engineering, The Cleveland Clinic Foundation, Cleveland, Ohio USA

Albert Folch, Dept. of Bioengineering, University of Washington, Seattle, Washington USA

Nobuyuki Futai, Dept. of Biomedical Engineering & Macromolecular Science & Engineering, University of Michigan, Ann Arbor, Michigan USA

Darren S. Gray, Dept. of Biomedical Engineering, Johns Hopkins University School of Medicine, Baltimore, Maryland USA

Jay T. Groves, Dept. of Chemistry, University of California, Berkeley, Berkely, California USA

Naomi J. Halas, Dept. of Bioengineering, Rice University, Houston, Texas USA

Leon R. Hirsch, Dept. of Bioengineering, Rice University, Houston, Texas USA

Allan S. Hoffman, Dept. of Bioengineering, University of Washington, Seattle, Washington USA

Dongeun Huh, Dept. of Biomedical Engineering & Macromolecular Science & Engineering, University of Michigan, Ann Arbor, Michigan USA

Laura J. Itle, Dept. of Chemical Engineering, The Pennsylvania State University, University Park, Pennsylvania USA

Anjana Jain, WHC Dept. of Biomedical Engineering, Georgia Institute of Technology/ Emory University, Atlanta, Georgia USA

Wen Jiang, Institute of Biomaterials and Biomedical Engineering, Toronto, Canada

Yoko Kamotani, Dept. of Biomedical Engineering & Macromolecular Science & Engineering, University of Michigan, Ann Arbor, Michigan USA

Won-Gun Koh, Dept. of Chemical Engineering, The Pennsylvania State University, University Park, Pennsylvania USA

Samarth Kulkarni, Dept. of Bioengineering, University of Washington, Seattle, Washington USA

Yang Yang Li, Dept. of Chemistry & Biochemistry, The University of California, San Diego, La Jolla, California USA

Wendy F. Liu, Dept. of Biomedical Engineering, Johns Hopkins University School of Medicine, Baltimore, Maryland USA

Noah Malmstadt, Dept. of Bioengineering, University of Washington, Seattle, Washington USA

Sawitri Mardyani, Institute of Biomaterials and Biomedical Engineering, Toronto, Canada

Kristie Melnik, IMEDD Inc., Columbus, Ohio USA

Samir Mitragotri, Dept. of Chemical Engineering, University of California, Santa Barbara, Santa Barbara, California USA

Michael V. Pishko, Dept. of Chemical Engineering & Materials Science, The Pennsylvania State University, University Park, Pennsylvania USA

Amy Pope-Harmon, Dept. of Internal Medicine, The Ohio State University, Columbus, Ohio USA

Arfaan Rampersaud, IMEDD Inc., Columbus, Ohio USA

Shuvo Roy, Dept. of Biomedical Engineering, The Cleveland Clinic Foundation, Cleveland, Ohio USA

Erkki Ruoslahti, The Burnham Institute, Cancer Research Center, La Jolla, California USA

Michael J. Sailor, Dept. of Chemistry & Biochemistry, The University of California, San Diego, La Jolla, California USA

John Shapiro, Department of Physiology and Biophysics, University of Illinois at Chicago, Chicago, IL

Sadhana Sharma, Department of Physiology and Biophysics, University of Illinois at Chicago, Chicago, IL

Anupam Singhal, Institute of Biomaterials and Biomedical Engineering, Toronto, Canada

Piyush M. Sinha,

Patrick S. Stayton, Dept. of Bioengineering, University of Washington, Seattle, Washington USA

Shuichi Takayama, Dept. of Biomedical Engineering & Macromolecular Science & Engineering, University of Michigan, Ann Arbor, Michigan USA

Sarah L. Tao, Dept. of Biomedical Engineering, Boston University, Boston, Massachusetts USA

Anna Tourovskaia, Dept. of Bioengineering, University of Washington, Seattle, Washington USA

Valerie Liu Tsang, Health Sciences and Technology, Massachusetts Institute of Technology, Cambridge, Massachusetts USA

Robbie J. Walczak, IMEDD Inc., Foster City, CA

Jennifer L. West, Dept. of Bioengineering, Rice University, Houston, Texas USA

Teri West, IMEDD Inc., Foster City, CA

Shuguang Zhang, Center for Bits and Atoms, Massachusetts Institute of Technology, Cambridge, Massachusetts USA

Xiaojun Zhao, Center for Biomedical Engineering, Massachusetts Institute of Technology, Cambridge, Massachusetts USA

Foreword

Less than twenty years ago photolithography and medicine were total strangers to one another. They had not yet met, and not even looking each other up in the classifieds. And then, nucleic acid chips, microfluidics and microarrays entered the scene, and rapidly these strangers became indispensable partners in biomedicine.

As recently as ten years ago the notion of applying nanotechnology to the fight against disease was dominantly the province of the fiction writers. Thoughts of nanoparticle-vehicled delivery of therapeuticals to diseased sites were an exercise in scientific solitude, and grounds for questioning one's ability to think "like an established scientist". And today we have nanoparticulate paclitaxel as the prime option against metastatic breast cancer, proteomic profiling diagnostic tools based on target surface nanotexturing, nanoparticle contrast agents for all radiological modalities, nanotechnologies embedded in high-distribution laboratory equipment, and no less than 152 novel nanomedical entities in the regulatory pipeline in the US alone.

This is a transforming impact, by any measure, with clear evidence of further acceleration, supported by very vigorous investments by the public and private sectors throughout the world. Even joining the dots in a most conservative, linear fashion, it is easy to envision scenarios of personalized medicine such as the following:

- patient-specific prevention supplanting gross, faceless intervention strategies;
- early detection protocols identifying signs of developing disease at the time when the disease is most easily subdued;
- personally tailored intervention strategies that are so routinely and inexpensively realized, that access to them can be secured by everyone;
- technologies allowing for long lives in the company of disease, as good neighbors, without impairment of the quality of life itself.

These visions will become reality. The contributions from the worlds of small-scale technologies are required to realize them. Invaluable progress towards them was recorded by the very scientists that have joined forces to accomplish the effort presented in this 4-volume collection. It has been a great privilege for me to be at their service, and at the service of the readership, in aiding with its assembly. May I take this opportunity to express my gratitude to all of the contributing Chapter Authors, for their inspired and thorough work. For many of them, writing about the history of their specialty fields of *BioMEMS and Biomedical Nanotechnology* has really been reporting about their personal, individual adventures through scientific discovery and innovation—a sort of family album, with equations, diagrams, bibliographies and charts replacing Holiday pictures . . . .

It has been a particular privilege to work with our Volume Editors: Sangeeta Bhatia, Rashid Bashir, Tejal Desai, Michael Heller, Abraham Lee, Jim Lee, Mihri Ozkan, and Steve Werely. They have been nothing short of outstanding in their dedication, scientific vision, and generosity. My gratitude goes to our Publisher, and in particular to Greg Franklin for his constant support and leadership, and to Angela De Pina for her assistance.

Most importantly, I wish to express my public gratitude in these pages to Paola, for her leadership, professional assistance throughout this effort, her support and her patience. To her, and our children Giacomo, Chiara, Kim, Ilaria and Federica, I dedicate my contribution to BioMEMS and Biomedical Nanotechnology.

With my very best wishes

Mauro Ferrari, Ph.D.
Professor, Brown Institute of Molecular Medicine Chairman
Department of Biomedical Engineering
University of Texas Health Science Center, Houston, TX

Professor of Experimental Therapeutics
University of Texas M.D. Anderson Cancer Center, Houston, TX

Professor of Bioengineering
Rice University, Houston, TX

Professor of Biochemistry and Molecular Biology
University of Texas Medical Branch, Galveston, TX

President, the Texas Alliance for NanoHealth
Houston, TX

Preface

The human body is composed of structures organized in a hierarchical fashion: from biomolecules assembled into polymers, to multimeric assemblies such as cellular organelles, to individual cells, to tissues, to organ systems working together in health and disease- each dominated by a characteristic length scale. Decades of science and engineering are now converging to provide tools that enable the orderly manipulation of biological systems at previously inaccessible, though critically important, length scales (<100 microns). Thus, the approaches described in this volume provide a snapshot of how micro- and nanotechnologies can enable the investigation, prevention, and treatment of human disease.

The volume is divided into three parts. The first part, *Cell-based therapeutics*; covers the merger of cells with micro- and nanosystems for applications in regenerative medicine spanning the development of novel nanobiomaterials, methods of tissue assembly with control over tissue microarchitecture, and methods to specify patterns of protein distribution that vary on the micro- and nanoscale for application in tissue regeneration (A), and therapeutic applications of integrating MEMS with cells and tissues including label-free microfluidic sorting of cells based on their function, using living cell arrays as biosensors, and micron-scale devices for surgical applications (B). The second part, *Drug Delivery*; covers intravascular delivery of nanoparticles such as semiconductor quantum dots and metal nanoshells in the context of vascular specialization or 'zip codes' (A) as well as non-vascular modes of delivery including implantation, oral, and inhalation using both encapsulated drugs as well as living cells that produce therapeutic products (B). Finally, the third part, *Molecular Surface Engineering for the Biological Interface*; covers platforms that provide enabling tools for fundamental investigations of cells in culture as they interact with biomolecular structures such as responsive biomaterials and lipid bilayers (A) as well as micropatterned adhesive and fluidic environments (B).

We would like to thank the contributing authors, our co-editors in this exciting compilation of volumes, and Dr. Mauro Ferrari for his tireless efforts to lead this endeavor. We hope the collected works will provide an excellent reference for an audience with a diversity of background and interests including industry, students, academic researchers, policy-makers, and enthusiasts.

Sangeeta N. Bhatia
Massachusetts Institute of Technology

Tejal Desai
University of California, San Francisco

Mauro Ferrari
Professor, Brown Institute of Molecular Medicine Chairman
Department of Biomedical Engineering
University of Texas Health Science Center, Houston, TX

Professor of Experimental Therapeutics
University of Texas M.D. Anderson Cancer Center, Houston, TX

Professor of Bioengineering, Rice University, Houston, TX

Professor of Biochemistry and Molecular Biology
University of Texas Medical Branch, Galveston, TX

President, the Texas Alliance for NanoHealth, Houston, TX

I

Cell-based Therapeutics

1

Nano- and Micro-Technology to Spatially and Temporally Control Proteins for Neural Regeneration

Anjana Jain and Ravi V. Bellamkonda

WHC Department of Biomedical Engineering, Georgia Institute of Technology/Emory University, Atlanta, Georgia 30332-0535

1.1. INTRODUCTION

Nano- and micro-technologies in the field of neural tissue engineering have implications in the pursuit of spatial and temporal control of protein and sugar cues at the site of injury and in the control over cellular response to these cues to promote regeneration and healing. The nervous system consists of two main components that are relevant from a regeneration and tissue engineering perspective. These components are the central nervous system (CNS), consisting of the cells and processes contained within the spinal and cranial cavities, and the peripheral nervous system (PNS), comprising of the nervous system outside of the CNS. Although a third component, the autologous nervous system, exists and is important physiologically, this component will not be the focus of this chapter for the sake of relevance and brevity.

The cellular and molecular events that follow injury in the CNS and PNS are different. These differences have implications for the kinds of nano- and micro- scale control necessary to stimulate regeneration or healing after injury. The nerves in the PNS have the ability to spontaneously regenerate if the gap between the two nerve ends is less than 10 mm. However, in the CNS, nerves cannot regenerate after injury and permanent functional loss occurs.

1.1.1. Response after Injury in CNS and PNS

A possible reason for the CNS's relative inability to regenerate compared to the PNS is due to the body's wound healing response. After injury to the CNS, a glial scar forms and there is a migration of microglia/macrophages to the wound site to remove the debris produced [23]. At the injury site, there are cells that up-regulate or expose inhibitory molecules that make the environment non-permissive for nerve regeneration. For example, astrocytes up-regulate chondroitin sulfate proteoglycans (CSPGs) and oligodendrocyte debris, produced during injury, expose myelin associated glycoprotein (MAG) [43, 48], NOGO [50], and oligodendrocyte myelin glycoprotein (OMpg) [35]. Other inhibitory molecules are also present in the glial scar; however, the non-regenerative environment has been mainly attributed to CSPGs, MAG, NOGO, and OMpg.

In the PNS, a slightly different series of events occur after injury. Macrophages infiltrate the injury site within three days and deliver various growth factors, as well as encourage Schwann cells to start producing NGF [63]. Other events that occur during the 3rd and 4th days after injury involve mast cells, axonal sprouts, and fibroblasts [81]. Similar to the macrophages, the mast cells also release important growth factors and other molecules, such as cytokines and interleukins. The presence of these cells peaks around the 4th day post injury and declines back to normal levels around 4 weeks after injury [63]. Another event that occurs in the two nerve ends is angiogenesis. The combination of these events makes it conducive for regeneration to take place in the PNS. The delivery of various growth factors by the macrophages and the mast cells along with the angiogenesis create a microenvironment allowing axonal outgrowth to occur from the proximal nerve end to the distal end.

Macrophages have a similar role in both the CNS and the PNS as they enter the lesion site to remove debris. However, the required time for the macrophages to enter the glial scar is much longer in the CNS than in the PNS [2]. One cause is the macrophages have problems entering the injury site due to the blood-brain barrier [69]. Also, in the distal end of the severed nerve in the CNS, the cell adhesion molecules are not up-regulated like they are in the PNS [59]. Although there are morphological and physiological differences between the PNS and CNS, the basic approach to aid in regeneration in both of these systems is similar.

Currently the options for therapeutic application in the CNS after injury are limited. Anti-inflammatory agents, such as methylprednisolone, are administered after injury; however, this does not provide any type of regenerative stimuli. In the PNS, there are two types of clinical treatments that can be applied to reattach the severed nerve. The type of treatment depends upon the size of the gap between the nerve ends. If the gap is short, then the two ends can be sutured together, restoring the axonal connections. However, if the gap is longer, then typically autografts or allografts are used [59]. Although clinically autografts are the state-of-the-art to regenerate nerves in the PNS, they are not the ideal solution because the donor supply is limited and obtaining a nerve graft from the patient means that the donor site has been denervated. Therefore, there is a great need to find a tissue engineered therapeutic solution to regenerate axons in the CNS and the PNS.

1.1.2. Nano- and Micro-scale Strategies to Promote Axonal Outgrowth
in the CNS and PNS

The current clinical challenges are in regenerating peripheral nerves across gaps greater than 20 mm and to obtain any regeneration with accompanying functional recovery after injury in the CNS. As alluded to earlier, the crux of the approaches currently under development to promote regeneration in the CNS and PNS involve either understanding or manipulating of nano- and micro-scale events in a controlled spatio-temporal sequence.

To regenerate nerves across a PNS or CNS nerve gap, there are three basic areas that are being explored. These strategies are 1) to provide permissive bioactive substrates for axonal outgrowth; (2) to deliver trophic factors in order to stimulate growth; and (3) to alleviate signaling due to the inhibitory entities present in the extracellular environment to allow axons to regenerate between the proximal and distal ends. First, nerves are anchorage dependent and the design of a substrate or bridge between the severed ends is an opportunity to present the correct spatial and temporal cues to both guide and stimulate axonal growth across the nerve gap. Second, promotion and stimulation of axonal growth involves the application of specific factors known as neurotrophic factors in a spatially and temporally controlled manner *in vivo*. Third, any inhibitory cellular responses and/or any inhibitory cues that may be generated after injury must be modulated such that they do not interfere in any putative regenerative attempt after injury. This third aspect is especially critical in the CNS where an extremely inhibitory glial scar is generated at the distal nerve segment, often leading to regenerative failure in the CNS. There are a series of technologies and challenges that are encountered in achieving this goal and they are discussed below.

1.1.2.1. Growth Permissive Substrates to Actively Support Growing Axons An important strategy used to promote axonal regeneration is to provide substrates for the outgrowth to occur. There are four types of substrates that have been investigated the most to provide an adequate scaffold; hydrogels, fibers, nerve guide conduits (NGCs), and transplanted cells. Proteins and oligopeptides are coupled onto the substrates to provide a more permissive surface that mimics the ECM for the axons to anchor up and extend through the nerve gap. Collagen and laminin (LN) are the more common proteins coupled to the substrates, as well as the oligopeptides, RGD, YIGSR, and IKVAV. The contribution of these proteins and oligopeptides will be discussed further later in this chapter. Transplanting cellular substrates is another approach to encourage axonal regeneration. Schwann cells, olfactory ensheathing glia (OEG), and astrocytes are examples of the cells that could be transplanted in the nerve gaps in the PNS and the CNS. These cells secrete proteins and growth factors that make the microenvironment less inhibitory for axonal outgrowth.

1.1.2.2. Stimulating Process Extension Using Trophic Factors Neurotrophic factors have an important role in the neural development and in adult life for axonal regeneration. Neurotrophins are a specific family of neurotrophic factors we are interested in to promote regeneration in the nervous system. The main focus in the PNS has been on two members of the neurotrohin family, nerve growth factor (NGF) and fibroblast growth factor (FGF). After injury, NGF is up-regulated. It was mentioned that Schwann cells and other cells that migrate to the injured area release trophic factors, one of them being NGF. FGF has also been shown to enhance axonal outgrowth [1] as well as angiogenesis [3]. In the CNS, brain-derived neurotrophic factor (BDNF) and neutrophin-3 (NT-3) have been investigated for their regenerative capabilities. Several studies, which will be discussed later, have shown that after administration of BDNF or NT-3, axonal regeneration was exhibited in spinal cord injured adult animals.

1.1.2.3. Alleviating Inhibitory Environment at the Site of Injury   Neurites extend from the neuronal body and have a growth cone at the tip. The role of the growth cone is to read the environmental cues and decide which direction the neurite will grow towards. The growth cone extends filopodia and lamellipodia to read the cues. In the glial scar the inhibitory molecules prevent the extension of the filopodia and lamellipodia and induce growth cone collapse. Therefore, it is necessary to mask or remove these negative components from the microenvironment in order to support axonal outgrowth. Protein transduction is one of the methods utilized to promote and stimulate axonal regeneration. By intracellularly modulating levels of protein and other chemical concentrations, the growth cone can be manipulated to extend and grow into an inhibitory environment. Modulating Rho GTPases is one of the ways to overcome glial scar inhibition through protein transduction [21, 31, 45, 70]. Rho GTPases are involved in actin cytoskeleton dynamics, specifically promoting filopodial and lamellipodial extension [51]. By modulating Rho GTPases' levels in the neurons and elevating the concentration, the inhibitory effects of the glial scar will be masked and the growth cone will lead the neurite towards the distal nerve ending.

Other molecules that have been used to encourage neurite outgrowth in the face of inhibitory signals is cAMP and calcium in the CNS [41]. It has been shown that the modulation of cAMP, using the active and inactive analogs, can encourage neurite outgrowth through inhibitory substrates [5]. The activation of the signal transduction pathway by cAMP shows that axonal regeneration can be stimulated and promoted [54]. *In vivo* studies have also shown that increasing cAMP levels will promote axonal outgrowth [19].

The last method that will be reviewed is removing the inhibitory effects that do not allow regeneration to occur in the PNS and the CNS. An example of removing the inhibition is the enzymatic degradation of chondroitin sulfate proteoglycans (CSPGs) in the CNS. After injury, astrocytes and oligodendrocyte precursor cells release CSPGs into the glial scar matrix [46]. It has been demonstrated that both components, the protein core and the glycosaminoglycan side chain, contribute to the inhibitory nature of the macromolecule [28, 46]. Consequently, studies have shown that by treating the CSPGs with chondroitinase ABC, which cleaves the glycosaminoglycan (GAG) side chain into dissacharide units, axonal regeneration could occur and extend through the inhibitory glial scar region into the distal nerve end [12, 42, 82, 83].

In order to achieve axonal regeneration in the CNS and PNS it is important to construct a scaffold that allows axons to extend through; has the mechanical integrity to support cell migration, such as Schwann cells, into the scaffold; allows the delivery of growth factors, such as neurotrophins, that encourage axonal outgrowth; and integrate the scaffold and extracellular matrix (ECM) [22].

1.2. SPATIALLY CONTROLLING PROTEINS

In order for the three strategies to promote nerve regeneration, proteins must be controlled spatially and temporally in the CNS and PNS. Typically, biomaterial scaffolds are used for the spatial control of proteins in three-dimension (3D). An important factor when developing scaffolds is that it must mimic the ECM in order to encourage axons to grow through the glial scar. Therefore, in order to control the location of the proteins, 3D scaffolds,

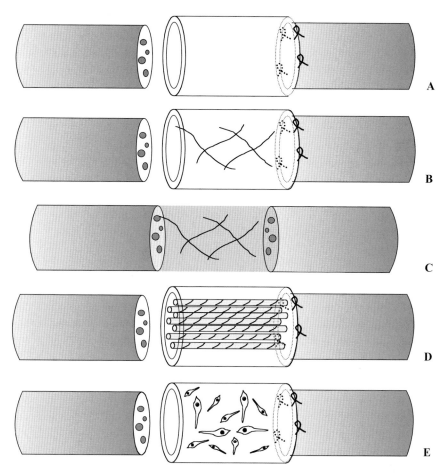

FIGURE 1.1. Spatial Control of Proteins. **A.** Nerve guide conduit (NGC) sutured to the proximal and distal ends of the nerve. **B.** Hydrogel filled NGC sutured to the nerve ends. **C.** Hydrogel scaffold injected between the nerve ends without the aid of the NGC. **D.** Fibers filled in the NGC to act as contact guidance cues for axonal regeneration. **E.** NGC containing transplanted cells, such as Schwann cells or olfactory ensheathing glia to aid in axonal regeneration.

such as hydrogels, fibers, NGCs, and cell transplantation can be utilized to promote axonal regeneration in the nervous system (Fig. 1.1).

1.2.1. Spatial Control: Permissive Bioactive Hydrogel Scaffolds for Enhanced Regeneration

The use of hydrogels provides a substrate for axonal outgrowth. Hydrogels are polymers that swell with the addition of water and are crosslinked. There are three main biomaterials belonging to the hydrogel family that have been used to provide a scaffold for axonal regeneration: (1) agarose, (2) alginate, and (3) collagen.

1.2.1.1. Agarose as a Scaffolding Material   Agarose, which is a thermoreversible copolymer of 1,4-linked 3,6-anhydro-α-L-galactose and 1,3-linked β-galactose, is derived from red algae. Agarose is a beneficial biomaterial to use as a scaffold for a few reasons. The hydrogel is biocompatible as it causes no adverse reaction when implanted *in vivo*. Its porosity and mechanical properties can be manipulated and optimized to maximize axonal growth [47]. Most of all, agarose is beneficial because it can be used to control proteins spatially by binding proteins to the agarose and it can be used to support cell migration [6, 10].

Agarose gel can be used to encourage axonal outgrowth by covalently coupling growth promoting molecules to the agarose hydrogel, which would embody the characteristics of the ECM allowing axonal outgrowth into the glial scar and reconnect with the distal nerve. *In vitro* studies have shown that covalently coupling a growth promoting ECM molecule, such as LN, to the agarose gel encouraged neurite outgrowth compared to a scaffold that did not have any modifications [6, 80]. Along with coupling whole proteins, such as LN and collagen, oligopeptides can be bound to hydrogels as well. The oligopeptides of interest are the ones that influence cell-matrix interactions, such as RGD, which is responsible for the interaction between fibronectin and an integrin receptor, and YIGSR, which is a peptide on the β1 chain of laminin aiding in cell attachment [10].

The application of the engineered scaffold, which was a polysulfone tube containing LN-bound agarose and a slow release system of NGF, *in vivo* in the peripheral nervous system demonstrated that the regenerated myelinated axons were comparable to the regeneration found in autografts [79].

1.2.2. Spatial Control: Chemical vs. Photochemical Crosslinkers for Immobilization of Bioactive Agents

There are different types of crosslinkers that can be used to couple the proteins to the gel. There are thermochemical bifunctional crosslinkers, such as 1,1'-carbonyldiimidazole (CDI), which can couple the protein to the agarose. Another class of crosslinkers that can be used is photocrosslinkers. These photocrosslinkers are activated by shining UV light onto the agarose gel that contains the crosslinker [40]. Free radicals are created, which then can be bound to the protein of interest. Photocrosslinkers can be used to covalently bind macromolecules to the agarose hydrogel. Using UV light to produce free radicals is beneficial because laser beams can be used to create patterns in the hydrogel. One such application used UV laser beams to create channels through the agarose gel, encouraging the neurons to extend their neurites down the channel, providing directional cues for the neurites [40]. In a study, both CDI and a photocrosslinker, benzophenone, were used to couple YIGSR to agarose hydrogel. Results from both *in vivo* and *in vitro* experiments have shown that DRG neurite outgrowth was enhanced when cultured in 0.5% agarose. Additionally, *in vivo* the number of myelinated axons was higher in the agarose coupled to the YIGSR peptide than plain agarose [10]. This study also concluded that the effectiveness of the gel was not determined by the type of crosslinker used to couple the oligopeptide. Although there is not a functional difference between the types of crosslinkers, the advantage of using the photocrosslinker is the ability to pattern the hydrogel to favor the axonal growth in a specific direction.

1.2.2.1. Alginate as a Scaffolding Material Alginate is another scaffold material, similar to agarose, which can be utilized to control proteins spatially in order to influence axonal regeneration. Alginate can be found in brown seaweed and is a copolymer formed from α-L-glucoronic acid and β-D-mannuronic acid. The studies that have used alginate have been predominantly for axonal outgrowth in the PNS. Therefore, the use of alginate scaffold should support the migration of Schwann cells so that the cells can deliver LN, neurotrophic factors and cytokines. It was demonstrated that alginate sponge could support axonal regeneration from the proximal end to the distal end. The alginate allowed Schwann cell infiltration and enhanced the promotion of axonal growth through the nerve gap [26]. Growth promoting molecules can also be added to alginate gels similar to the agarose gels. Although soluble fibronectin added to the alginate gel did not exhibit any positive significance on axonal growth compared to empty poly-3-hydroxybutyrate conduits, the addition of soluble fibronectin in the alginate along with Schwann cells demonstrated a significant effect on axonal regeneration *in vivo* and shows promise as a possible method to attempt regeneration over long nerve gaps [47]. A study comparing the potency of implanted alginate with and without the use of a conduit demonstrated that in a 50 mm sciatic nerve gap made in cats, the alginate without the aid of a conduit was able to provide a microenvironment that allowed axonal outgrowth across the gap and the number of myelinated and unmyelinated axons that regenerated were the same in both conditions [62]. Regeneration in the PNS has mostly focused on using nerve conduits that are filled with a form of hydrogel. This study has suggested a possible alternative in case it is not possible to use the conduits. In addition, alignate sponge was investigated as a potential scaffold to promote regeneration in the CNS after the spinal cord was transected in rats. It was shown that regenerating axons infiltrated the alginate gel significantly higher compared to collagen gels [32]. It was also suggested that the formation of glial scar could be reduced by the alginate gel due to little infiltration of the connective tissue.

1.2.2.2. Collagen as a Growth Permissive Scaffold for Nerve Regeneration The third type of hydrogel applied as a scaffold for nerve regeneration is collagen, more specifically type I collagen. Collagen is found in the ECM and helps promote axonal outgrowth and cell adhesion. Comparisons among different types of gel matrices, collagen, methylcellulose, and Biomatrix, were conducted and showed that collagen along with methylcellulose had the best results in regenerating axons across a peripheral nerve gap [68]. It has been reported that filling tubes with ECM molecules, such as collagen, LN, and fibronectin, improves axonal regeneration. The affect of collagen and LN gels that are magnetically aligned improves the distance of axonal outgrowth compared to collagen added without any alterations [16, 66]. The collagen hydrogel has also been used in combination with GAG, and inserted into an NGC in the PNS. The combination of the collagen-GAG matrix exhibited a greater number of myelinated axons compared to unfilled conduits [17]. It is suggested that the collagen-GAG matrix elicited axonal outgrowth because it provided an adequate scaffold needed for attachment and cell migration. Collagen gels have also been inserted into lesions after dorsal transections in rat spinal cords. Although axonal regeneration did not occur through the entire lesion area, the collagen gel along with the neurotrophin encouraged outgrowth into the matrix [29]. The study also showed minimal glial scar formation, which would provide a more promoting microenvironment for axonal regeneration.

1.2.3. Other Hydrogel Scaffolds

There are other hydrogels that can be used as scaffolds besides the three main ones discussed above. Some of the other hydrogels are Matrigel, NeuroGel™, and Biomatrix. Matrigel is made out of a mixture of ECM proteins, such as LN and collagen. *In vivo* studies have shown that Matrigel alone is not adequate scaffold to promote axonal outgrowth [25, 65]. However, when the Matrigel is used in conjunction with Schwann Cells, axonal outgrowth is significantly noticeable. NeuroGel™ is a crosslinked copolymer hydrogel made of N-2-(hydroxypropyl) methacrylamide. When this hydrogel was inserted into the thoracic region of the spinal cord after a contusion injury, it was observed that the rats that had implanted NeuroGel™ in the lesioned cavity had an improved locomotion according to the BBB test and there was evidence of axonal fibers infiltrating the hydrogel, thereby crossing the tissue-implant interface [71]. NeuroGel™ also demonstrated the capability to hinder glial scar formation when it was implanted in the lesion of spinal cords in adult rats [72]. Biomatrix is a hydrogel, similar to Matrigel, made of ECM proteins, such as LN. However, Biomatrix does not appear to have as adequate regenerative capabilities as collagen and other hydrogels [68].

1.2.4. Spatial Control: Contact Guidance as a Strategy to Promote Regeneration

It was previously mentioned that besides the use of hydrogels as a scaffold, fibers could also be utilized to direct axonal growth from the proximal to distal ends of the nerve. This is another strategic technique to gain spatial control of proteins using a substrate. Tubes are inserted between the nerve gaps and then the nerve ends are sutured to the tubes with the fibers placed through the length of the tube (Fig. 1.1D). Due to the fibers being oriented longitudinally through the tube, it provides the orientation for the axons to grow from the proximal to distal end of the gap. Fibers are used to encourage the occurrence of two events in order to obtain successful myelinated axonal regeneration in the PNS. The first event is the formation of the fibrin matrix, which will have the same orientation as the filaments. The second event that needs to occur is the infiltration of Schwann cells. The goal is to have the Schwann cells adhere to the filaments and travel along the entire length of the filaments, which is the length of the nerve gap. This would then provide an environment which would encourage axonal outgrowth. Poly (L-Lactide) (PLLA) is another material that is commonly used to make filaments. In an *in vitro* study, it was demonstrated that if the PLLA was coated with LN, then the neurite outgrowth was significantly greater than neurite outgrowth on uncoated PLLA surface or the poly-L-lysine coated filaments [55]. Tubes are generally used to encapsulate the filaments and provide an environment for axonal growth along the filaments. However, in a study conducted in the PNS, collagen filaments were sutured to the proximal and distal ends of the nerve without the aid of tubes *in vivo*. The study showed that the number of myelinated axons that regenerated was greater to that found in the group that received the autograft, although it was not significantly greater [77]. This is the only study that did not use a conduit for the filaments or any neurotrophic factors, however, the regeneration was abundant and demonstrated that perhaps these two components are not completely necessary if the proper conditions are provided for axonal growth. Another variable that needs to be considered in the application of fibers is the number of fibers that should be inserted between the nerve ends. In studies conducted by Yoshii et al., collagen filaments were sutured to the sciatic nerve ends without the aid of a tube, two thousand filaments were connected at the ends to keep them joined over a 20 mm and 30 mm gap [77, 78]. The myelinated axon regeneration was comparable to the results observed with autografts for the 20 mm gap [77]. However, in the case of the 30 mm gap, the axonal regeneration was significantly less. These studies suggest that a large number of filaments would aid in axonal outgrowth. However, in another study, which inserted PLLA filaments inside silicone tubes, demonstrated that a lower packing density of filaments elicited the greatest number of myelinated axons [49].

Although filaments are predominantly used in the PNS, studies have been performed where filaments were inserted in CNS to promote axonal outgrowth. Carbon filaments were implanted in the lesion of a fully transected rat spinal cord. The carbon filaments allowed a scaffold for axons to advance through the lesion [33]. This study was taken further, where 10,000 carbon filaments were cultured with fetal tissue and implanted into the spinal cord lesion. This condition exhibited an improvement in electrical conduction through the injured axons [39]. A study conducted by the same group who inserted 2000 filaments into a nerve gap in the PNS, utilized the collagen filaments to encourage axonal regeneration in the CNS after spinal cord injury (SCI) [78]. Four thousand collagen fibers were inserted between the two nerve ends parallel to the spinal cord. It was demonstrated that the collagen fibers provided an adequate scaffold to bridge the nerve ends and allow axons to extend across the gap.

It was previously mentioned that proteins and oligopeptides could be coupled to hydrogels. A similar method was used to couple peptides to fibers that could potentially be implanted as a scaffold in the CNS. Two laminin peptides, YIGSR and IKVAV, were coupled to poly(tetrafluoroethylene) (PTFE) fibers and DRGs were cultured to observe neurite extension [60]. The peptide surface modified fibers encouraged neurite outgrowth; however, the neurites could not extend along unmodified PTFE fibers. To have successful axonal regeneration using fibers as the scaffold, it is important to either use a biomaterial that encourages fibrin matrix formation and Schwann cell infiltration or to coat the fibers with a protein that does those things. Current research has demonstrated that fibers made out of collagen, coated with proteins, such as collagen or laminin, or oligopeptides have produced the most significant axonal regeneration. Controlling proteins spatially through fiber scaffolds allows a surface for axons to adhere, as well as orient the direction of growth.

1.2.5. Spatial Control: Nerve Guide Conduits Provide an Environment for Axonal Regeneration

The use of nerve guide conduits has greatly influenced axonal regeneration. They aid in providing a scaffold to promote axonal regeneration and have the potential to both spatially and temporally control the protein environment at the site of injury. Importantly, the conduit serves as a physical barrier to prevent proteins and other molecules from inhibiting axonal regeneration. When NGCs were first being used, it was believed that the best material for the tube was silicone due to its mechanical properties. However, silicone NGCs are non-absorbable, non-semipermeable and require a second surgery to remove the conduit, otherwise it could cause chronic tissue response, such as scar formation, as well as nerve compression [18]. Most NGCs in use today are semi-permeable and even biodegradable. However, as NGCs have been extensively reviewed elsewhere [8, 18, 30, 59], we choose to concentrate this chapter on approaches where the NGCs are used as carriers for other bioactive agents to enhance their functionality.

1.2.6. Spatial Control: Cell-scaffold Constructs as a Way of Combining Permissive Substrates with Stimuli for Regeneration

Cell transplantation techniques are an elegant way to combine two promising strategies to elicit regeneration: permissive substrates and spatio-temporally controlled delivery of trophic factors at the site of injury. This strategy has been explored both in the CNS and the PNS and is described below. Typically, NGCs are used as carriers for the delivery of these cells to the site of injury in the PNS or the CNS (Fig. 1.1E).

Schwann cells and OEG are two cell types commonly used to promote regeneration in the CNS, while Schwann cells are typically the cells of choice in the PNS. These cells provide both trophic cues, as well as physical, contact guidance type cues in promoting regeneration as described below. The use of cells, such as these glia, utilizes the strategy that modulates intrinsic mechanisms to promote axonal outgrowth. The transplantation of Schwann cells and OEG allows for spatial control of growth factors and other proteins, which are secreted by the cells.

Schwann cells have been shown to enhance peripheral nerve regeneration. It was mentioned previously that infiltration by endogenous Schwann cells increased axonal regeneration [24]. Schwann cells were embedded in a scaffold, such as Matrigel, and transplanted into an NGC implanted between two nerve ends, myelinated and unmyelinated axons are regenerated [25]. It was believed that by implanting Schwann cells already present throughout the conduit, the pace of regeneration could be increased. Schwann cells align along the tube and arrange themselves so that they are end to end, which is called Bungner bands. It was demonstrated that syngeneic Schwann cells elicited a better axonal regeneration than heterologous Schwann cells, which elicited an immune response [25]. It was also shown that as the Schwann cell density increased in the NGC, the axonal regeneration improved along similar lines to nerve autografts. Schwann cells myelinate peripheral nerves and it has been established that transplantation of these cells encourages the outgrowth of myelinated and unmyelinated axons.

Schwann cells have also shown to promote regeneration in the CNS. In studies that transected rat spinal cords and then implanted grafts containing Schwann cells and Matrigel, it was demonstrated that the number of myelinated and unmyelinated axons was greater compared to grafts containing only Matrigel and the myelinated axons formed fascicles through the conduit [74, 75]. In another study that transplanted Schwann cells into the spinal cord, it was shown that Schwann cells that released increased amounts of NGF had significantly more axons growing into the graft compared to Schwann cells that were not modified to release increased amounts of NGF [67]. It was also demonstrated that these Schwann cells expressed the same phenotype and myelinated axons in the CNS as in the PNS. The combination of NGF and Schwann cells allows for the outgrowth of axons into the grafts due to the presence of NGF and then the Schwann cells provides direction for axonal growth due to the Bunger bands [67]. It was mentioned previously, cAMP has been investigated to promote axonal regeneration. In a study cAMP and Schwann cells were both inserted into the spinal cord to observe whether there was a synergistic effect [53]. The results demonstrated that by implanting Schwann cells and elevating cAMP, the number of myelinated axons increased and functional recovery was observed compared to the transplantation of only Schwann cells.

Unlike Schwann cells, which can be transplanted in both the PNS and CNS, OEG is primarily transplanted in the CNS to promote axonal regeneration. OEG ensheath olfactory axons and shield the axons from inhibitory molecules exposed in the environment, thus allowing the axons to regenerate throughout adult life [58]. OEG demonstrates a promising method to ensheath the axons in other areas of the CNS that are injured and aid in regeneration. The olfactory bulb is the main supplier for OEG and one of the main benefits of using this source for OEG is because the glia can migrate into other regions of the CNS and integrate with other CNS glia [58].

Comparisons have been made between Schwann cells and OEG for their effectiveness in promoting axonal regeneration in the CNS. In a study that was comparing the response of astrocytes and CSPG expression after OEG or Schwann cell transplantation in the CNS, it was demonstrated that OEG elicited less of an astrocytic response and lower expression of CSPG compared to Schwann cells [37]. Although OEG do not induce as severe a response as Schwann cells do, Schwann cells have shown more promising results in improving locomotor performance compared to OEG after adult rats have suffered from contused thoracic SCI [64].

It was mentioned earlier that astrocytes can also be used as a substrate for axonal outgrowth. These studies were performed *in vitro*. It was demonstrated that uniformly orienting the astrocytes and organizing the ECM and cell adhesion molecules in order to culture neurons on the astrocytes lead to the enhancement of neurites extending in a direction parallel to the astrocytes [9]. The use of glial cells, such as astrocytes, as a substrate can be combined with a biomatrix to enhance neurite extension in a specific direction [20]. Glial cells were cultured on the biodegradable poly(D,L)-lactide matrices to orient the cells in a specific direction. Although this substrate did not enhance either the number of extended neurites or the length of the neurites, the cultured cortical neurons extended neurites along the orientation of the glial cells/biomatrix substrate.

1.3. TEMPORALLY CONTROLLING THE RELEASE OF PROTEINS

As important as it is to control the proteins spatially, it is equally imperative to control the amount of protein delivered over a period of time. Regeneration over long nerve gaps requires several months. Therefore, for axonal outgrowth to occur during this time period, the microenvironment must be actively supportive over this time scale. If proteins, such as Rho GTPases and neurotrophic factors, are only administered as a single dose at the time of implantation of the scaffold, then some of the protein will be taken up intracellularly, diffuse into the surrounding tissue, and degrade. Then there will not be a therapeutic level of protein to promote axonal outgrowth over the time necessary to have complete regeneration. For example, it was concluded that after local administration of NGF into the brain, the half-life of NGF was 30 minutes [36]. Once the effective concentration for the proteins is known, then it can be delivered and sustained. Sustaining the presence of proteins at the effective concentration can be achieved through a controlled slow release delivery system. There are currently four main techniques that are being investigated for controlling protein concentration at the site of injury over time: (1) osmotic pumps, (2) embedded

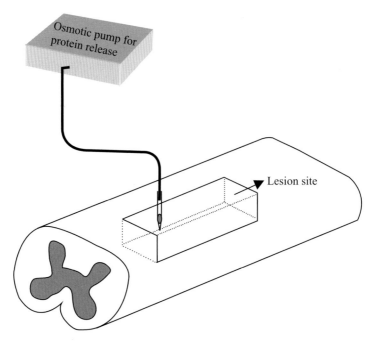

FIGURE 1.2. Osmotic Pumps for Temporal Control of Proteins. An external osmotic pump provides a reservoir of protein that is delivered via a catheter implanted near or at the lesion area.

microspheres, (3) microtubules and (4) enzyme dependent demand-driven trophic factor release.

1.3.1. Temporal Control: Osmotic Pumps Release Protein to Encourage Axonal Outgrowth

Osmotic pumps can be used to deliver proteins, such as neurotrophic factors, to promote axonal regeneration. Osmotic pumps are mostly utilized to deliver the proteins in the CNS. There are two parts to this delivery system, one component is the infusion pump that is usually implanted under the skin on the back of the animal, and other component is the catheter that is inserted in the lesion of the nerve (Fig. 1.2).

1.3.1.1. Temporal Control: Using Osmotic Pumps to Stimulate Process Extension by Sustained, Local Trophic Factor Delivery   Several studies have investigated the benefits of continuous infusion of the neurotrophic factors BDNF and NT-3 after SCI. Typically, after SCI, methylprednisolone (MP) is administered to the patient. It has been demonstrated that the levels of BDNF and NT-3 decrease after the administration of MP. In a study, after treatment of MP, it was concluded that if BDNF was continuously delivered, then the rats locomotor function improved [34]. In a study that delivered both BDNF and NT-3 over a short time period (2 weeks) and a longer time period (8 weeks), it was shown that only the rats treated with BDNF and NT-3 over the 8 week time period allowed for the survival of the rubrospinal neurons [52]. However, rubrospinal axonal regeneration was not observed.

In another study that delivered either NT-3 or BDNF for 4 weeks into the spinal cord after it was crushed, the rats treated with BDNF did not exhibit any axonal regeneration. However, fiber sprouting was observed into and through the lesion in the rats that had NT-3 administered to the spinal cord lesion [11]. In a study that infused only BDNF for two weeks into the rat motor cortex after SCI, sprouting of corticospinal fibers was observed; however, axonal regeneration did not occur into the peripheral nerve transplant that was placed in the lesion [27]. The constant release of neurotrophic factors using the osmotic pump appears to exhibit therapeutic results. The site of administration seems to affect the response of axonal regeneration and fiber sprouting. The only disadvantage of utilizing the osmotic pump is the different locations of its components.

1.3.1.2. Temporal Control: Alleviation of Inhibitory Environments by Using Osmotic Pumps It was mentioned above that osmotic pumps can be used to deliver neurotrophic factors to the CNS to modulate intra-neuronal mechanisms. Osmotic pumps have also been utilized to infuse IN-1 antibody that neutralizes NOGO-A, an isoform of NOGO that is one of the main inhibitory molecules located in the glial scar [13]. It was observed that after 2 weeks of IN-1 delivery, regenerating fibers were observed through the lesion in the thoracic region into the lumbar region of the spinal cord. Therefore, the use of osmotic pumps can also be used to deliver proteins that can neutralize the inhibitory environment of the glial scar.

Other than the use of osmotic pumps to deliver proteins, Gelfoam, an insoluble gelatin sponge, was used to deliver chondroitinase ABC into the spinal cord lesion. The animals treated with chondroitinase ABC filled Gelfoam displayed axonal regeneration of the Clarke's neurons through the lesion area and it was exhibited that CSPG was digested by the chondroitinase ABC [76].

1.3.2. Temporal Control: Slow Release of Trophic Factors Using Microspheres

Microspheres, used in drug delivery applications, are being investigated to deliver protein to the PNS and CNS in order to encourage axonal outgrowth (Fig. 1.3). Microspheres have an advantage over osmotic pumps because a single administration is needed to release the protein over time. The size of the microspheres depends upon the application. The size of the microparticles in the studies that use microspheres to promote axonal outgrowth is around 12-16 μm. The materials that are used to make the microsphere are typically biodegradable polymers. The use of copolymers and altering the ratio of the polymers can affect the biodegradation profiles because the polymeric characteristics, such as glass transition temperature and hydrophilicities, change [61]. The polymeric materials mostly used for the microspheres are poly(lactic acid) (PLA), the copolymer poly(lactic-co-glycolic acid) (PLGA) and polyphosoesters. When investigating a specific polymer or another biomaterial, it is important to make sure that when the material degrades it does not denature the protein due to the possible immunogeneic response it can cause, thus altering the release profile and bioactivity [61].

1.3.2.1. Temporal Control: Use of Microspheres to Stimulate Process Extension in the PNS and CNS Most of the research, currently, focuses on delivering NGF loaded microspheres to regenerate nerves in the PNS. In a study performed by Xu et al., NGF

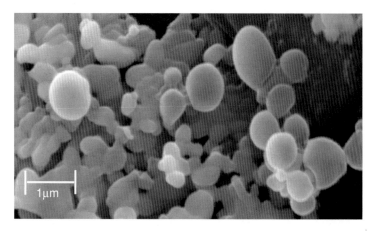

FIGURE 1.3. SEM Image of Microspheres. Microspheres can be used to encapsulate protein that will be slowly released as the microsphere degrades. Scale bar = 1 μm. Figure courtesy of YT Kim and RV Bellamkonda, Department of Biomedical Engineering, Georgia Institute of Technology.

was loaded into poly(phosphoester) (PPE) microspheres. First, in *in vitro* studies, it was determined that the microspheres released bioactive NGF up to 10 weeks. The NGF loaded PPE microspheres in a saline solution were loaded into PPE NGCs. When these constructs were implanted into rat sciatic nerves, it was observed that treatment with NGF loaded microspheres in the NGC had a cable that bridged the entire 10 mm gap between the nerve ends. Also, compared to the controls, there were more myelinated axons, higher fiber density, and thicker myelin sheath [73].

In the CNS, one of the first studies conducted using microspheres to deliver protein to the CNS was by Camarata et al. In order to combat neurodegenerative disease, they inserted microspheres loaded with NGF that could be released *in vivo* for 4 to 5 weeks [14]. In another *in vitro* study, the number of days NGF was released was increased to 91 days. Various ratio of PLGA were tested to determine the release characteristics, as well as poly(ε-caparolactone) (PCL) [15]. The surface morphology of the microspheres that are loaded versus unloaded ones is different. The surface of protein loaded microspheres is rougher, whereas the unloaded microspheres have a smoother surface. The smaller the microsphere, the greater the surface area, thus increasing the degradation rate of the microsphere and release of the protein.

1.3.3. Temporal Control: Lipid Microtubules for Sustained Release of Stimulatory Trophic Factors

Another method to slowly release protein in the CNS and PNS is the use of lipid microtubules, also referred to as microcylinders (Fig. 1.4). These microtubules are hollow cylinders with a diameter of 0.5 μm [44]. The length of the microtubules varies based on the time period in which the protein, DNA, or other desired molecule needs to be released. The molecule is released at the ends of the microtubules, which is the reason why the length of the microcylinders controls the release profile of the protein. In a study previously mentioned, to aid axonal regeneration in the PNS, a two-step slow release system was developed. The first step was NGF loaded microtubules, which had a length of 40 μm, and

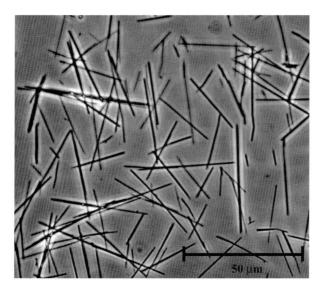

FIGURE 1.4. Micrograph Image of Lipid Microtubules. The image depicts the lipid microtubules being on average 40 μm in length. The microtubules release the protein from the ends. Scale bar = 50 μm.

the second step was the loaded microtubules embedded in agarose hydrogel [79]. The two step release system was thus, first the diffusion of the NGF from the microtubules into the agarose and then the release of the NGF from the agarose into the gap between the two nerve ends. This slow release system allows the NGF to last longer in the nerve gap and prevents degradation or dilution by macrophages and other fluids. Two months post-implantation, a cable formed, the number of myelinated axons was statistically similar to the autograft condition, and the density of myelinated axons was similar to that of the autograft and a normal sciatic nerve.

1.3.4. Temporal Control: Demand Driven Release of Trophic Factors

Another form of controlled release of a protein is the fibrin matrix, which was initially developed for wound healing. Cells that migrate to the area degrade the matrix through proteolysis, thereby releasing the contained protein (Fig. 1.5) [57]. A fibrin matrix covalently coupled to heparin that interacted with neurotrophins, NGF, BDNF, and NT-3 was developed. It was demonstrated *in vitro* that the neurite outgrowth was enhanced when the neurotrophins were released using this delivery system compared when soluble neurotrophins were added to the fibrin matrix [56]. When the heparin immobilized fibrin matrix was implanted in a nerve gap in the PNS, fiber sprouting was observed through the conduit to the distal end [38].

1.4. CONCLUSION

The advancement in CNS and PNS regeneration has been due to the utilization of nano- and micro-technologies. Most of the technology that has been developed has been geared

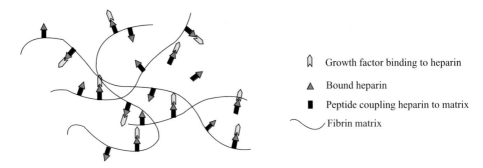

FIGURE 1.5. Schematic of Fibrin Matrix Releasing Protein. The peptides bind the heparin to the fibrin matrix. The growth factor or protein is then able to bind the heparin, thus attaching the growth factor to the matrix. The cells migrating to the area will then degrade the matrix releasing the growth factor. Figure adapted from Ref. [57].

towards controlling proteins spatially and temporally. There are three main strategies used to elicit axonal outgrowth after injury, which allows spatial and temporal control of proteins. The three strategies mentioned are to 1) provide permissive bioactive substrates for the axonal outgrowth; (2) use trophic factors to stimulate growth; and (3) alleviate signaling due to the inhibitory entities present in the extracellular environment to allow axons to regenerate between the proximal and distal ends.

This chapter briefly describes studies that have incorporated various nano- and micro-technologies using biomaterials based design. While, for analytical convenience we divide this chapter into sections with various strategies, it is becoming evident that a coordinated, multiple component strategy may be required for successful regeneration. For example, one approach is to design a substrate that is coupled to proteins, contains either Schwann cells or OEG, and has a delivery vehicle slowly releasing proteins. The key combination remains elusive and is the focus of active, ongoing investigation.

REFERENCES

[1] P. Aebischer, A.N. Salessiotis, and S.R. Winn. Basic fibroblast growth factor released from synthetic guidance channels facilitates peripheral nerve regeneration across long nerve gaps. *J. Neurosci. Res.*, 23(3):282–289, 1989.

[2] A.M. Avellino, D. Hart, A.T. Dailey, M. MacKinnon, D. Ellegala, and M. Kliot. Differential macrophage responses in the peripheral and central nervous system during wallerian degeneration of axons. *Exp. Neurol.*, 136(2):183–198, 1995.

[3] A. Baird, P.A. Walicke. Fibroblast growth factors. *Br. Med. Bull.*, 45(2):438–452, 1989.

[4] A.P. Balgude, X. Yu, A. Szymanski, and R.V. Bellamkonda. Agarose gel stiffness determines rate of DRG neurite extension in 3D cultures. *Biomaterials*, 22(10):1077–1084, 2001.

[5] C.E. Bandtlow. Regeneration in the central nervous system. *Exp. Gerontol.*, 38(1–2):79–86, 2003.

[6] R. Bellamkonda, J.P. Ranieri, and P. Aebischer. Laminin oligopeptide derivatized agarose gels allow three-dimensional neurite extension in vitro. *J. Neurosci. Res.*, 41(4):501–509, 1995a.

[7] R. Bellamkonda, J.P. Ranieri, N. Bouche, and P. Aebischer. Hydrogel-based three-dimensional matrix for neural cells. *J. Biomed. Mater. Res.*, 29(5):663–671, 1995b.

[8] R. Bellamkonda and P. Aebischer. Review: Tissue Engineering in the Nervous System. *Biotech. Bioeng.*, 43:543–1994, 1993.

[9] R. Biran, M.D. Noble, and P.A. Tresco. Directed nerve outgrowth is enhanced by engineered glial substrates. *Exp. Neurol.*, 184(1):141–152, 2003.

[10] M. Borkenhagen, J.F. Clemence, H. Sigrist, and P. Aebischer. Three-dimensional extracellular matrix engineering in the nervous system. *J. Biomed. Mater. Res.*, 40(3):392–400, 1998.

[11] E.J. Bradbury, S. Khemani, R. Von King, J.V. Priestley, and S.B. McMahon. NT-3 promotes growth of lesioned adult rat sensory axons ascending in the dorsal columns of the spinal cord. *Eur. J. Neurosci.*, 11(11):3873–3883, 1999.

[12] E.J. Bradbury, L.D. Moon, R.J. Popat, V.R. King, G.S. Bennett, P.N. Patel, J.W. Fawcett, and S.B. McMahon. Chondroitinase ABC promotes functional recovery after spinal cord injury. *Nature*, 416(6881):636–640, 2002.

[13] C. Brosamle, A.B. Huber, M. Fiedler, A. Skerra, and M.E. Schwab. Regeneration of lesioned corticospinal tract fibers in the adult rat induced by a recombinant, humanized IN-1 antibody fragment. *J. Neurosci.*, 20(21):8061–8068, 2000.

[14] P.J. Camarata, R. Suryanarayanan, D.A. Turner, R.G. Parker, and T.J. Ebner. Sustained release of nerve growth factor from biodegradable polymer microspheres. *Neurosurgery*, 30(3):313–319, 1992.

[15] X. Cao and M.S. Schoichet. Delivering neuroactive molecules from biodegradable microspheres for application in central nervous system disorders. *Biomaterials*, 20(4):329–339, 1999.

[16] D. Ceballos, X. Navarro, N. Dubey, G. Wendelschafer-Crabb, W.R. Kennedy, and R.T. Tranquillo. Magnetically aligned collagen gel filling a collagen nerve guide improves peripheral nerve regeneration. *Exp. Neurol.*, 158(2):290–300, 1999.

[17] L.J. Chamberlain, I.V. Yannas, H.P. Hsu, G. Strichartz, and M. Spector. Collagen-GAG substrate enhances the quality of nerve regeneration through collagen tubes up to level of autograft. *Exp. Neurol.*, 154(2):315–329, 1998.

[18] L.B. Dahlin and G. Lundborg. Use of tubes in peripheral nerve repair. *Neurosurg. Clin. N. Am.*, 12(2):341–352, 2001.

[19] S. David and S. Lacroix. Molecular Approaches to Spinal Cord *Repair. Annu. Rev. Neurosci.*, 2003.

[20] R. Deumens, G.C. Koopmans, C.G. Den Bakker, V. Maquet, S. Blacher, W.M. Honig, R. Jerome, J.P. Pirard, H.W. Steinbusch, and E.A. Joosten. Alignment of glial cells stimulates directional neurite growth of CNS neurons in vitro. *Neuroscience*, 125(3):591–604, 2004.

[21] C.I. Dubreuil, M.J. Winton, and L. McKerracher. Rho activation patterns after spinal cord injury and the role of activated Rho in apoptosis in the central nervous system. *J. Cell. Biol.*, 162(2):233–243, 2003.

[22] G.R. Evans. Peripheral nerve injury: a review and approach to tissue engineered constructs. *Anat. Rec.*, 263(4):396–404, 2001.

[23] J.W. Fawcett and R.A. Asher. The glial scar and central nervous system repair. *Brain. Res. Bull.*, 49(6):377–391, 1999.

[24] S.P. Frostick, Q. Yin, and G.J. Kemp. Schwann cells, neurotrophic factors, and peripheral nerve regeneration. *Microsurgery*, 18(7):397–405, 1998.

[25] V. Guenard, N. Kleitman, T.K. Morrissey, R.P. Bunge, and P. Aebischer. Syngeneic Schwann cells derived from adult nerves seeded in semipermeable guidance channels enhance peripheral nerve regeneration. *J. Neurosci.*, 12(9):3310–3320, 1992.

[26] T. Hashimoto, Y. Suzuki, M. Kitada, K. Kataoka, S. Wu, K. Suzuki, K. Endo, Y. Nishimura, and C. Ide. Peripheral nerve regeneration through alginate gel: analysis of early outgrowth and late increase in diameter of regenerating axons. *Exp. Brain. Res.*, 146(3):356–368, 2002.

[27] G.W. Hiebert, K. Khodarahmi, J. McGraw, J.D. Steeves, and W. Tetzlaff. Brain-derived neurotrophic factor applied to the motor cortex promotes sprouting of corticospinal fibers but not regeneration into a peripheral nerve transplant. *J. Neurosci. Res.*, 69(2):160–168, 2002.

[28] A. Hoke and J. Silver. Proteoglycans and other repulsive molecules in glial boundaries during development and regeneration of the nervous system. *Prog. Brain. Res.*, 108:149–163, 1996.

[29] D.A. Houweling, A.J. Lankhorst, W.H. Gispen, P.R. Bar, and E.A. Joosten. Collagen containing neurotrophin-3 (NT-3) attracts regrowing injured corticospinal axons in the adult rat spinal cord and promotes partial functional recovery. *Exp. Neurol.*, 153(1):49–59, 1998.

[30] T.W. Hudson, G.R. Evans, and C.E. Schmidt. Engineering strategies for peripheral nerve repair. *Orthop. Clin. North. Am.*, 31(3):485–498, 2000.

[31] A. Jain, S.M. Brady-Kalnay, and R.V. Bellamkonda. Modulation of Rho GTPase activity alleviates chondroitin sulfate proteoglycan-dependent inhibition of neurite extension. *J. Neurosci. Res.*, 77(2):299–307, 2004.

[32] K. Kataoka, Y. Suzuki, M. Kitada, T. Hashimoto, H. Chou, H. Bai, M. Ohta, S. Wu, K. Suzuki, and C. Ide. Alginate enhances elongation of early regenerating axons in spinal cord of young rats. *Tissue. Eng.*, 10(3-4):493–504, 2004.

[33] T. Khan, M. Dauzvardis, and S. Sayers. Carbon filament implants promote axonal growth across the transected rat spinal cord. *Brain. Res.*, 541(1):139–145, 1991.

[34] D.H. Kim and T.A. Jahng. Continuous brain-derived neurotrophic factor (BDNF) infusion after methylprednisolone treatment in severe spinal cord injury. *J. Korean. Med. Sci.*, 19(1):113–122, 2004.

[35] V. Kottis, P. Thibault, D. Mikol, Z.C. Xiao, R. Zhang, P. Dergham, P.E. Braun. Oligodendrocyte-myelin glycoprotein (OMgp) is an inhibitor of neurite outgrowth. *J. Neurochem.*, 82(6):1566–1569, 2002.

[36] C.E. Krewson, M.L. Klarman, and W.M. Saltzman. Distribution of nerve growth factor following direct delivery to brain interstitium. *Brain. Res.*, 680(1–2):196–206, 1995.

[37] A. Lakatos, S.C. Barnett, R.J. Franklin. Olfactory ensheathing cells induce less host astrocyte response and chondroitin sulphate proteoglycan expression than Schwann cells following transplantation into adult CNS white matter. *Exp. Neurol.*, 184(1):237–246, 2003.

[38] A.C. Lee, V.M. Yu, J.B. Lowe, 3rd, M.J. Brenner, D.A. Hunter, S.E. Mackinnon, S.E. Sakiyama-Elbert. Controlled release of nerve growth factor enhances sciatic nerve regeneration. *Exp. Neurol.*, 184(1):295–303, 2003.

[39] L.S. Liu, T. Khan, S.T. Sayers, M.F. Dauzvardis, and C.L. Trausch. Electrophysiological improvement after co-implantation of carbon filaments and fetal tissue in the contused rat spinal cord. *Neurosci. Lett.*, 200(3):199–202, 1995.

[40] Y. Luo and M.S. Shoichet. A photolabile hydrogel for guided three-dimensional cell growth and migration. *Nat. Mater.*, 3(4):249–253, 2004.

[41] M.P. Mattson, A. Taylor-Hunter, and S.B. Kater. Neurite outgrowth in individual neurons of a neuronal population is differentially regulated by calcium and cyclic AMP. *J. Neurosci.*, 8(5):1704–1711, 1988.

[42] R.J. McKeon, A. Hoke, and J. Silver. Injury-induced proteoglycans inhibit the potential for laminin-mediated axon growth on astrocytic scars. *Exp. Neurol.*, 136(1):32–43, 1995.

[43] L. McKerracher, S. David, D.L. Jackson, V. Kottis, R.J. Dunn, and P.E. Braun. Identification of myelin-associated glycoprotein as a major myelin-derived inhibitor of neurite growth. *Neuron*, 13(4):805–811, 1994.

[44] N.J. Meilander, X. Yu, N.P. Ziats, and R.V. Bellamkonda. Lipid-based microtubular drug delivery vehicles. *J. Control. Release.*, 71(1):141–152, 2001.

[45] P.P. Monnier, A. Sierra, J.M. Schwab, S. Henke-Fahle, and B.K. Mueller. The Rho/ROCK pathway mediates neurite growth-inhibitory activity associated with the chondroitin sulfate proteoglycans of the CNS glial scar. *Mol. Cell. Neurosci.*, 22(3):319–330, 2003.

[46] D.A. Morgenstern, R.A. Asher, and J.W. Fawcett. Chondroitin sulphate proteoglycans in the CNS injury response. *Prog. Brain. Res.*, 137:313–332, 2002.

[47] A. Mosahebi, M. Wiberg, and G. Terenghi. Addition of fibronectin to alginate matrix improves peripheral nerve regeneration in tissue-engineered conduits. *Tissue. Eng.*, 9(2):209–218, 2003.

[48] G. Mukhopadhyay, P. Doherty, F.S. Walsh, P.R. Crocker, and M.T. Filbin. A novel role for myelin-associated glycoprotein as an inhibitor of axonal regeneration. *Neuron*, 13(3):757–767, 1994.

[49] T.T. Ngo, P.J. Waggoner, A.A. Romero, K.D. Nelson, R.C. Eberhart, and G.M. Smith. Poly(L-Lactide) microfilaments enhance peripheral nerve regeneration across extended nerve lesions. *J. Neurosci. Res.*, 72(2):227–238, 2003.

[50] B. Niederost, T. Oertle, J. Fritsche, R.A. McKinney, and C.E. Bandtlow. Nogo-A and myelin-associated glycoprotein mediate neurite growth inhibition by antagonistic regulation of RhoA and Rac1. *J. Neurosci.*, 22(23):10368–10376, 2002.

[51] C.D. Nobes and A. Hall. Rho, rac, and cdc42 GTPases regulate the assembly of multimolecular focal complexes associated with actin stress fibers, lamellipodia, and filopodia. *Cell*, 81(1):53–62, 1995.

[52] L.N. Novikova, L.N. Novikov, and J.O. Kellerth. Differential effects of neurotrophins on neuronal survival and axonal regeneration after spinal cord injury in adult rats. *J. Comp. Neurol.*, 452(3):255–263, 2002.

[53] D.D. Pearse, F.C. Pereira, A.E. Marcillo, M.L. Bates, Y.A. Berrocal, M.T. Filbin, and M.B. Bunge. cAMP and Schwann cells promote axonal growth and functional recovery after spinal cord injury. *Nat. Med.*, 10(6):610–616, 2004.

[54] J. Qiu, D. Cai, H. Dai, M. McAtee, P.N. Hoffman, B.S. Bregman, and M.T. Filbin. Spinal axon regeneration induced by elevation of cyclic AMP. *Neuron*, 34(6):895–903, 2002.

[55] N. Rangappa, A. Romero, K.D. Nelson, R.C. Eberhart, and G.M. Smith. Laminin-coated poly(L-lactide) filaments induce robust neurite growth while providing directional orientation. *J. Biomed. Mater. Res.*, 51(4):625–634, 2000.

[56] S.E. Sakiyama-Elbert and J.A. Hubbell. Controlled release of nerve growth factor from a heparin-containing fibrin-based cell ingrowth matrix. *J. Control. Release*, 69(1):149–158, 2000a.

[57] S.E. Sakiyama-Elbert and J.A. Hubbell. Development of fibrin derivatives for controlled release of heparin-binding growth factors. *J. Control Release*, 65(3):389–402, 2000b.

[58] F.F. Santos-Benito and A. Ramon-Cueto. Olfactory ensheathing glia transplantation: a therapy to promote repair in the mammalian central nervous system. *Anat. Rec.*, 271B(1):77–85, 2003.

[59] C.E. Schmidt and J.B. Leach. Neural tissue engineering: strategies for repair and regeneration. *Annu. Rev. Biomed. Eng.*, 5:293–347, 2003.

[60] D. Shaw and M.S. Shoichet. Toward spinal cord injury repair strategies: peptide surface modification of expanded poly(tetrafluoroethylene) fibers for guided neurite outgrowth in vitro. *J. Craniofac. Surg.* 14(3):308–316, 2003.

[61] V.R. Sinha and A. Trehan. Biodegradable microspheres for protein delivery. *J. Control. Release*, 90(3):261–280, 2003.

[62] W. Sufan, Y. Suzuki, M. Tanihara, K. Ohnishi, K. Suzuki, K. Endo, and Y. Nishimura. Sciatic nerve regeneration through alginate with tubulation or nontubulation repair in cat. *J. Neurotrauma.*, 18(3):329–338, 2001.

[63] S. Sunderland, Sir. *Nerve Injuries and Their Repair: A Critical Appraisal*. Edinburgh: Churchill Livingstone, 1991.

[64] T. Takami, M. Oudega, M.L. Bates, P.M. Wood, N. Kleitman, and M.B. Bunge. Schwann cell but not olfactory ensheathing glia transplants improve hindlimb locomotor performance in the moderately contused adult rat thoracic spinal cord. *J. Neurosci.*, 22(15):6670–6681, 2002.

[65] R.F. Valentini, P. Aebischer, S.R. Winn, P.M. Galletti. Collagen- and laminin-containing gels impede peripheral nerve regeneration through semipermeable nerve guidance channels. *Exp. Neurol.*, 98(2):350–356, 1987.

[66] E. Verdu, R.O. Labrador, F.J. Rodriguez, D. Ceballos, J. Fores, and X. Navarro. Alignment of collagen and laminin-containing gels improve nerve regeneration within silicone tubes. *Restor. Neurol. Neurosci.*, 20(5):169–179, 2002.

[67] N. Weidner, A. Blesch, R.J. Grill, and M.H. Tuszynski. Nerve growth factor-hypersecreting Schwann cell grafts augment and guide spinal cord axonal growth and remyelinate central nervous system axons in a phenotypically appropriate manner that correlates with expression of L1. *J. Comp. Neurol.*, 413(4):495–506, 1999.

[68] M.R. Wells, K. Kraus, D.K. Batter, D.G. Blunt, J. Weremowitz, S.E. Lynch, H.N. Antoniades, and H.A. Hansson. Gel matrix vehicles for growth factor application in nerve gap injuries repaired with tubes: a comparison of biomatrix, collagen, and methylcellulose. *Exp. Neurol.*, 146(2):395–402, 1997.

[69] W.D. Whetstone, J.Y. Hsu, M. Eisenberg, Z. Werb, and L.J. Noble-Haeusslein. Blood-spinal cord barrier after spinal cord injury: relation to revascularization and wound healing. *J. Neurosci. Res.*, 74(2):227–239, 2003.

[70] M.J. Winton, C.I. Dubreuil, D. Lasko, N. Leclerc, and L. McKerracher. Characterization of new cell permeable C3-like proteins that inactivate Rho and stimulate neurite outgrowth on inhibitory substrates. *J. Biol. Chem.*, 277(36):32820–32829, 2002.

[71] S. Woerly, V.D. Doan, F. Evans-Martin, C.G. Paramore, and J.D. Peduzzi. Spinal cord reconstruction using NeuroGel implants and functional recovery after chronic injury. *J. Neurosci. Res.*, 66(6):1187–1197, 2001.

[72] S. Woerly, V.D. Doan, N. Sosa, J. de Vellis, A. Espinosa-Jeffrey. Prevention of gliotic scar formation by NeuroGel allows partial endogenous repair of transected cat spinal cord. *J. Neurosci. Res.*, 75(2):262–272, 2004.

[73] X. Xu, W.C. Yee, P.Y. Hwang, H. Yu, A.C. Wan, S. Gao, K.L. Boon, H.Q. Mao, K.W. Leong, and S. Wang. Peripheral nerve regeneration with sustained release of poly(phosphoester) microencapsulated nerve growth factor within nerve guide conduits. *Biomaterials*, 24(13):2405–2412, 2003.

[74] X.M. Xu, A. Chen, V. Guenard, N. Kleitman, and M.B. Bunge. Bridging Schwann cell transplants promote axonal regeneration from both the rostral and caudal stumps of transected adult rat spinal cord. *J. Neurocytol.*, 26(1):1–16, 1997.

[75] X.M. Xu, S.X. Zhang, H. Li, P. Aebischer, and M.B. Bunge. Regrowth of axons into the distal spinal cord through a Schwann-cell-seeded mini-channel implanted into hemisected adult rat spinal cord. *Eur. J. Neurosci.*, 11(5):1723–1740, 1999.

[76] L.W. Yick, P.T. Cheung, K.F. So, and W. Wu. Axonal regeneration of Clarke's neurons beyond the spinal cord injury scar after treatment with chondroitinase ABC. *Exp. Neurol.*, 182(1):160–168, 2003.

[77] S. Yoshii and M. Oka. Collagen filaments as a scaffold for nerve regeneration. *J. Biomed. Mater. Res.*, 56(3):400–405, 2001.

[78] S. Yoshii, M. Oka, M. Shima, A. Taniguchi, and M. Akagi. Bridging a 30-mm nerve defect using collagen filaments. *J. Biomed. Mater. Res.*, 67A(2):467–474, 2003.

[79] X. Yu and R.V. Bellamkonda. Tissue-engineered scaffolds are effective alternatives to autografts for bridging peripheral nerve gaps. *Tissue Eng.*, 9(3):421–430, 2003.

[80] X. Yu, G.P. Dillon, and R.B. Bellamkonda. A laminin and nerve growth factor-laden three-dimensional scaffold for enhanced neurite extension. *Tissue Eng.*, 5(4):291–304, 1999.

[81] D.W. Zochodne. The microenvironment of injured and regenerating peripheral nerves. *Muscle Nerve Suppl.*, 9:S33–38, 2000.

[82] J. Zuo, D. Neubauer, K. Dyess, T.A. Ferguson, and D. Muir. Degradation of chondroitin sulfate proteoglycan enhances the neurite-promoting potential of spinal cord tissue. *Exp. Neurol.*, 154(2):654–662, 1998.

[83] J. Zuo, D. Neubauer, J. Graham, C.A. Krekoski, T.A. Ferguson, and D. Muir. Regeneration of axons after nerve transection repair is enhanced by degradation of chondroitin sulfate proteoglycan. *Exp. Neurol.*, 176(1):221–228, 2002.

2

3-D Fabrication Technology for Tissue Engineering

Alice A. Chen, Valerie Liu Tsang, Dirk R. Albrecht, and Sangeeta N. Bhatia

Harvard- MIT Division of Health Sciences and Technology (HST), Electrical Engineering and Computer Science, Massachusetts Institute of Technology; Department of Medicine, Brigham & Women's Hospital

2.1. INTRODUCTION

Tissue engineering typically involves the combination of cells and biomaterials to form tissues with the goal of replacing or restoring physiological functions lost in diseased organs. The biomaterial scaffolds are designed to provide mechanical support for the cells; however, in practice, the simple addition of cells to porous scaffolds often does not recapitulate sufficient tissue function. Scaffold design previously focused on the incorporation of macroscale features such as interconnected pores for nutrient transport and tissue remodeling. One strategy to further augment the function of tissue-engineered constructs is to mimic the in vivo tissue microarchitecture and cellular microenvirnment. Tissues in the body are divided into repeating functional units (e.g., nephron, islet) [1], whose 3-D architecture coordinates the processes of multiple types of specialized cells. Further, the local environment of these cells presents biochemical and physical stimuli that specifically modulate both cellular functions, e.g. biosynthesis and metabolism, and cellular fate processes such as differentiation, proliferation, apoptosis and migration. Thus, the fabrication of functional 3-D tissue constructs that incorporate both microscale features for appropriate cell functions and macroscale mechanical and transport properties demands control over chemistry and architecture over multiple length scales.

Tissue engineering scaffolds that mimic the complex architecture of native tissues have been more difficult to produce than conventional porous polymer scaffolds that support undirected cell adhesion and spreading within homogeneous and relatively large (millimeter scale) constructs [2]. Recently, computer-controlled rapid-prototyping technologies have been adapted toward the fabrication of 3-D scaffolds with precise geometric control at the macro- and micro-scale. These 3-D fabrication approaches offer numerous opportunities with great potential for tissue engineering. For example, the function of complex tissue units is expected to rely on the independent control of macro- and micro-scale features. The incorporation of vascular beds would allow for larger constructs than could be supported by nutrient diffusion alone. In addition, the combination of clinical imaging data with CAD-based freeform techniques allows the fabrication of replacement tissues that are customized to the shape of a particular defect. Finally, the large-scale production of identical functional tissue units may find use in cell-based assays for drug discovery or for fundamental biological studies. Recently, microfabrication tools have been applied to the study of cell-cell and cell-matrix interactions within a two-dimensional cell culture context [1]. Extending these studies to three dimensional cellular control may provide further insight on cellular interactions and structure/function relationships within a tissue.

In this review, we describe various three-dimensional technologies used in tissue design and fabrication and compare their modes of assembly, spatial resolution, development stage, and feasibility for tissue engineering. Specifically, our discussion focuses on three general approaches (Fig. 2.1): acellular polymer scaffold fabrication, cellular assembly techniques, and hybrid cell/scaffold strategies.

2.2. FABRICATION OF ACELLULAR CONSTRUCTS

Traditional scaffold fabrication methods, including solvent-casting/particulate-leaching, gas foaming, fiber bonding, phase separation, and emulsion freeze drying, allow for limited control of pore size and shape but lack the sensitivity to precisely determine scaffold architecture [3]. In contrast, CAD-based rapid prototyping methods provide excellent spatial control over polymer architecture and have recently been applied to the fabrication of 3-D tissue engineering scaffolds. In Figure 2.1, various methods for creating acellular scaffolds are categorized according to their modes of fabrication, using heat, light, adhesives, or molding. These techniques are presented below, along with recent applications and advances.

2.2.1. Heat-Mediated 3D Fabrication

Fabrication by heat energy combines pre-fabricated polymer layers into simple three-dimensional structures by raising the polymer above its glass transition temperature and fusing the softened layers together with applied pressure [4]. In sheet lamination fabrication, laser-cut polymer sheets are sequentially bonded by the application of heat and pressure. Currently, scaffolds created with this method have very low void volume and are generally too dense for the construction of tissues with high cellularity.

Lamination techniques can also be used to fabricate more intricate scaffolds that contain small, well-defined pores to increase void volume. For example, biodegradable polyester polymers such as poly(DL-lactic-co-glycolic) acid (PLGA), have been micropatterned by various techniques and laminated into three-dimensional structures. Borenstein and colleagues constructed thin biodegradable films containing small trenches by casting

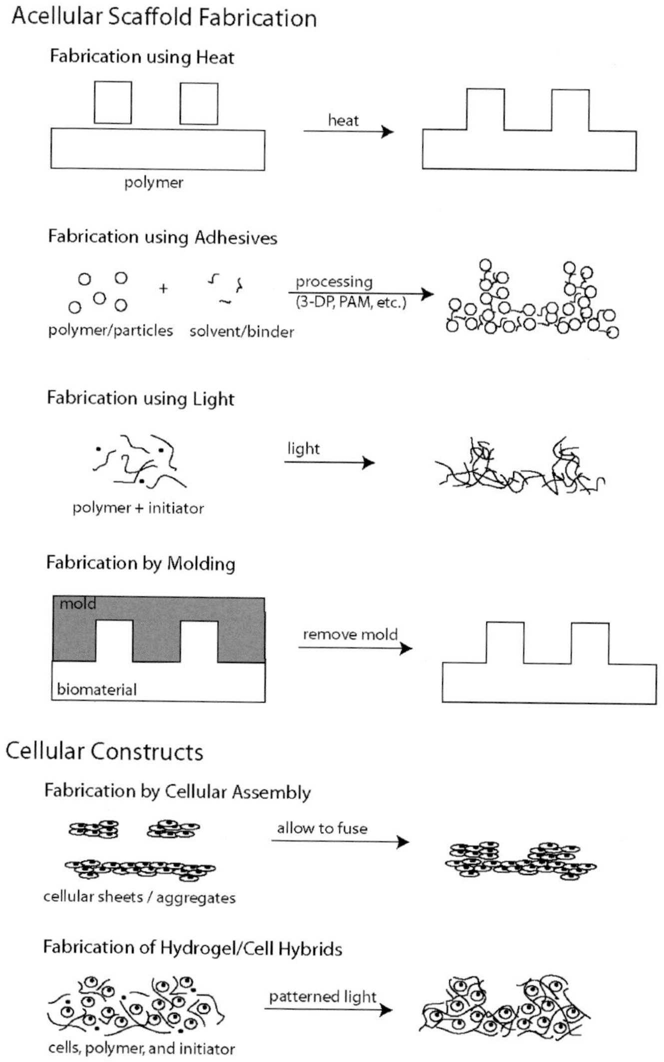

Acellular Scaffold Fabrication

Fabrication using Heat

heat →

polymer

Fabrication using Adhesives

O O + ʃ ʃ → processing (3-DP, PAM, etc.)

polymer/particles   solvent/binder

Fabrication using Light

light →

polymer + initiator

Fabrication by Molding

mold

remove mold →

biomaterial

Cellular Constructs

Fabrication by Cellular Assembly

allow to fuse →

cellular sheets / aggregates

Fabrication of Hydrogel/Cell Hybrids

patterned light →

cells, polymer, and initiator

FIGURE 2.1. Summary of 3-D Scaffold Fabrication Methods. Acellular scaffolds can be fabricated using various techniques, such as heat (FDM), chemicals (3-DP), light (SLA), and molding. Cells themselves can be incorporated in the fabrication process by cellular addition or by photopatterning of hydrogels.

PLGA onto microfabricated silicon masters. When laminated together, these patterned films formed a vascular tissue engineering scaffold with 20 μm diameter channels between layers (Fig. 2.2a) [5]. Researchers later developed similar scaffolds with soft lithography techniques that utilize inexpensive elastomeric polydimethylsiloxane (PDMS) molds cast from silicon masters [6]. By introducing a PLGA solution into the mold and heating, Bhatia and colleagues created polymer layers that exhibited microstructures similar in shape and resolution (20–30 μm) to those on the silicon master and could be fused together (Fig. 2.2b) [7]. To further increase the scaffold surface area for cell attachment and proliferation,

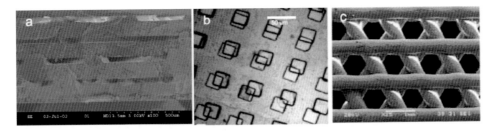

FIGURE 2.2. Fabrication using Heat. (a–b) Molded Lamination. Membranes of the biodegradable polymer PLGA are cast from silicon (a) or PDMS (b) molds and then laminated to create 3-D scaffolds. In (a), layers of PLGA are fused together to form microfluidic channels for vascular tissue engineering (c) Fused Deposition Molding. Molten biomaterials are extruded through a nozzle to build 3-D scaffolds layer by layer. (Photo courtesy of Jeff Borenstein and Kevin King, Draper Laboratory).

micropores can be incorporated into the patterned PLGA membranes by solvent casting and particulate leaching strategies.

Selective laser sintering (SLS) is a heat-based fabrication technique that uses laser energy to combine powdered polymeric materials into defined shapes. A laser beam directed across a powder bed locally increases polymer temperature to fuse with the surrounding material and form a layer of patterned structures [4]. Three-dimensional SLS scaffolds are created sequentially with fresh powder deposited over each patterned layer. Unfused powder released from the scaffold yields high porosity and surface area while retaining mechanical integrity. The pattern resolution of SLS is limited by the diameter of the laser beam diameter to about 400 um [4], and maximum pore size is about 50 um due to the powder particle size [8]. Lee and Barlow first utilized SLS with polymer-coated calcium phosphate powders to fabricate oral implants and demonstrated extensive bone tissue ingrowth in dog models [9]. Since then, Leong and others have broadened SLS utilization for various biopolymer applications [8].

Fused deposition modeling (FDM) combines heat and extrusion techniques to create 3-D scaffolds layer by layer. A nozzle directs a stream of molten plastic or ceramic onto a previously deposited layer of material. By altering the direction of material deposition with each layer, scaffolds with complex internal organization can be formed (Fig. 2.2c). Zein and Hutmacher used this method to produce biodegradable poly(ε-caprolactone) (PCL) scaffolds exhibiting various honeycomb geometries with finely tuned pore and channel dimensions of 250–700 μm [10]. Primary human fibroblasts cultured in these scaffolds proliferated and produced extracellular matrix [11], and scaffolds composed of other bio-compatible polymers and composites have demonstrated utility for various tissue engineering applications [12–14]. While FDM exhibits high pattern resolution in the xy-plane, it is limited in the z-direction by the diameter of the extruded polymer filament that defines layer thickness and corresponding pore height. Further, high processing temperatures limit the biomaterials that are compatible with the method. However, FDM capabilities are expanding with new developments such as multi-phase jet solidification (MJS), a technique that allows simultaneous extrusion of multiple melted materials [15].

3-D plotting is a similar heat-based extrusion technology that is not limited to synthetic polymers that must withstand high temperatures while retaining their desired properties such as degradation and biocompatibility. Instead, fabrication is based on a sol-gel phase

FIGURE 2.3. 3-D Plotting. Heated liquid agar solidifies into a 3-D hydrogel scaffold when deposited into a cooled medium. (from [16], reprinted with permission of Elsevier).

transition that occurs at lower temperatures. This strategy has been demonstrated with natural hydrogel biomaterials that are substantially more versatile for tissue engineering applications. For example, Mulhaupt and coworkers deposited agar and gelatin solutions heated to 90°C into a cooled plotting medium, resulting in a 3-D hydrogel scaffold (Figure 2.3) [16]. Similarly, Ang and colleagues used robotic dispensing to form chitosan and chitosan-hydroxyapatite scaffolds [17]. Following fibrin treatment, these scaffolds supported the adhesion of human osteosarcoma cells or mouse fibroblasts.

2.2.2. Light-Mediated Fabrication

Light energy can also be used to fabricate structured 3-D polymer scaffolds. Photopolymerization uses light to initiate a chain reaction that solidifies a liquid polymer solution. Stereolithography (SLA) is a photopolymerization method that utilizes a deflected UV laser beam to irradiate and solidify exposed polymer regions at the surface of a vat of photosensitive polymer (Fig. 2.4). Multiple layers are formed sequentially by lowering the stage and repeating the laser illumination. While SLA machines are traditionally used to build prototypes and molds for implants, Cooke et al. fabricated biodegradable 3-D polymer scaffolds for bony tissue consisting of diethyl fumarate, poly(propylene fumarate) and the photoinitiator bisacylphosphine oxide [18]. Similarly, a photocurable ceramic acrylate suspension formed cancellous bone [19] and hydroxyapatite bone tissue scaffolds [20], with overall dimensions suitable for healing critical-sized (4-mm thickness, 50-mm diameter) bone defects. As with SLS, stereolithography is limited in resolution by laser beam diameter to approximately 250 μm, although small-spot laser systems have demonstrated the production of smaller (70 μm) features [4].

Light energy can also be used to photopolymerize hydrogel polymer scaffolds that are less rigid than conventional stereolithography materials. Hydrogels are crosslinked networks of insoluble hydrophilic polymers that swell with water. Their increasing popularity as tissue

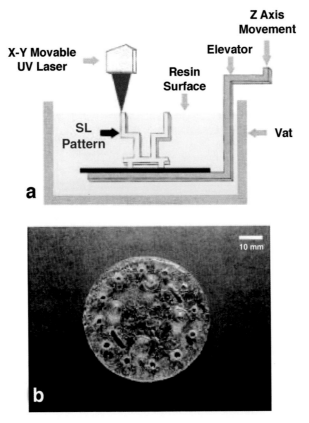

FIGURE 2.4. Stereolithography. (a) UV light is used to crosslink the material in specific regions of a layer. The elevator is then lowered to reveal a new layer of polymer, and the process is repeated to create the desired shape. (b) A prototype scaffold designed using SLA (from [18], reprinted by permission of John Wiley & Sons, Inc.).

engineering biomaterials reflects mechanical properties and high water content analogous to those of natural tissue. Yu and colleagues demonstrated a photolithographic method of patterning layers of dried 2-hydroxyethyl methacrylate that were subsequently rehydrated and seeded with cells [21]. However, the resolution of hydrogel scaffold fabrication may be compromised during rehydration of the polymer. Instead, Matsuda et al. later created scaffolds with improved strength and limited swelling using combinations of vinylated polysaccharides and diacrylated polyethylene glycol [22]. Additionally, the photopatterning of hydrogels has recently been extended to incorporate living cells into hybrid constructs, as discussed in a later section.

2.2.3. Adhesive-Mediated Fabrication

Scaffolds fabricated by binding polymers with solvents or adhesives, rather than by heat or light, circumvent biomaterial limitations for thermostable polymers or for biocompatible photoinitiators. Three-dimensional printing (3-DP), for instance, utilizes an ink jet printer to deposit a binder solution onto a polymer powder bed. Multiple layers can be fabricated and stacked with dimensions on the scale of polymer particle size (approximately 200–300 μm)

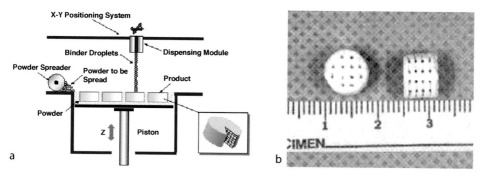

FIGURE 2.5. 3-D printing. Ink jet technology is used to print a binder solution onto a bed of polymer powder. An additional layer of powder is then deposited, and the process is repeated to form 3-D scaffolds (a) from Therics, website, with permission; (b) from [24], reprinted with permission of Leppincott Williams & Wilkins.

(Fig. 2.5) [23]. Scaffolds composed of natural biopolymers such as starch, dextran, and gelatin can be formed using aqueous solvents, and further can incorporate micropores by particle leaching. Griffith at al. explored porous scaffolds of PLGA for liver tissue engineering and demonstrated rat hepatocyte attachment [24]. Others have extended this technique to examine the effects of pore size on the attachment, growth, and matrix deposition of different cell types [25].

Pressure assisted microsyringe (PAM) fabrication is another adhesion-based technique that uses a solvent to bind polymers in a layer by layer format. A stage controlled microsyringe delivery system deposits a stream of polymer dissolved in solvent through a 10–20 μm glass capillary needle [7]. The polymer stream thickness can be modified by varying the solution viscosity, syringe-tip diameter, syringe pressure, and stage motor speed, to generate structures that range in dimension from 5 μm to 600 μm. This method is similar to FDM scaffold fabrication, but is capable of high resolution features and does not require heat. However, the limited size of the syringe-needle system prohibits the use of particulate leaching to increase microporosity and scaffold surface area.

2.2.4. Indirect Fabrication by Molding

In addition to the methods described above that directly fabricate 3-D scaffolds, scaffolds can also be cast from microstructured molds formed using the same methods. This indirect fabrication strategy is advantageous for sensitive biomaterials that are incompatible with fabrication conditions, since only the mold itself is subjected to the processing environment. Further, the resulting scaffold represents an inverse of the mold, thereby extending the 3-D design possibilities. For example, Orton et al. casted a hydroxyapatite/acrylate suspension onto a negative epoxy mandible mold made by stereolithography (Fig. 2.6a) [26]. After heat-curing the polymer, the mold and acrylate binder were incinerated. The resulting hydroxyapatite scaffolds contained different internal channel architectures and resulted in bone ingrowth in minipigs up to nine weeks post-implantation [27]. Others have created molds for indirect scaffolds using 3-DP by depositing wax or other low melting point compounds that can be later removed with elevated temperature or solvents. This method has been combined with particulate leaching to indirectly fabricate porous scaffolds composed of hypoxyapatite, poly(L)lactide, and polyglycolide [28]. Sachlos et al. similarly used ink

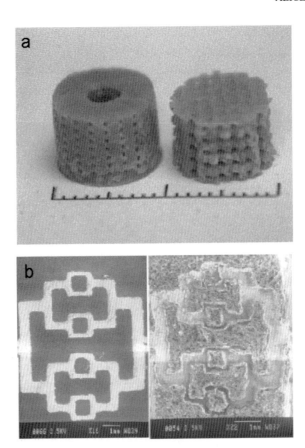

FIGURE 2.6. Molded Scaffolds. (a) Hydroxyapatite was cast into a negative epoxy mold (manufactured using stereolithography) and then cured by heat. The scaffold was then placed in a furnace to burn out the mold. (b) The extracellular matrix compound collagen was cast onto a negative mold that was printed using ink jet technology. The mold was then dissolved away with ethanol, leaving a patterned collagen scaffold (from [29], reprinted with permission of Elsevier).

jet printing to manufacture molds for casting collagen microstructures with 200 um feature size (Fig. 2.6b) [29]. The molds were then dissolved with ethanol to form scaffolds with predefined internal morphology. Furthermore, these scaffolds would present a component of extracellular matrix specifically recognized by cells to enhance attachment. However, implantation of such a construct may also be problematic in that other host cells may also react and respond in a nonspecific manner.

2.3. FABRICATION OF CELLULAR CONSTRUCTS

Although acellular assembly techniques have proven useful for defining the macroscopic and microscopic features of a 3-D scaffold, these methods are generally limited by inefficient and inhomogenous incorporation of cells. The direct assembly of cultured layers of living cells is an alternative being pursued by several groups. This strategy was demonstrated for myocardial tissue by culturing cardiomyocytes on dishes selectively grafted with poly(N-iso-propylacrylamide) (PIPAAm) to form a substrate with temperature-sensitive adhesive properties. At a reduced temperature, the grafted polymer hydrated and promoted the release of a cellular sheet that retained cell-cell junctions and extracellular matrix and could be layered with additional sheets [30]. In a similar manner, Okano and colleagues created corneal epithelial sheets from corneal stem cells without the use of polymer scaffolds, and transplanted sheets remained stable for up to six months in rabbit models [31]. Auger and colleagues have also adapted this technique to vascular tissue engineering. To mimic the structure of blood vessels, they wrapped smooth muscle cell sheets around a tubular mandrel and subsequently seeded endothelial cells within the lumen of the cylinder [32]. After culture with pulsatile flow, the multilayer engineered blood vessels demonstrated excellent mechanical properties and exhibited cellular markers that resembled native vessels.

These cell layering techniques have limited capability for the formation of complex 3-D patterned structures. Instead of using cell monolayers that fuse into a sheet, some groups have achieved greater complexity of tissue construction by the selective delivery and spontaneous fusion of living cells into 3-D structures. For example, Mironov et al. used a jet-based printer to position cell aggregates and embryonic heart mesenchymal fragments that fused together within biocompatible gels of varying chemical and mechanical properties [33]. In the future, this 'organ-printing' technology may allow for precise 3-D cell positioning that could be scaled up to larger tissue engineered constructs [34, 35]. Odde et al. have explored the use of laser-based optical forces to precisely deliver a stream of living cells and 'write' them into arbitrary positions on a substrate [36]. While single cells can be positioned in this manner, the serial deposition process may pose problems for scaling up to tissue dimensions. Further characterization of these constructs will be required to determine which tissues are amenable to fabrication by these emerging cellular assembly techniques.

2.4. FABRICATION OF HYBRID CELL/SCAFFOLD CONSTRUCTS

One disadvantage of direct cellular assembly is that constructs may not possess adequate mechanical stability for tissue engineering applications. Conversely, acellular scaffolds have excellent mechanical strength but may be difficult to populate with cells. Hydrogel polymers have the ability to provide both structural support and high tissue density while maintaining an *in vivo*-like environment for cells [37]. Many of these water-swollen polymers can be formed in mild conditions compatible with living cells, allowing the formation of hybrid cell/scaffold constructs. A significant advantage of this strategy is that the construct can then encapsulate a homogeneous cell population. The construct shape is determined either by the mold or container used during crosslinking, or by spatial photopatterning using selective light exposure.

2.4.1. Cell-laden Hydrogel Scaffolds by Molding

Hybrid cell/hydrogel constructs conform to the dimensions of the mold or container used. Initial studies generated constructs containing a random distribution of cells in the

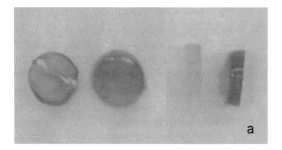
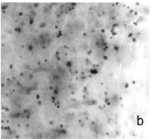

FIGURE 2.7. PEG-based hydrogels containing cells. (a) PEG-based hydrogels are crosslinked to form the shape of the container (dye added for clarity). (b) Living cells are suspended within the crosslinked hydrogel (MTT stain for viability). (Photos courtesy Jennifer Elisseef, Johns Hopkins University)

shape of thin sheets or disks by polymerization within tissue culture wells or tubes (Fig. 2.7). In this manner, Hubbell and coworkers explored the encapsulation of cells in various biological hydrogels, such as collagen and fibrin, that were functionalized with genetically engineered bioactive sites to enhance cell adhesion and proteolytic remodeling [38]. Desai et al. extended these studies by using microfluidic molding technologies to deposit micropatterned collagen gel structures containing living cells [39]. While complex biological microstructures containing a few layers were fabricated in this manner, microfluidic molding is typically constrained to flat surfaces and may be difficult to generalize to 3-D tissue architectures.

Synthetic polymer hydrogels offer several advantages over biological hydrogels for producing scaffolds containing homogeneously dispersed cells. In particular, poly(ethylene glycol) (PEG)-based hydrogels are increasingly used for cell encapsulation because of their biocompatibility, hydrophilicity, and customizable mechanical and transport properties by changing the monomer chain length and polymer fraction. Photopolymerizable PEG hydrogels have been shown in numerous studies to support the viability and function of various immobilized cell types, including osteoblasts [40], chondrocytes [41, 42], vascular smooth muscle cells [43], and fibroblasts [44]. Additionally, these hydrogels can be chemically functionalized by incorporating biologically relevant molecules [45]. For example, extracellular matrix protein domains were shown to enhance cell attachment and incorporation [43, 46–49], tethered growth factors modulated cell functions [46], and degradable/cleavable linkages within the polymer backbone allowed cell proliferation and migration [43, 44, 46, 47, 50–53]. Thus, photopolymerized synthetic hydrogels are particularly appealing for tissue engineering because of their controllable biochemical and mechanical properties, gentle crosslinking processes that are compatible with living cells, and hydrated, tissue-like 3-D environment.

2.4.2. Cell-laden Hydrogel Scaffolds by Photopatterning

Although molding techniques can form hybrid cell/hydrogel constructs with micropatterned external features, they are not amenable to patterning internal structure of complex engineered tissues. However, recent developments have exploited the ability to localize light exposure, and therefore hydrogel crosslinking, in defined micropatterns, potentially allowing the buildup of complex microstructures in a layer by layer manner. Outside the tissue engineering arena, hydrogel microstructures have been formed by photolithographic patterning with applications such as microfluidic valves [54] and cell-laden microstructures on silicon [55].

Liu and Bhatia recently adapted photolithographic techniques to existing PEG-based cell-encapsulation chemistry and created organized, three-dimensional cell/hydrogel networks (Fig. 2.8) [56]. In this method, living cells were suspended in a polymer solution and exposed to UV light through a photolithographic mask to form multiple cellular domains

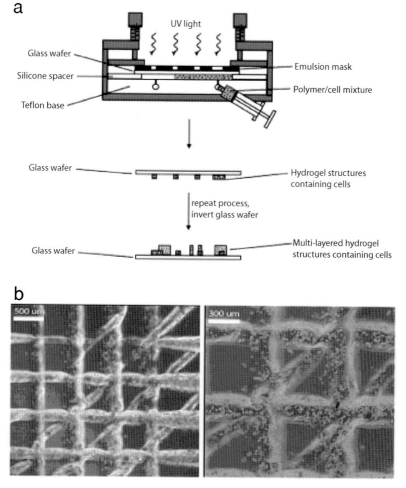

FIGURE 2.8. 3-D Photopatterning of Hydrogels. (a) Photopatterning Method. Polymer solution and cells are introduced into a chamber. The unit is exposed to 365 nm light through an emulsion mask, causing crosslinking of the polymer in the exposed areas and trapping the cells within these regions. The uncrosslinked polymer solution and cells are then washed away, and the process is repeated with thicker spacers and a new mask to create 3-D cellular hydrogel structures. Each layer may contain the same type of polymer/cell mixture, or can be composed of different polymer properties or different cell types. (b) Three layered hydrogel structure containing cells (from [56]).

with controlled hydrogel architecture. Uncrosslinked polymer and cells were then rinsed away and additional domains could be photopatterned with different cell types, polymer formulations, and exposure patterns. During each exposure cycle, the newly crosslinked polymer fused with existing hydrogel domains. Furthermore, 3-D multilayer constructs with complex internal structure were formed by increasing the height of the photocrosslinking chamber between exposure steps. In this manner, a three layered hydrogel construct was fabricated with raised protrusions containing a high cell density (Fig. 2.8b). To date, microstructure feature size has approached 50 μm, thus enabling the patterning of cells on the scale of functional tissue units. While these structures contain randomly dispersed cells, Albrecht et al. have developed a complimentary technology capable of defining the organization of encapsulated cells within a hydrogel to a resolution of <10 μm [57, 58]. This method utilizes electromagnetic fields to specify the position of cells in the liquid polymer solution prior to photocrosslinking. In conjunction with bioactive hydrogel technologies being explored by numerous groups, the photopatterning of hydrogels containing homogeneous or organized patterns of living cells may lead to the development of improved tissue engineered constructs with customized spatial, physical, and chemical properties. The flexibility of these hydrogel systems shows great promise for the fabrication of 3-D tissues that mimic the structural, multicellular, and biochemical complexity found in many organs in the body.

2.5. FUTURE DIRECTIONS

Novel scaffold fabrication methods, often based on technologies borrowed from the manufacturing industry, have led to rapid progress in the development of complex 3-D tissue engineered constructs in recent years. The various acellular, cellular, or hybrid cell/scaffold fabrication strategies described in this review are summarized in Table 2.1 with regard to their spatial resolution, advantages, and limitations. The utility of each fabrication method will ultimately depend on design criteria specific to each tissue engineering application, including chemical composition, mechanical strength, degradation profile, nutrient transport, and cellular organization.

The field of tissue engineering has advanced significantly from the initial examples of seeding living cells into synthetic polymer scaffolds to the development of tissue constructs with physical and biochemical complexity. As researchers develop a greater understanding of the biology underlying fundamental structure-function relationships, factors that influence cell fate (e.g., growth, morphogenesis, apoptosis) and function (gene expression, biosynthesis) can be incorporated into the design of tissue engineering strategies. These factors include signals from the cellular microenvironment that are sensed in a three-dimensional context, such as cell-cell and cell-matrix interactions, soluble signals, and mechanical forces. The ability to control the presentation of microenviromental cues at the microscale (cell and functional subunit, 10–100 μm) as well as bulk properties at the macroscale (tissue implant, 1–100 mm) will be enabled by leveraging these emerging 3-D fabrication technologies. While the goal of engineering complex tissues remains a difficult challenge, continued interaction between interdisciplinary fields of cell and molecular biology, chemistry, biomaterials, and medicine, will ensure continued progress toward substantial improvements in human health.

TABLE 2.1. Comparison of 3-D scaffolding methods.

	Resolution (μm)	Advantages	Disadvantages
ACELLULAR 3-D SCAFFOLDS		use of well-established fabrication methods, usually automated	must seed cells post-processing, less control in cell placement and distribution
Heat-Mediated Fabrication			
Micro Molding [5, 7]	20–30	simple; reusable molds	limited to thin membranes, each layer must be contiguous structure, manual alignment required
Selective Laser Sintering [4, 8, 9]	400	high porosity, automated	high temperatures during process, powder may be trapped
Fused Deposition Modeling [8, 10, 11]	250–700	no trapped particles or solvents, automated	high temperatures during processing
3-D Plotting [16]	1000	use of hydrogel materials (agar, gelatin), automated	limited resolution
Light-Mediated Fabrication			
Stereolithography [4, 18]	70–250	ease of use, easy to achieve small features, automated	limited choice of materials- must be photosensitive and biocompatible; exposure of material to laser
Adhesive-Mediated Fabrication			
3-D Printing [23–25]	200–500	versatile; high porosity, automated	limited choice of materials (e.g. organic solvents as binders); difficult to reduce resolution below polymer particle size
Pressure Assisted Microsyringe [7]	10	high resolution, not subject to heat, automated	viscosity dependent, no inclusion of particles
Indirect Fabrication by Molding			
Matrix Molding [29]	200	use of biological matrix materials (collagen), mold fabrication can use automated methods (above)	features must be interconnected, weaker mechanical properties
CELLULAR 3-D SCAFFOLDS		Precise placement of cells throughout construct, ability to place multiple cell types arbitrarily	limited fabrication conditions (sterility, temperature, pH), still in earlier phases of development
Cellular Assembly			
Organ Printing [33, 35]	100	incorporation of cell aggregates or tissue explants, precise cell placement, automated	lack of structural support, dependence on self assembly
Laser-Guided Deposition [36]	<1	precise single cell placement, automated	has yet to be extended to 3-D structures, lack of structural support

(cont.)

TABLE 2.1. *Continued*

	Resolution (μm)	Advantages	Disadvantages
Hybrid Cell/Scaffold Assembly			
Hydrogel Photopatterning [56]	100	incorporation of living cells within scaffold, leverages existing hydrogel chemistry (incorporation of peptides, degradation domains), versatile	not yet automated, exposure of cells to ultraviolet light, diffusion of large molecules limited by hydrogel pore size

ACKNOWLEDGEMENTS

We would like to thank the Whitaker Foundation (V.L.T., D.A.), American Association of University Women (V.L.T.), NIH NIDDK, NSF CAREER, David and Lucile Packard Foundation, and NASA for their generous support.

REFERENCES

[1] S.N. Bhatia and C.S. Chen. Tissue Engineering at the Micro-Scale. *Biomedical. Microdevices.*, 2:131–144, 1999.

[2] R. Langer and J.P. Vacanti. Tissue engineering. *Science*, 260(5110):920–926, 1993.

[3] S. Yang et al. The design of scaffolds for use in tissue engineering. Part I. Traditional factors. *Tissue Engineering*, 7(6):679–689, 2001.

[4] S. Yang et al. The design of scaffolds for use in tissue engineering. Part II. Rapid prototyping techniques. *Tissue Engineering*, 8:1–11, 2002.

[5] J.T. Borenstein et al. Microfabrication technology for vascularized tissue engineering. *Biomedical Microdevices*, 4:167–175, 2002.

[6] C.S. Chen et al. Geometric control of cell life and death. *Science*, 276(5317):1425–1428, 1997.

[7] G. Vozzi et al. Fabrication of PLGA scaffolds using soft lithography and microsyringe deposition. *Biomaterials*, 24:2533–2540, 2003.

[8] K.F. Leong, C.M. Cheah, and C.K. Chua. Solid freeform fabrication of three-dimensional scaffolds for engineering replacement tissues and organs. *Biomaterials*, 24(13):2363–2678, 2003.

[9] G. Lee et al. *Biocompatibility of SLS-formed Calcium Phosphate Implants*. Proceedings of Solid Freeform Fabrication Symposium, pp. 15–22, 1996.

[10] I. Zein et al. Fused deposition modeling of novel scaffold architectures for tissue engineering applications. *Biomaterials*, 23:1169–1185, 2002.

[11] D.W. Hutmacher. Scaffold design and fabrication technologies for engineering tissues–state of the art and future perspectives. *J. Biomater. Sci. Polym. Ed.*, 12(1):107–124, 2001.

[12] M. Endres et al. Osteogenic induction of human bone marrow-derived mesenchymal progenitor cells in novel synthetic polymer-hydrogel matrices. *Tissue Eng.*, 9(4):689–702, 2003.

[13] B. Rai et al. The effect of rhBMP-2 on canine osteoblasts seeded onto 3D bioactive polycaprolactone scaffolds. *Biomaterials*, 25(24):5499–5506, 2004.

[14] T.B. Woodfield et al. Design of porous scaffolds for cartilage tissue engineering using a three-dimensional fiber-deposition technique. *Biomaterials*, 25(18):4149–4161, 2004.

[15] D.W. Hutmacher, M. Sittinger, and M.V. Risbud. Scaffold-based tissue engineering: rationale for computer-aided design and solid free-form fabrication systems. *Trends Biotechnol*, 22(7):354–362, 2004.

[16] R. Landers et al. Rapid prototyping of scaffolds derived from thermoreversible hydrogels and tailored for applications in tissue engineering. *Biomaterials*, 23(23):4437–4447, 2002.
[17] T.H. Ang et al. Fabrication of 3D chitosan-hydroxyapatite scaffolds using a robotic dispensing system. *Mat. Sci. Eng.*, C20:35–42, 2002.
[18] M.N. Cooke et al. Use of stereolithography to manufacture critical-sized 3D biodegradable scaffolds for bone ingrowth. *J. Biomed. Mater. Res.*, 64B(2):65–69, 2003.
[19] C.M. Langton et al. Development of a cancellous bone structural model by stereolithography for ultrasound characterisation of the calcaneus. *Med. Eng. Phys.*, 19(7):599–604, 1997.
[20] G. Chu et al. *Solid Freeform and Additive Fabrication*. Materials Research Society Symposium Proceedings, Warrendale, PA., pp. 119–123, 1999.
[21] T. Yu and C.K. Ober. Methods for the topographical patterning and patterned surface modification of hydrogels based on hydroxyethyl methacrylate. *Biomacromolecules*, 4(5):1126–1131, 2003.
[22] T. Matsuda and T. Magoshi. Preparation of vinylated polysaccharides and photofabrication of tubular scaffolds as potential use in tissue engineering. *Biomacromolecules*, 3(5):942–950, 2002.
[23] A. Park, B. Wu, and L.G. Griffith. Integration of surface modification and 3D fabrication techniques to prepare patterned poly(L-lactide) substrates allowing regionally selective cell adhesion. *J. Biomater. Sci. Polym. Ed.*, 9(2):89–110, 1998.
[24] S.S. Kim et al. Survival and function of hepatocytes on a novel three-dimensional synthetic biodegradable polymer scaffold with an intrinsic network of channels. *Ann. Surg.*, 228(1):8–13, 1998.
[25] J. Zeltinger et al. Effect of pore size and void fraction on cellular adhesion, proliferation, and matrix deposition. *Tissue Eng.*, 7(5):557–572, 2001.
[26] T.M. Chu et al. Manufacturing and characterization of 3-d hydroxyapatite bone tissue engineering scaffolds. *Ann. NY Acad. Sci.*, 961:114–117, 2002.
[27] T.M. Chu et al. Mechanical and in vivo performance of hydroxyapatite implants with controlled architectures. *Biomaterials*, 23(5):1283–1293, 2002.
[28] J.M. Taboas et al. Indirect solid free form fabrication of local and global porous, biomimetic and composite 3D polymer-ceramic scaffolds. *Biomaterials*, 24(1):181–194, 2003.
[29] E. Sachlos et al. Novel collagen scaffolds with predefined internal morphology made by solid freeform fabrication. *Biomaterials*, 24(8):1487–1497, 2003.
[30] T. Shimizu et al. Cell sheet engineering for myocardial tissue reconstruction. *Biomaterials*, 24(13):2309–2316, 2003.
[31] K. Nishida et al. Functional bioengineered corneal epithelial sheet grafts from corneal stem cells expanded ex vivo on a temperature-responsive cell culture surface. *Transplantation*, 77(3):379–385, 2004.
[32] N. L'Heureux et al. A completely biological tissue-engineered human blood vessel. *Faseb. J.*, 12(1):47–56, 1998.
[33] V. Mironov et al. Organ printing: computer-aided jet-based 3D tissue engineering. *Trends Biotechnol.*, 21(4):157–161, 2003.
[34] K. Jakab et al. Engineering biological structures of prescribed shape using self-assembling multicellular systems. *Proc. Natl. Acad. Sci. U.S.A.*, 101(9):2864–2869, 2004.
[35] T. Xu et al. Inkjet printing of viable mammalian cells. *Biomaterials*, 26(1):93–99, 2005.
[36] D.J. Odde and M.J. Renn. Laser-guided direct writing of living cells. *Biotechnol. Bioeng.*, 67(3):312–318, 2000.
[37] K.T. Nguyen and J.L. West. Photopolymerizable hydrogels for tissue engineering applications. *Biomaterials*, 23(22):4307–4314, 2002.
[38] S.E. Sakiyama, J.C. Schense, and J.A. Hubbell. Incorporation of heparin-binding peptides into fibrin gels enhances neurite extension: an example of designer matrices in tissue engineering. *Faseb. J.*, 13:2214–2224, 1999.
[39] W. Tan and T.A. Desai. Microfluidic patterning of cells in extracellular matrix biopolymers: effects of channel size, cell type, and matrix composition on pattern integrity. *Tissue Eng.*, 9(2):255–267, 2003.
[40] E. Behravesh, K. Zygourakis, and A.G. Mikos. Adhesion and migration of marrow-derived osteoblasts on injectable in situ crosslinkable poly(propylene fumarate-co-ethylene glycol)-based hydrogels with a covalently linked RGDS peptide. *J. Biomed. Mater. Res.*, 65A(2):260–270, 2003.
[41] J. Elisseeff et al. Photoencapsulation of chondrocytes in poly(ethylene oxide)-based semi-interpenetrating networks. *J. Biomed. Mater. Res.*, 51(2):164–171, 2000.

[42] S.J. Bryant and K.S. Anseth. Hydrogel properties influence ECM production by chondrocytes photoencapsulated in poly(ethylene glycol) hydrogels. *J. Biomed. Mater. Res.*, 59(1):63–72, 2002.

[43] B.K. Mann et al. Smooth muscle cell growth in photopolymerized hydrogels with cell adhesive and proteolytically degradable domains: synthetic ECM analogs for tissue engineering. *Biomaterials*, 22(22):3045–3051, 2001.

[44] A.S. Gobin and J.L. West Cell migration through defined, synthetic ECM analogs. *Faseb. J.*, 16(7):751–753, 2002.

[45] N.A. Peppas et al. Hydrogels in pharmaceutical formulations. *Eur. J. Pharm. Biopharm.*, 50(1):27–46, 2000.

[46] E. Alsberg et al. Engineering growing tissues. *Proc. Natl. Acad. Sci. U.S.A.*, 99(19):12025–12030, 2002.

[47] R.H. Schmedlen, K.S. Masters, and J.L. West. Photocrosslinkable polyvinyl alcohol hydrogels that can be modified with cell adhesion peptides for use in tissue engineering. *Biomaterials*, 23(22):4325–4332, 2002.

[48] W.J. Kao and J.A. Hubbell. Murine macrophage behavior on peptide-grafted polyethyleneglycol-containing networks. *Biotechnol. Bioeng.*, 59(1):2–9, 1998.

[49] L.Y. Koo et al. Co-regulation of cell adhesion by nanoscale RGD organization and mechanical stimulus. *J. Cell. Sci.*, 115(Pt 7):1423–1433, 2002.

[50] S. Halstenberg et al. Biologically engineered protein-graft-poly(ethylene glycol) hydrogels: a cell adhesive and plasmin-degradable biosynthetic material for tissue repair. *Biomacromolecules*, 3(4):710–723, 2002.

[51] M.P. Lutolf et al. Synthetic matrix metalloproteinase-sensitive hydrogels for the conduction of tissue regeneration: engineering cell-invasion characteristics. *Proc. Natl. Acad. Sci. U.S.A.*, 100(9):5413–5418, 2003.

[52] A.S. Sawhney et al. Optimization of photopolymerized bioerodible hydrogel properties for adhesion prevention. *J. Biomed. Mater. Res.*, 28(7):831–838, 1994.

[53] C.R. Nuttelman, S.M. Henry, and K.S. Anseth. Synthesis and characterization of photocrosslinkable, degradable poly(vinyl alcohol)-based tissue engineering scaffolds. *Biomaterials*, 23(17):3617–3626, 2002.

[54] D.J. Beebe et al. Functional hydrogel structures for autonomous flow control inside microfluidic channels. *Nature*, 404(6778):588–590, 2000.

[55] W.G. Koh, A. Revzin, and M.V. Pishko. Poly(ethylene glycol) hydrogel microstructures encapsulating living cells. *Langmuir*, 18(7):2459–2462, 2002.

[56] V.A. Liu and S.N. Bhatia. Three-dimensional photopatterning of hydrogels containing living cells. *Biomedical Microdevices*, 4:257–266, 2002.

[57] D.R. Albrecht, R.L. Sah, and S.N. Bhatia Dielectrophoretic Cell Patterning within Tissue Engineering Scaffolds. *Second Joint EMBS-BMES Conference. Annual International Conference of the Engineering in Medicine and Biology Society. Annual Fall Meeting of the Biomedical Engineering Society. IEEE*, 2:1708–1709, 2002.

[58] D.R. Albrecht et al. Photo- and Electropatterning of Hydrogel-Encapsulated Living Cell Arrays. *Lab. Chip.*, In Press.

3

Designed Self-assembling Peptide Nanobiomaterials

Shuguang Zhang[1,2] and Xiaojun Zhao[2]

[2] *Center for Biomedical Engineering, NE47-379,*
[1] *Center for Bits & Atoms, Massachusetts Institute of Technology, 77 Massachusetts Avenue*
Cambridge, MA 02139-4307

There are two complementary technologies that can be used in the production of biomaterials. In the 'top-down' approach, biomaterials are produced by polymerizing homogeneous monomers into covalently linked microfibers, sheet, coatings and other structures. This is in sharp contrast from tailor-made approach, where biomaterials are assembled heterogeneous population of molecules to produce supra-structures and architectures. The latter approach is likely to become an integral part of biomaterials manufacture in the coming years. This approach requires a deep understanding of individual molecular building blocks, their structures, assembling properties and dynamic behaviors. Two key elements in molecular nanobiomaterial production are chemical complementarity and structural compatibility, both of which require the weak and noncovalent interactions that bring building blocks together during self-assembly. Significant advances have been made at the interface of materials chemistry and biology, including the design of helical ribbons, peptide nanofiber scaffolds for three-dimensional cell cultures and tissue engineering, peptide surfactants, peptide detergents for solubilizing, stabilizing and crystallizing diverse types membrane proteins and their complexes and molecular ink peptides for arbitrary printing and coating surfaces. These designed self-assembling peptides have far reaching implications in broad spectrum of applications and some of which are beyond our current imaginations.

3.1. INTRODUCTION

Manufacture of new molecular scale biomaterials has become increasingly important for next generation of nanobiomaterials and nanobiotechnology, namely, the design, synthesis and fabrication of nano-devices at the molecular scale from bottom up. Nature has already produced numerous and diverse building units through billions of years molecular selection and evolution [1, 8]. Basic engineering principles for micro- and nano-fabrication can now be learned through understanding of the molecular self-assembly and programmed assembly phenomena. Self- and programmed assembly phenomena are ubiquitous in nature. The key elements in molecular self-assembly are chemical complementarity and structural compatibility through noncovalent weak interactions.

Several self-assembling peptide systems have been developed ranging from models for studying protein folding and protein conformational diseases, to molecular materials to produce peptide nanofibers, peptide scaffolds, peptide surfactants and peptide ink [20, 21]. These self-assembling peptide systems are simple, versatile, affordable and easy to produce in large scale to spur a new industry. These self-assembly peptide systems represent a significant advance in molecular engineering for diverse nanobiomaterial innovations. We here only focus research activities from our laboratory from the last decade. Those who are interested in a broad range of biomaterials are referred to elsewhere in this encyclopedia.

Molecular self-assembly is ubiquitous in nature and has recently emerged as a new approach in chemical synthesis, nanotechnology, polymer science, materials and engineering. Molecular self-assembly systems lie at the interface between molecular biology, protein science, biochemistry, polymer science, materials science and engineering [13, 14]. Many self-assembling systems have been developed. Molecular self-assembly systems represent a significant advance in the molecular engineering of simple molecular building blocks useful for a wide range of applications. This field is growing at an accelerating pace riding on the tide of nanobiotechnology.

3.2. PEPTIDE AS BIOLOGICAL MATERIAL CONSTRUCTION UNITS

Similar to the construction of a house, doors, windows, and many other parts of house can be prefabricated and programmed assembled according to architectural plans. If we shrink the construction units many orders of magnitude into nanoscale, we can apply similar principles to construct molecular materials and devices, through molecular self-assembly and programmed molecular assembly. Given the growing interest but limited space, only three self-assembling construction units are summarized here. They include: "Lego peptide" that form well-ordered nanofiber scaffolds for 3-D cell culture and for regenerative medicine; "surfactant/detergent peptides" for drug, protein and gene deliveries as well as for solubilizing and stabilizing membrane proteins; "peptide ink" for surface biological engineering. These designed construction peptide units are structurally simple and versatile for a wide spectrum of applications as nanobiomaterials and beyond.

Three distinct classes of self-assembling peptide construction units are described. The first class belongs to amphiphilic peptides that form well-ordered nanofibers. These peptides have two distinctive sides, one hydrophobic and the other hydrophilic. The hydrophobic

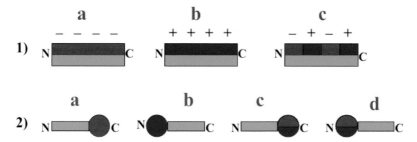

FIGURE 3.1. Two distinct classes of self-assembling peptides construction units. The first class "Lego peptides" belongs to amphiphilic peptides that form well-ordered nanofibers. These peptides have two sides, one hydrophobic (green) and another hydrophilic (red and blue). The hydrophobic side forms a double sheet inside of the fiber and hydrophilic side forms the outside of the nanofibers that interact with water molecules that they can form extremely high water content hydrogel, containing as high as 99.9% water. At least three types of molecules can be made, with $-$, $+$, $-/+$ on the hydrophilic side. The second class of the self-assembling peptide belongs to a surfactant-like molecule. These peptides have a hydrophilic head and a hydrophobic tail, much like lipids or detergents. They sequester their hydrophobic tail inside of micelle, vesicles or nanotube structures and their hydrophilic heads expose to water. At least three kinds molecules can be made, with $-$, $+$, $-/+$ heads.

side forms a double sheet inside of the fiber and hydrophilic side forms the outside of the nanofibers that interact with water molecules that they can form extremely high water content hydrogel, containing as high as 99.9% water. At least three types of molecules can be made, with $-$, $+$, $-/+$ on the hydrophilic side.

The second class of the self-assembling peptide belongs to a surfactant-like molecule. These peptides have a hydrophilic head and a hydrophobic tail, much like lipids or detergents. They sequester their hydrophobic tail inside of micelle, vesicles or nanotube structures and their hydrophilic heads expose to water. At least three kinds molecules can be made, with $-$, $+$, $-/+$ heads. The third class of the self-assembling peptide belongs to molecular ink, which has 3 distinct segments, the ligand, the linker and the anchor. These molecular ink peptides have been used as ink to directly print arbitrary patterns on surfaces.

3.2.1. Lego Peptide

Molecular-designed "Lego Peptide", at the nanometer scale, resembles the Lego bricks that have both pegs and holes in a precisely determined manner and can be programmed to assemble in well-formed structures. This class of "Lego peptide" can spontaneously assemble into well-formed nanofibers [16]. The first member of the Lego peptide was serendipitously discovered from a segment in a left-handed Z-DNA binding protein in yeast, Zuotin (Zuo means Left in Chinese, tin means protein in biology) [15].

These peptides form beta-sheet structures in aqueous solution thus they form two distinct surfaces, one hydrophilic, the other hydrophobic, like the pegs and holes in Lego bricks. The hydrophobic sides shield themselves from water thus facilitate them to self-assemble in water, similar to that as seen in the case of protein folding. The unique structural feature of these "peptide Lego" is that they form complementary ionic bonds with regular repeats on the hydrophilic surface (Fig. 3.2). The complementary ionic sides have been classified into several moduli, i.e. modulus I, II, III, IV, etc., and mixed moduli. This

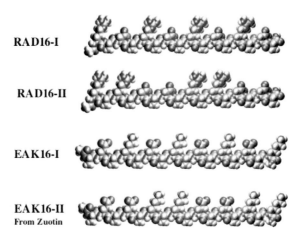

RAD16-I

RAD16-II

EAK16-I

EAK16-II
From Zuotin

FIGURE 3.2. Molecular models of several self-assembling peptides, RAD16-I, RAD16-II, EAK16-I and EAK16-II. Each molecule is ~5nm in length with 8 alanines on one side and 4 negative and 4 positive charge amino acids in an alternating arrangement on the other side.

classification is based on the hydrophilic surface of the molecules that have alternating + and − charged amino acid residues, either alternating by 1, 2, 3, 4 and so on. For example, charge arrangements are for modulus I, − + − + − + −+; modulus II, − − + + − − ++; modulus III, − − − + ++; and modulus IV, − − − − + + ++. The charge orientation can also be designed in the reverse orientation that can yield entirely different molecules (Fig. 3.2). These well-defined Lego peptides undergo ordered self-assembly, resembling some polymer assemblies.

The Lego peptide molecules can undergo self-assembly in aqueous solutions to form well-ordered nanofibers that further associate to form nanofiber scaffolds. One of them, RADA16-I, is called PuraMatrix, because of its purity as a designed biological scaffold in contrast to other biologically derived scaffolds from animal collagen and Matrigel which contain unspecified components in addition to known materials.

Since these nanofiber scaffolds contain 5–200 nm pores and have an extremely high water content (>99.5% or 1–5 mg/ml), they have been used as three-dimensional (3-D) cell-culture media. The scaffolds closely mimic the porosity and gross structure of extracellular matrices, allowing cells to reside and migrate in a 3-D environment and molecules, such as growth factors and nutrients, to diffuse in and out very slowly. These peptide scaffolds have been used for 3-D cell culture, controlled cell differentiation, tissue engineering and regenerative medicine applications.

3.2.2. Surfactant/detergent Peptides

We designed a second class of surfactant/detergent peptides with hydrophobic tails and hydrophilic heads, taking advantage of their self-assembly properties in water [10–12]. Several surfactant peptides have been designed using nature's lipid as a guide. These peptides have a hydrophobic tail with various degrees of hydrophobicity and a hydrophilic

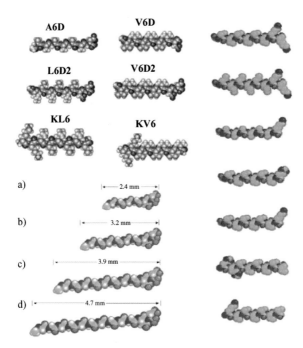

FIGURE 3.3. Molecular models of peptide surfactants/detergents. Left upper panel. A_6D, V_6D, V_6D_2 and L_6D_2. KKL_6, KKV_6. D (Aspartic acid) bears negative charges and A (alanine), V (valine) and L (leucine) constitute the hydrophobic tails with increasing hydrophobicity. Each peptide is about 2–3 nm in length, similar to biological phospholipids. K (lysine). Left lower panel. Molecular structures of individual glycine tail-based surfactant peptides. G_4D_2, G_6D_2, G_8D_2 and $G_{10}D_2$. The tail length of glycines varies depending on the number of glycine residues. The lengths of these molecules in the extended conformation range from 2.4 nm of G_4D_2 to 4.7 nm of $G_{10}D_2$. Color code: carbon, green; hydrogen, white; red, oxygen; and blue, nitrogen.

head, either negatively charged aspartic and glutamic acids or positively charged lysine or histidine (Fig. 3.3). These peptide monomers contain 7 to 8 amino acid residues and have a hydrophilic head composed of aspartic acid and a tail of hydrophobic amino acids such as alanine, valine or leucine. The length of each peptide is approximately 2 nm, similar to that of biological phospholipids [10–12]. The length can also be varied by adding more amino acids, one at a time to a desired length as shown in figure 3.3.

Although individually these peptide surfactants/detergents have completely different composition and sequences, these peptides share a common feature: the hydrophilic heads have 1–2 charged amino acids and the hydrophobic tails have four or more consecutive hydrophobic amino acids. For example, A_6D (AAAAAAD), V_6D (VVVVVVD) peptide has six hydrophobic alanine or valine residues from the N-terminus followed by a negatively charged aspartic acid residue, thus having two negative charges, one from the side chain and the other from the C terminus; likewise, G_8DD (GGGGGGGGDD), has eight glycines following by two asparatic acids with three negative charges. In contrast, K_2V_6 (KKVVVVVV) has two positively charged lysines as the hydrophilic head, following by six valines as the hydrophobic tail [10–12].

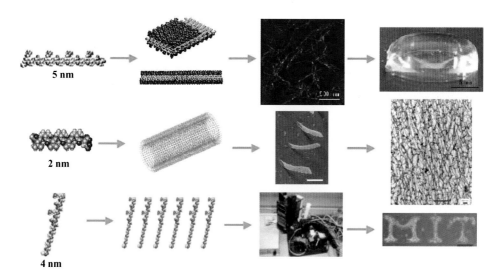

FIGURE 3.4. Design of various peptide materials. **(a) Lego Peptide**, also called ionic self-complementary peptide has 16 amino acids, ~ 5 nm in size, with an alternating polar and nonpolar pattern. They form stable □-strand and □-sheet structures, thus the side chains partition into two sides, one polar and the other nonpolar. They undergo self-assembly to form nanofibers with the nonpolar residues inside (green) and + (blue) and − (red) charged residues form complementary ionic interactions, like a checkerboard. These nanofibers form interwoven matrices that further form a scaffold hydrogel with very high water content, >99.5% water (images courtesy of Hidenori Yokoi). **(b) Surfactant/detergent peptides**, ~2 nm in size, which has a distinct head group, either positively charged or negatively charged, and a hydrophobic tail consisting of six hydrophobic amino acids. They can self-assemble into nanotube and nanovesicles with a diameter of ~30–50 nm (image courtesy Steve Santoso and Sylvain Vauthey). These nanotubes go on to form an inter-connected network, which has similar been observed in other nanotubes. **(c) Ink peptide**, This type of peptide has three distinct segments: a functional segment where it interacts with other proteins and cells; a linker segment that can not only be flexible or stiff, but also sets the distance from the surface, and an anchor for covalent attachment to the surface. These peptides can be used as ink for an inkjet printer to directly print on a surface, instantly creating any arbitrary pattern, as shown here. Neural cells from rat hippocampal tissue form defined patterns. (Images courtesy S. Fuller & N. Sanjana).

These peptides undergo self-assembly in water to form nanotubes and nanovesicles having an average diameter of 30–50 nm [10–12]. The tails consisting of alanines and valines produce more homogeneous and stable structures than those of glycines, isoleucine and leucine. This property may be due to their hydrophobic and hydrophilic ratios. These monomer surfactant peptides were used for molecular modeling. The negatively charged aspartic acid is modeled as red and positively lysine is blue with the green as the hydrophobic tails. They form tubular and vesicle structures.

Quick-freeze/deep-etch sample preparation where the sample was flash-frozen at −190°C instantly produced a 3-D structure with minimal structural disturbance. It revealed a network of open-ended nanotubes with a helical twist using transmission electron microscopy (Fig. 3.5) [10–12]. Likewise, A_6K cationic peptide also exhibited similar nanotube structures with the opening ends clearly visible.

It is interesting that these simple surfactant peptides can produce remarkable complex and dynamic structures. This is another example of building biological materials from the bottom up.

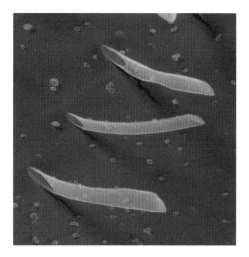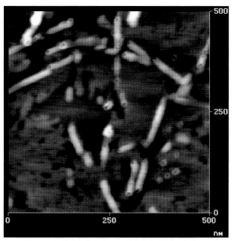

FIGURE 3.5. Quick-freeze/deep-etch TEM image of V_6D dissolved in water (4.3 mM at pH 7) at high-resolution and AFM image of A_6K. A) The images show the dimensions, 30–50 nm in diameters with openings of nanotube ends. Note opening ends of the peptide nanotube may be cut vertically. The strong contrast shadow of the platinum coat also suggests the hollow tubular structure. There are openings at the ends with the other ends possibly buried. The diameter of the V_6D nanobutes is ~30–50 nm. B) The nanotubes of A_6K peptide surfactant/detergent. The openings are clearly visible in the AFM image. It should be pointed out that V_6D and other anionic peptide surfactants cannot be imaged on negatively charged mica surface with AFM because they do not adhere to mica surface well.

How could these simple surfactant peptides form such well-ordered nanostructures? There are molecular and chemical similarities between lipids and the peptides since both have a hydrophilic head and a hydrophobic tail. The packing between lipids and peptides is however likely to be quite different. In lipids, the hydrophobic tails pack tightly against each other to completely displace water, precluding the formation of hydrogen bonds. On the other hand, in addition to hydrophobic tail packing between the amino acid side chains, surfactant peptides also interact through intermolecular hydrogen bonds along the backbone.

These peptide surfactants/detergents have been found to be excellent materials for solubilizing, stabilizing and crystallizing several diverse membrane proteins. These simply designed peptide detergents may now open a new avenue to overcome one of the biggest challenges in biology–to obtain high resolution structures of membrane proteins. Study of membrane proteins will not only enrich and deepen our knowledge of how cell communicate with their surroundings, thus all living system response to their environments, but these membrane proteins can also be used to fabricate the most advanced molecular devices, from energy harness devices, extremely sensitive sensors to medical detection devices, we cannot now even imagine.

Furthermore, several cationic peptide surfactants have been tested for their ability to encapsulate DNA and deliver DNA into cells [12].

3.2.3. Molecular Ink Peptides

Whitesides and collaborators developed a microcontact printing technology that combines semi-conducting industry fabrication, chemistry, and polymer science to produce

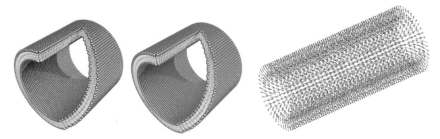

FIGURE 3.6. Molecular models of peptide surfactant/detergent nanobutes. Color code: hydrophobic tail, green; red, negatively charged head (V_6D); and blue, positively charged head (V_6K, or KV_6).

defined features on a surface down to the micrometer or nanometer scale [2, 7, 13]. Following microcontact printing, a surface can be functionalized with different molecules using a variety of methods, such as covalent coupling, surface adhesion, coordination chemistry. Surfaces have now been modified with a variety of chemical compounds. Furthermore peptides and proteins as inks have also been printed onto surfaces. This development has spurred new research into the control of molecular and cellular patterning, cell morphology, and cellular interactions and fueled new technology development.

Molecular surface assembly can be targeted to alter the chemical and physical properties of a material's surface. Surface coatings instantly alter a material's texture, color, compatibility with, and responsiveness to the environment.

Molecular ink peptides have been directly printed on surfaces and to allow adhesion molecules to interact with cells and adhere to the surface (Fig. 3.7). These peptides have three general regions along their lengths: a ligand for specific cell recognition and attachment, a linker for physical separation from the surface, and an anchor for covalent attachment to the surface [9, 18]. The ligands may be of the RGD (arginine-glycine-aspartic acid) motif that is known to promote cell adhesion, or other sequences for specific molecular recognition, or specific cell interactions. The linker is usually a string of hydrophobic amino acids such as alanine or valine. The anchor can be a cysteine residue for gold surfaces, asparatic acid linking on amine surface, or lysine linking on carboxylic surface.

This approach may facilitate research into cell–cell communication. Recently, we have moved one step further: using proteins or peptide as ink, we have directly microprinted specific features onto the non-adhesive surface of polyethylene glycol to write any arbitrary patterns rapidly without preparing the mask or stamps (Fig. 3.7). The process is similar to using an ink pen for writing—here, the printing device is the pen and the biological substances are the inks [9].

This simple and rapid printing technology permitted us to design arbitrary patterns to address questions in neurobiology that could not have been possible.

Since understanding of correct complex neuronal connections is absolutely central for comprehension of our own consciousness, human beings are always interested in finding ways to probes the question deeper. However, the neuronal connections are exceedingly complex, we must dissect the complex neuronal connections into smaller and manageable units to study them in a well-controlled manner through systematic biomedical engineering approaches. Therefore, nerve fiber guidance and connections can now be studied on special engineered pattern surfaces that are printed with protein and peptide materials. The surface

A)

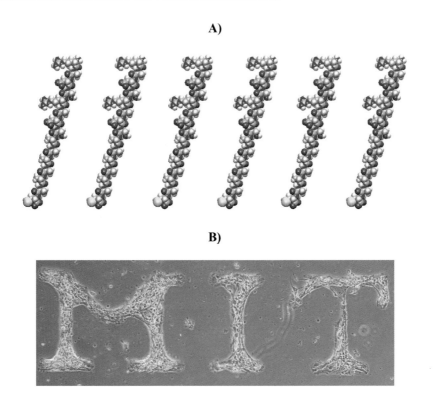

B)

FIGURE 3.7. Molecular structure of molecular ink peptide. A) This class of peptide ink has three general regions along their lengths: a ligand for specific cell recognition and attachment, a linker for physical separation from the surface, and an anchor for covalent attachment to the surface. Color code: carbon, green; hydrogen, white; red, oxygen; blue, nitrogen; yellow, thiol group. B) Cells adhere to printed patterns. The protein was printed onto a uniform PEG inhibitory background. Cells adhered to patterns after 8–10 days in culture spelling MIT. (Cell images courtesy Sawyer Fuller & Neville Sanjana).

substrate can be either from purified native proteins or from self-assembling peptide containing cell adhesion motifs, prepared through systematic molecular engineering of amino acids. Previous studies have shown that nerve fibers attach and outgrow extensively on the self-assembling peptide matrices [3]. We are interested in studying in detail of the nerve fiber navigation on designed pattern surfaces. Studies of nerve fiber navigation and nerve cell connections will undoubtedly enhance our general understanding of the fundamental aspects of neuronal activities in the human brain and brain-body connections. It will also likely have applications in screening neuropeptides and drugs that stimulate or inhibit nerve fiber navigation and nerve cell connections.

3.3. PEPTIDE NANOFIBER SCAFFOLD FOR 3-D CELL CULTURE, TISSUE ENGINEERING AND REGENERATIVE MEDICINE

3.3.1. Ideal Synthetic Biological Scaffolds

The ideal biological scaffold [22] and its building blocks should meet several important criteria: 1). derived from chemically-defined, synthetic sources which are present in native tissue; 2). amenable to design and modification to customize specific functional requirements; 3). allow cell attachment, migration, cell-cell, cell-substrate interactions, and recovery of cells from the scaffold; 4). exhibit no cytotoxicity or biocompatibility problems while chemically compatible with aqueous solutions, cell culture and physiological conditions; 5). compatible with microscopy, molecular and cell biology analyses, flow cytometry; 6). sterile and stable for long shelf life at room temperature, transportation, bioproduction and closed system cell therapy culture; 7). economically viable and scaleable material production, purification and processing; 8). exhibit a controlled rate of material biodegradation *in vivo* with non-detectable immune responses and inflammation; 9). foster cell migration and angiogenesis to rapidly integrate with tissues in the body. 10). injectable along with cells, compatible with cell delivery and surgical tools.

3.3.2. *Peptide Scaffolds*

One aspect of molecular self-assembly for production of nanobiomaterials is to fabricate three-dimensional peptide scaffolds by exposing the self-assembling peptide to a salt solution or to physiological media that accelerate the formation of macroscopic structures [3, 16, 17, 22]. Scanning electron microscopy (SEM), transmission EM (TEM), and atomic force microscopy (AFM) [6] reveal that the matrices formed are made of interwoven nanofibers having a diameter of $\sim$10 nm and pores of $\sim$5–200 nm in size. If the alanines are changed to more hydrophobic residues, such as valine, leucine, isoleucine, phenylalanine or tyrosine, the molecules have a greater tendency to self-assemble and form peptide matrices. These simple, defined and tailor-made self-assembling peptides have provided the first *de novo* designed scaffolds for three-dimensional cell culture, with potential implications for basic studies of cell growth and applied studies in tissue engineering and ultimately regenerative medicine.

Since these nanofiber scaffolds contain 5–200 nm pores and have an extremely high water content ($>$99.5% or 1–5 mg/ml), they are of potential utility in three-dimensional cell-culture media. The scaffolds closely mimic the porosity and gross structure of extracellular matrices, allowing cells to reside in a three-dimensional environment and molecules, such as growth factors and nutrients, to diffuse in and out very slowly.

We have shown that a variety of tissue cells encapsulated and grown in three-dimensional peptide scaffolds exhibit interesting functional cellular behaviors, including proliferation, functional differentiation, active migration and extensive production of their own extracellular matrices (Fig. 3.8). When primary rat neuron cells are allowed to attach to the peptide scaffolds, the neuron cells not only project lengthy axons that follow the contours of the scaffold surface, but also form active and functional synaptic connections (Fig. 3.8a). Furthermore, some neuronal progenitor cells migrate long distances into the three-dimensional peptide scaffold (Fig. 3.8b). Recently, the same peptide scaffold has been used for animal brain lesion repair where the severed brain section could be re-patched (Fig. 3.8c) (Ellis-Behnke, R., personal communication).

Similarly, when young and adult bovine chondrocytes are encapsulated in the peptide scaffold, they not only maintain their differentiated state, but also produce abundant type II and type XI collagen as well as glycosaminoglycans (Fig. 3.8d). Moreover, adult liver

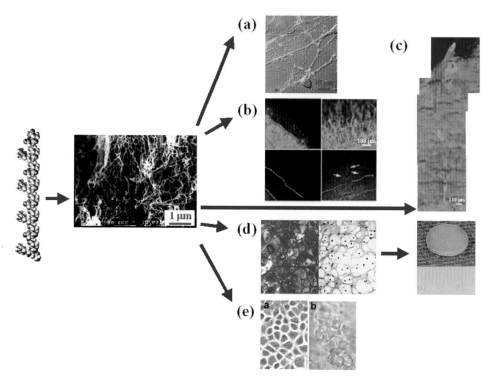

FIGURE 3.8. Self-assembling peptides form a three-dimensional scaffold woven from nanofibers ~10 nm in diameter. The scaffolds have been applied in several three-dimensional cell culture studies and in tissue engineering applications. **(a)** Rat hippocampal neurons form active nerve connections, each green dot represents a single synapsis. **(b)** Neural cells from rat hippocampal tissue slide migrate on the three-dimensional peptide scaffold. Cells on the polymer membrane (left) and on the peptide scaffold (right) are shown. Both glia cells (green) and neural progenitors (red) migrate into the three-dimensional peptide scaffold (image courtesy of C. Semino). **(c)** Brain damage repair in hamster. The peptide scaffold was injected into the optical nerve area of brain that was first severed with a knife. The cut was sealed by the migrating cells after two days. A great number of neurons form synapses (image courtesy of R. Ellis-Behnke). **(d)** Chondrocytes from young and adult bovine encapsulated in the peptide scaffold. These cells not only produce a large amount of glycosaminoglycans (purple) and type II collagen (yellow), characteristic materials found in the cartilage, but also a cartilage-like tissue *in vitro* (images courtesy of J. Kisiday & A. Grodzinsky). **(e)** Adult rat liver progenitor cells encapsulated in the peptide scaffold. The cells on the two-dimensional dish did not produce cytochrome P450 type enzymes (left-panel). However, cells in three-dimensional scaffolds exhibited cytochrome P450 activity (right panel). (image courtesy of C. Semino).

progenitor cells proliferate and differentiate into cells that express enzyme activities that metabolize toxic substances (Fig. 3.8e).

3.3.3. *PuraMatrix* in vitro *Cell Culture Examples*

One of the Lego peptides, RADA16-I now called PuraMatrix for its single component purity, is now commercially available for research (BD Bioscience). PuraMatrix has been used to culture diverse types of tissue cells including stem and progenitor cells, as well as

TABLE 3.1. A diversity of tissue cells and tissues cultured on PuraMatrix.

Mouse fibroblast	Bovine calf & adult chondrocytes
Chicken embryo fibroblast	Bovine endothelial cells
Chinese hamster ovary	Rat adult liver progenitor cells
Rat pheochromocytoma	Rat cardiac myocytes
Rat neural stem cells	Rat hippocampal neural tissue slice
Mouse embryonic stem cells	Mouse neural colony stem cells
Mouse cerebellum granule cells	Mouse & rat hippocampal cells
Bovine osteoblasts	Hamster pancreas cells
Human cervical carcinoma	Human osteosarcoma
Human hepato-cellular carcinoma	Human neuroblastoma
Human embryonic Kidney	Human foreskin fibroblast
Human epidermal keratinocytes	Human neural stem cells
Human hepatocytes	Human embryonic stem (ES) cells

These cells include stable cell lines, primary isolated cells from animals, progenitor and stem cells.

differentiated cell types and organotypic tissue slices ([17] summarized in [22]). Examples of cell types cultured in PuraMatrix are listed in (Table 3.1).

3.3.4. Extensive Neurite Outgrowth and Active Synapse Formation on PuraMatrix

PuraMatrix serves as a substrate and portable membrane media to support both neuronal attachment and differentiation in the form of extensive neurite outgrowth. Functional synapse formation also occurs between attached neurons when the cells are grown on PuraMatrix (Fig. 3.8) [3]. Neurite outgrowth from primary neuronal cultures was also tested by using several nerve cell types, including primary dissociated neurons from the mouse cerebellum and rat hippocampus. Cerebellar granule neurons undergo postnatal development and are morphologically distinguishable from other cerebellar cells. PuraMatrix scaffolds support extensive neurite out-growth from cerebellar granule neurons prepared from 7-day-old mice and the neurites were readily visualized in two different focal planes, suggesting that the neurites closely follow the contours of the matrices. The primary cerebellar neuronal cultures on the PuraMatrix scaffolds were maintained for up to 4 weeks. Dissociated mouse and rat hippocampal neurons also attach and project neurites on the scaffolds (Table 3.2).

TABLE 3.2. Neurtite outgrowth on PuraMatrix membrane surface.

Cell type	Length of process (micron)	Cell sources
NGF-treated Rat PC12	400–500	Cultured cell line
NGF-preprimed PC12	400–500	Cultured cell line
Human SY5Y neuroblastoma	400–500	Cultured cell line
Mouse cerebellar granule neurons	200–300	[‡]Primary cells
Mouse hippocampal neurons	100–200	[‡]Primary cells
Rat hippocampal neurons	200–300	[§]Primary cells

Cells were seeded onto PuraMatrix scaffold coating. The cell-bearing PuraMatrix was transferred to dishes with fresh medium. Maximum neurite length was estimated visually with scale bars 3–7 days after cell attachment for primary cultures and 10–14 days after matrix attachment for the cultured cell lines. [‡]Seven-day-old mouse. [§]One-day-old rat.

3.3.5. Compatible with Bioproduction and Clinical Application

PuraMatrix fulfills both traditional biological materials and a nanoporous/nanofiber hydrogel to enable proper 3-D cell growth and the creation of microenvironments surrounding cell colonies. For cells that prefer culture on a 2-D surface, better results were obtained by culturing cells on PuraMatrix floating sheets, allowing a 3-D nutrient bath within the 2-D growth context.

PuraMatrix mimics important aspects of the in vivo environment, while eliminating complicating variables traditionally experienced from animal-derived materials. Unlike most other scaffolds, PuraMatrix can be sterilized through UV radiation or filtration, and has proven itself shelf-stable at room temperature for over 10 years. Not only lowering the handling, processing, transportation, and inventory costs, but also meeting the stringent bioproduction requirements through its synthetic composition, free of the undesired factors and adventitious agents from animal product. PuraMatrix not only provides the 3-D context for cell research, but they are forward-compatible with bioproduction and clinical requirements necessary for eventual tissue engineering and stem cell based therapies. Furthermore, PuraMatrix not only did not elicit detectable immune responses in many kinds of animals, but it also did not cause appreciable inflammatory reactions in animals nor gliosis when injected into several kinds of animal brains [22].

3.3.6. Synthetic Origin and Clinical-Grade Quality

PuraMatrix is synthetic and sterile, hence suitable for bioproduction and can be readily used in bioproduction and clinical settings, unlike many animal derived materials (bovine, pig or others) that lack the consistent quality control, thus complicating clinical reproducibility and risking the introduction of undesirable contaminants (cell signaling factors, prions, adventitious agents, etc.). PuraMatrix is currently manufactured in large-scale quantities, some under GMP-grade processes and commercially available. Additionally, PuraMatrix can be used for closed, sterile system culture *in vitro* and can also be injected *in vivo* without eliciting immune responses, unlike collagens and alginates. Upon injection or pipetting into solution, the PuraMatrix volume does not swell, retaining a consistent volume since it already has >99% water content.

3.3.7. Tailor-Made PuraMatrix

In order to fabricate tailor-made PuraMatrix, it is crucial to understand the finest detail of peptide and protein structures, their influence on the nanofiber structural formation and stability. Since there is a vast array of possibilities to form countless structures, a firm understanding of all available amino acids, their properties, the peptide and protein secondary structures is an absolute prerequisite for further advance fabrication of peptide and protein materials. We are moving in that direction and will further accelerate new scaffold development.

PuraMatrix so far used in diverse cell and tissue systems from a variety of sources demonstrated a promising prospect in further improvement for specific needs since tissues are known to reside in different microenvironments. The PuraMatrix used thus far are general peptide nanofiber scaffolds and not tailor-made for specific tissue environment.

We are designing 2nd generation PuraMatrix scaffolds and to show that these tailor-made PuraMatrix incorporating specific functional motifs will perform as superior scaffolds in specific applications. They may not only create a fine-tuned microenvironment for 3-D tissue cell cultures, but also may enhance cell-materials interactions, cell proliferation, migration, differentiation and performing their biological function. The ultimate goal is to produce tailor-made scaffolds for particular tissue engineering, regenerative and reparative medical therapies.

3.4. PEPTIDE SURFACTANTS/DETERGENTS STABILIZE MEMBRANE PROTEINS

Although membrane proteins are composed of at approximately one-third of total cellular proteins and carry out some of the most important functions in cells, only few dozens of membrane protein structures have been elucidated. This is in sharp contrast to ~23,000 non-membrane proteins structures that have been solved [5]. The main reason for this delay is due to difficulty to purify and crystallize membrane proteins because removal of lipids from membrane proteins affects protein solubility and conformation stability. Despite a variety of detergents and lipids as surfactants have been used to facilitate solubilize, stabilize, purify, crystallize and manipulate the membrane proteins for over several decades, how surfactants interact with membrane protein to impact its structure and functions and how to choose good surfactants for the right membrane proteins remain largely unknown. This is partly due to complexity of membrane protein-detergent-lipid interactions and lack of a "magic material" surfactant/detergents. Therefore, the need to develop new membrane compatible material is acute.

Several of our detergent peptides have been shown to complement other chemical detergents to solubilize, stabilize and crystallize membrane proteins. These findings are very important to overcome the barrier of working on membrane proteins. Furthermore, these stabilized membrane proteins could be extremely important for fabrication of array of ultra-sensitive detection devices, not only for biomedical research and clinical diagnosis, but for utilities far outside traditional applications and beyond our current imagination.

3.5. PERSPECTIVE AND REMARKS

One of the emerging fields in nanobiomaterials is the development of new biologically inspired materials. These materials will often broaden the questions we can address therefore deepen our understanding of seemingly intractable biological phenomena. The versatile self-assembling peptide units will create a new class of molecular materials that will likely have a high impact in many fields.

We have encountered many surprises since we started our serendipitous journey of working on various self-assembling peptide systems: from developing a class of pure peptide nanofiber scaffolds for 3-D tissue culture and for tissue engineering, studying of the model system of protein conformational diseases, designing peptide/protein inks for surface printing to finding peptide surfactant/detergents that solubilize and stabilize membrane proteins.

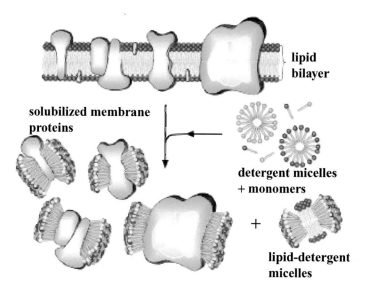

lipid
bilayer

solubilized membrane
proteins

detergent micelles
+ monomers

+

lipid-detergent
micelles

FIGURE 3.9. These simple surfactant/detergent peptides can be used to solubilize, stabilize and crystallize membrane proteins. These peptides have a hydrophilic head and a hydrophobic tail, much like lipids or detergents. They use their tail to sequester the hydrophobic part of membrane proteins, but the hydrophilic heads expose to water. Thus, they make membrane protein soluble and stable out their lipid milieu.

We believed that application of these simple and versatile molecular self-assembly systems will provide us with new opportunities to study some complex and previously intractable biological phenomena. Molecular engineering through molecular design of self-assembling peptides is an enabling technology that will likely play an increasingly important role in the future materials chemistry and will change our lives in the coming decades.

ACKNOWLEDGEMENTS

We would like to thank members of the lab, past and present, for making discoveries and conducting exciting research. We gratefully acknowledge the supports by grants from ARO, ONR, DARPA (BioComputing), DARPA/Naval Research Labs, DARPA/AFOSR, MURI/AFO, NIH, NSF-MIT BPEC and NSF CCR-0122419 to MIT Media Lab's Center for Bits & Atoms, the Whitaker Foundation, Du Pont-MIT Alliance, and Menicon, Ltd, Japan. We also acknowledge the Intel Corporation educational donation of computing cluster to the Center for Biomedical Engineering at MIT.

REFERENCES

[1] C.-I. Branden and J. Tooze. *Introduction to Protein Structure*. 2nd Ed. Garland Publishing, New York, N.Y. (1999).
[2] C.S. Chen, M. Mrksich, S. Huang, G.M. Whitesides, and D.E. Ingber. *Science*, 276:1425–1428, 1997.
[3] T. Holmes, S. Delacalle, X. Su, A. Rich, and S. Zhang. *Proc. Natl. Acad. Sci. U.S.A.*, 97:6728–6733, 2000.

[4] J. Kisiday, M. Jin, B. Kurz, H. Hung, C. Semino, S. Zhang, and A.J. Grodzinsky *Proc. Natl. Acad. Sci. U.S.A.*, 99:9996–10001, 2002.

[5] P.J. Loll. *J. Struct. Biol.*, 142:144–153, 2003.

[6] D. Marini, W. Hwang, D.A. Lauffenburger, S. Zhang, and R.D. Kamm. *NanoLetters*, 2:295–299, 2002.

[7] M. Mrksich and G.M. Whitesides. *Annu. Rev. Biophys. Biomol. Struct.*, 25:55–78, 1996.

[8] G.A. Petsko and D. Ringe. *Protein Structure and Function*. New Science Press Ltd. London, U.K., 2003.

[9] N.E. Sanjana and S.B. Fuller. A fast flexible ink-jet printing method for patterning dissociated neurons in culture. *J. Neurosci. Methods*, 136:151–163, 2004.

[10] S. Santoso, W. Hwang, H. Hartman, and S. Zhang. *NanoLetters*, 2:687–691, 2002.

[11] S. Vauthey, S. Santoso, H. Gong, N. Watson, and S. Zhang. *Proc. Natl. Acad. Sci. U.S.A.*, 99:5355–5360, 2002.

[12] G. von Maltzahn, S. Vauthey, S. Santoso, and S. Zhang. *Langmuir*, 19:4332–4337, 2003.

[13] G.M. Whitesides, J.P. Mathias, and C.T. Seto. *Science*, 254:1312–1319, 1991.

[14] G.M. Whitesides and B. Grzybowski. *Science*, 295:2418–2421, 2002.

[15] S. Zhang, C. Lockshin, A. Herbert, E. Winter, and A. Rich, *EMBO. J.*, 11:3787–3796, 1992.

[16] S. Zhang, T. Holmes, C. Lockshin, and A. Rich. *Proc. Natl. Acad. Sci. U.S.A.*, 90:3334–3338, 1993.

[17] S. Zhang, T. Holmes, M. DiPersio, R.O. Hynes, X. Su, and A. Rich. *Biomaterials*, 16:1385–1393, 1995.

[18] S. Zhang, L. Yan, M. Altman, M. Lässle, H. Nugent, F. Frankel, D. Lauffenburger, G.M. Whitesides, and A. Rich. *Biomaterials*, 20:1213–1220, 1999.

[19] S. Zhang, D. Marini, W. Hwang, and S. Santoso. *Current opinion in Chemical Biology*, 6:865–871, 2002.

[20] S. Zhang. *Biotechnology Advances*, 20:321–339, 2002.

[21] S. Zhang. *Nature Biotechnology*, 21:1171–1178, 2003.

[22] S. Zhang. *Nature Biotechnology*, 22:151–152, 2004.

[23] S. Zhang, X. Zhao, and L. Spirio. PuraMatrix: Self-assembling peptide nanofiber scaffolds. In *Scaffolding in Tissue Engineering*. (**In the press**) (2004).

4

At the Interface: Advanced Microfluidic Assays for Study of Cell Function

Yoko Kamotani, Dongeun Huh, Nobuyuki Futai, and Shuichi Takayama
Department of Biomedical Engineering and Macromolecular Science and Engineering, University of Michigan, Ann Arbor, Michigan 48109-2099

4.1. INTRODUCTION

Understanding basic biology and disease mechanisms, testing drug safety and efficacy, engineering tissues and cell-based biosensors all require methods to study and manipulate mammalian cell function. A convenient method that has been developed over the past century for these purposes is *in vitro* cell culture where cells are taken out of their normal physiological environment inside the body and kept alive in a dish. Although *in vitro* cell culture is powerful, there is increasing evidence that cellular responses in culture dishes do not necessarily reflect how cells may behave *in vivo*. This discrepancy is due, at least in part, to the fact that much of what happens in living organisms is affected by microscale patterns and heterogeneity which are not well controlled in traditional macroscopic culture systems. Cells, which are micron-sized, determine their behavior based on cues from their microenvironment. Cellular behaviors such as subcellular signaling, chemotaxis (directed migration towards a chemical), growth, differentiation, and death, for example, are determined by subcellular stimulation, microscale chemical gradients, and adhesive micro- & nanopatterns. Microfluidics and other microscale phenomena dominate at this level making microtechnology crucial to the understanding of the cellular basis of life.

The power of microtechnology in biology—as represented by DNA microarrays—is well appreciated for its ability to do more with less and in obtaining larger quantities of data more rapidly than is possible with macroscopic tools. We refer to this type of use of microtechnology as being "traditional". In the study of cell function, there is an additional incentive to the use of microtechnology, in that knowledge can be gained that is also qualitatively different and richer. In this chapter, we refer to this type of use of microtechnology as being "advanced". The chapter is illustrative rather than comprehensive and deals mainly with microfluidic technologies for subcellular biology, cell sorting, flow cytometry, and immunoassays. A focus is placed on the use of microfluidic systems to generate interfaces between two or more chemicals, fluids, cell types, or adhesive patterns and in studying how signals, cells, and molecules respond to these boundaries.

Traditional microscale *in vitro* assays generally involve the seeding of cells in static, flat, and homogeneous multi-well plates and observing their behavior. Tissues in the body, however, have active transport of molecules and fluids, are three-dimensional (3-D), and are composed of multiple cell types. In other words, the tissue has multiple boundaries and interfaces that create a complex environment. Cells in the body are influenced by the resulting dynamic (that is, it changes with time) three-dimensional adhesive, fluidic, and mechanical organization of their surroundings. Cells recognize and respond to microscale spatiotemporal variations such as the concentrations of molecules in the extracellular medium, the extracellular matrix, or the surfaces of adjacent cells (membrane receptors). A solution that can provide the convenience and versatility of *in vitro* assays as well as provide an added degree of physiological relevance is to use microtechnology to recreate aspects of biological interfaces seen *in vivo*. These advanced microscale bioassay systems allow micrometer-level design of factors such as the biochemical composition and topology of the substrate, the composition of the cell culture medium, and the interactions with other cells in its vicinity [4, 22, 72, 89]. Use of microfluidic interfaces also provides new opportunities in biochemical separations and analysis, such as flow cytometry and immunoassays.

The chapter is organized as follows: first, microfabrication technologies utilized in creating advanced bioassay systems are described with an emphasis on soft lithography. Second, phenomena encountered at the microscale are discussed. Third, several examples of microfluidic advanced bioassay systems are described.

4.2. MICROFABRICATION

Microfabrication is a prerequisite for creating and controlling cell microenvironments. Most modern microfabrication technologies are based on photolithography, or the patterning of a layer of photosensitive polymer by means of beams of light, electrons, or ions [51]. In this process photoresist layers are selectively exposed to ultraviolet (UV) light through a photomask that contains the desired pattern in opaque features. Photoresist is usually spincoated onto a flat substrate and dried to form uniform layers. The exposed or unexposed sections of the photoresist are then dissolved in a developer solution, thus generating the micropatterns. Although it is able to produce accurate patterns, photolithography is not always conducive to biological applications for the following reasons: (i) multiple steps are needed to generate the patterns, during which the system is exposed to organic solvents, plasmas, or other harsh chemical conditions that are toxic to cells and denature biomolecules; (ii) photolithography must be carried out in clean-room facilities containing expensive equipment such as photomasks and aligners; (iii) the substrate is restricted to flat and rigid materials (e.g. silicon wafers); (iv) specialized steps such as anodic bonding are needed for bonding and packaging; (v) and photolithography can only pattern a limited set of photosensitive materials making it difficult to pattern proteins onto surfaces or performing multiple chemical reactions on the same chip [72]. A set of microfabrication and pattern generation techniques called soft lithography that uses soft elastomeric materials, overcomes many of these limitations and offers its own advantages.

4.2.1. Soft Lithography

Soft lithography most commonly utilizes the elastomer poly(dimethylsiloxane) (PDMS) which offers many advantages over silicon and glass including its low cost, flexibility, and ease of handling and polymerizing. PDMS can replicate features with submicron resolution in molding procedures [26] and can easily be peeled off from most substrates after curing without leaving a residue [17]. It is optically transparent down to 230 nm and compatible with biological studies because it is impermeable to water, nontoxic to cells, and permeable to gases. PDMS can also be attached reversibly through contact with a variety of surfaces and bonded irreversibly to other surfaces such as PDMS, glass, silicon oxide, and oxidized polystyrene after brief exposure to oxygen plasma [21, 52, 73]. Two PDMS pieces that have asymmetric reagent excess may also be sealed together simply by additional curing [85].

To fabricate PDMS structures, a master mold is needed, which is the structure against which the PDMS prepolymer is poured and cured. The design of the master mold is created using a computer-aided design (CAD) program. The patterns are then printed onto transparencies which are used as a photomask in UV-photolithography to generate the master mold. This step is often the most expensive and time intensive step in soft lithography. Lateral resolutions of 25 μm can be achieved with transparencies (at a resolution of 5080 dots/in or 5 μm/dot) and can be extended to 8 μm using a resolution of 20,000 dots/in. with photoplotted silver halide films [49]. For features beyond 8 μm, chrome masks can be used, however they take have a longer turnaround time and are more expensive. Features less than 1 μm can be achieved using commercially available microscopes and projecting the patterns of the transparency onto a substrate [50].

To create the PDMS mold, the liquid PDMS prepolymer (a mixture to 10:1 base polymer:curing agent) is cast onto the surface of the silicon/photoresist master after it has been treated with fluorinated silanes as a mold release agent to prevent irreversible bonding to the PDMS. The prepolymer is then cured into solid PDMS at an elevated temperature of up to 150°C in an oven before being peeled off the master. The PDMS now has a replica of the desired microstructures. Holes can be cut into the PDMS for inlets, outlets, and reservoirs and after a brief plasma oxidation irreversibly sealed to another surface. The procedure is outlined in Figure 4.1.

PDMS consists of repeating $-OSi(CH_3)_2-$ units that make the surface hydrophobic. Any microchannels are susceptible to trapping air bubbles and the surface is prone to nonspecific adsorption of proteins and cells. [73] Surface oxidation produces hydroxyl groups on the surface and changes its property to being hydrophilic and silica-like. The surface can remain hydrophilic if it is in contact with water [52]. This is useful for microfluidic applications when it is necessary to draw fluid into channels with capillary force [16].

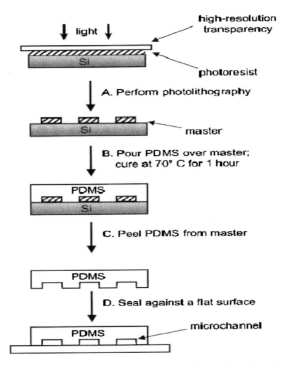

FIGURE 4.1. Scheme for making PDMS microchannels. The microchannel design is made using a CAD program and printed onto a high-resolution transparency. The transparency is then used as a photomask (A) to produce a master. Liquid PDMS is poured over the master and cured (B). The PDMS is then peeled from the master (C) and sealed against a flat surface to enclose the channels (D) [53]. (Reprinted with permission from McDonald JC 2002. Copyright 2002 American Chemical Society)

Particularly for cellular studies, PDMS is advantageous due to the following properties: (1) it is nontoxic to cells, (2) it is optically transparent allowing for the easy visualization of cells, (3) the surface properties of PDMS can be altered to allow for the patterned adsorption of proteins, (4) it is permeable to gases such as oxygen that cells require, and (5) depending on the photomask used, PDMS can be cured to form a variety of microstructures to resemble the *in vivo* environment of cells [90].

4.3. MICROSCALE PHENOMENA

The ability of microtechnologies to create structures and patterns on micron and smaller length scales provides a powerful means to manipulate cells and their microenvironments. Miniaturized bioanalytical systems can handle small reagent volumes with shorter reaction times and control biochemical composition and topology of substrates at a cellular/subcellular level. Since length scale is a fundamental quantity that dictates the type of forces governing physical phenomena, however, successful microengineering of cellular environments requires an understanding of physical effects that dominate in microscale. In this section, we briefly describe intrinsic small-scale phenomena that affect fluid transport and the preparation of patterned surfaces in microfabricated devices.

4.3.1. Scaling Effects

A general design paradigm in the development of microsystems is "scale-down". As a characteristic length becomes smaller and approaches the sub-millimeter scales, forces and physical effects that are often insignificant at macroscopic scales become very important, offering new possibilities of tailoring them to produce functional microdevices. Firstly, surface forces (e.g. surface tension, van der Waals forces, electrostatic forces) scaled to the first or second power of the length become dominant over body forces (e.g. gravity, inertial forces) which depend on the third power of the length scale. As a result, the movement and spatial configuration of fluids and biological molecules in microscale geometries can be manipulated effectively by changing physicochemical properties of surfaces such as wettability [25, 32, 46, 92], temperature [14, 15], charge distribution [78], topology [78, 81], etc. Secondly, the ratio of surface area to volume that is inversely proportional to the characteristic length scale becomes very large and thus, a much greater portion of fluids or sample molecules is permitted to interact with surfaces. The high surface-to-volume ratios attainable in microfabricated structures allow separations or intensive chemical reactions with higher yields due to increased heat and mass transfer rates [13, 34, 39]. Thirdly, a balance between inertial and viscous forces in fluid transport, described by the Reynolds number $Re = $ (inertial effect)/(viscous effect) $= \rho U L / \mu (\rho = $ density of fluid, $U = $ average flow speed, $L = $ characteristic length scale, $\mu = $ fluid viscosity) changes. In most cases, viscous effects dominate flow behaviors and the Reynolds number becomes much smaller than 1 mainly due to small length scales and slow flow speeds. The most important consequence of low-Reynolds-number flow is that fluid streams are driven parallel to the local orientation of microchannel walls without random fluctuations of the flow in time (laminar flow) and that multiple streams flow stably in contact with each other without significant mixing except due to diffusion at their interface [74, 77]. Finally, diffusion time scales that vary to the square power of a length scale can be much shorter in microchannels and transport phenomena mediated by diffusion become useful. In combination with convective fluid motions, diffusion serves as an important mode of fluidic mixing [38, 42], molecular separation [31, 87, 88], and the homogenization of surface compositions [28, 29] in microsystems.

4.3.2. Laminar Flow

Laminar flow is the definitive feature of microfluidics and is characterized by a smooth, orderly motion of fluids without random temporal variations and strong dissipations of momentum and energy. It allows generation of an interface between two different fluids. A drastic reduction in inertial effects in microscale laminar flows (low Reynolds number flows) also allows for tractable theoretical analysis and an accurate prediction of flow fields, making it easy to control flow patterns, positions of interfaces, and transport phenomena. In general, fluid motions in microfluidic systems operating at low values of Reynolds number are driven by externally applied pressure gradients [65], electroosmosis [5], surface wetting [32, 47, 59], and interfacial tension gradients [1]. It is most common in microfluidics that two or more miscible laminar streams are prompted to flow adjacent to each other in the same direction. This flow configuration renders two unique features: (i) Each laminar stream retains most of its chemical composition and physical contents over the short length of microchannels. This enables selective delivery of chemicals [17, 81, 83], proteins [16, 93], lipids [90], cells [80], or polymers [2] to specific locations to form spatially directed patterns on channel surfaces. This technique is collectively known as "laminar flow patterning" and has been found useful in studying cell-cell, cell-surface, cell-medium interactions [22] as well as engineering surface-directed flows [46, 91, 93]. (ii) In the absence of transverse convective fluid motions, mixing between liquid laminar streams occurs only via diffusion at the interface. This characteristic poses both challenges and advantages in microfluidics. It has been demonstrated that the diffusive transport and chemical reaction confined to an area in the vicinity of the interface of two miscible liquids provide novel tools for microelectrode fabrication [43], size-dependent separation of biological molecules [31, 87, 88], and immunoassays [30]. The cross-stream-wise diffusion localized at the interface, however, becomes a rate-limiting factor when it comes to mixing in microfluidics because mixing purely by diffusion is too slow. A rapid and efficient mixing in microfluidics is often achieved by either generation of chaotic convective fluid motions transverse to main laminar flows [76] or simple splitting and recombination of fluid streams to create an increased fluid interface and short diffusion path [39].

4.3.3. Surface Tension

When immiscible fluids (e.g. liquid and air, two immiscible liquids such water and oil) are present, molecules within the liquid are attracted equally from all directions, but those near the interface experience unequal attractions and thus, are drawn towards the center of liquid mass. At the microscale, surface tension originating from this cohesion between liquid molecules at the interface of different fluids becomes a dominant surface force and plays a crucial role in manipulating transport phenomena. Spatial gradients in surface tension that are often created by variations of temperature [14, 15], electric field [32, 47, 59], prepatterned surface energies [46, 92, 93], electrochemical reactions [25] on the walls of microchannels give rise to driving forces for the movement of liquid droplets and thin liquid films or for stable flows of immiscible fluids without any need for mechanical moving parts. In such multiphase microfluidic systems, a Reynolds number that is generally small, itself, does not usually serve as a characteristic system parameter. Instead, geometrical features and ratios of different forces become important in indicating the details of flows. The relative importance between surface tension and other forces is estimated by the following dimensionless parameters.

$$\text{Ca (capillary number)} = \frac{\mu_L U_L}{\sigma} \sim \frac{\text{viscous effect}}{\text{surface tension effect}}$$

$$\text{Bo (Bond number)} = \frac{\rho_L g L^2}{\sigma} \sim \frac{\text{gravitational effect}}{\text{surface tension effect}}$$

$$\text{We (Weber number)} = \frac{\rho_L U_L^2 L}{\sigma} \sim \frac{\text{inertial effect}}{\text{surface tension effect}}$$

Where μ_L is liquid viscosity, U_L is average velocity of liquid, σ is surface tension, ρ_L is density of liquid, g is gravitational acceleration, and L is a characteristic length. For example, assuming that there is continuous movement of water ($\mu_L = 0.001$ kg/ms^2, $\rho_L = 1000$ kg/m^3) droplets or flow of water through air at the average speed of 1 cm/s in a 100 μm-thick polymeric microchannel ($\sigma = 30$ dynes/cm), it can be shown that Ca $\sim 3.3 \times 10^{-4}$, Bo $\sim 3.2 \times 10^{-3}$, We $\sim 3.3 \times 10^{-4}$, implying the dominance of surface tension effects.

Generally, the manipulation of surface tension to change surface wettability in microfluidics is performed in two different formats depending on the reversibility of surface energy changes: (i) In the first case, electric potentials or thermal energies applied to microchannel walls through embedded electrodes enable reversible changes of surface tension. In this arrangement, it has been demonstrated that liquid droplets surrounded by immiscible fluids can be transported quickly to desired locations over two-dimensional arrays of microelectrodes [12, 20, 47]. Such routing of independent droplets and reversible motion of fluids provide novel methods to devise basic building blocks of microfluidic systems such as pumps, valves, and mixers [12, 57]. Also, a precise handling of pico- or nanoliter quantities of reagents and analytes isolated in discrete droplets becomes useful for biochemical assays [67]. In addition to droplet-based configurations, the alteration of surface tension by electric potentials has proven useful for reversibly switching flow patterns of continuous air-liquid laminar streams [35] or for transporting thin continuous liquid films [25] as well. (ii) In the second case, surface patterning by laminar flows or UV light can be used to irreversibly generate surface tension gradients. This method is based on the preparation of surfaces patterned with hydrophobic or hydrophilic self-assembled monolayers achieved by laminar flow- or UV patterning [46, 92, 93]. The predefined surface free energies facilitate the confinement of liquid samples in hydrophilic regions and assist in guiding continuous liquid streams without requiring external forces to drive the fluid motions. Selective wetting of liquid samples over patterned hydrophilic surfaces enables a nanoscopic patterning of biomolecules such as DNA [70]. Also, this system has been used to realize microfluidic gas-liquid chemical reactions and the microfabrication of thin polymeric membranes [91, 93, 94].

4.4. EXAMPLES OF ADVANCED MICROFLUIDIC CELLULAR BIOASSAYS

Inexpensive rapid microfabrication techniques combined with knowledge of microscale phenomena (e.g. laminar flows and dominance of surface tension) allow development of new advanced microfluidic cellular bioassays. Microfluidic devices for these applications typically have cross-sectional capillary dimensions of 10–500 µm. This size scale is similar to those of biological cells, so that these devices are well suited to the transport, manipulation, and chemical or biochemical treatment of single cells or small number of cells [48]. This section describes several specific systems that take advantage of microfluidics to create controlled microenvironments to study subcellular phenomena, to sort cells by function such as motility, and to analyze cells. These advanced systems allow for the microscale control of interfaces that have not been exploited in traditional systems.

4.4.1. Patterning with Individual Microfluidic Channels

The most straightforward application of microfluidics is the use of different channels to pattern different cells or molecules in different regions on the microscale. The small size of microfluidic channels and the ability to fabricate multiple channels in parallel provides a platform for simultaneously patterning different biomolecules onto a surface.

Delamarche et al. used microfluidic networks for the spatially controlled deposition of biomolecules onto various substrates. The networks are constructed using a PDMS mold

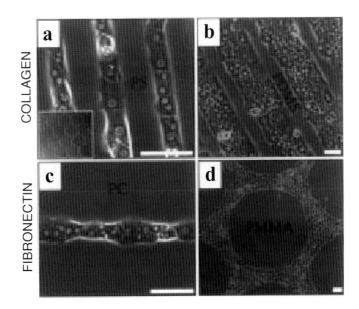

FIGURE 4.2. Phase-contrast optical micrographs of hepatocytes patterned onto various polymers (A: Polystyrene Petri dish, B: PDMS sheet, C: Polycarbonate Petri dish, and D: Poly(methyl methacrylate) disk) 12–24 hours after seeding. Hepatocytes adhere only to regions of collagen (A and B) or fibronectin (C and D). Scale bar is 50 μm. [23] (Reprinted with permission from Folch 1998. Copyright 1998 American Chemical Society)

sealed to a surface to form a conduit. The capillaries are filled with solutions containing the biomolecules of interest, allowed to deposit onto the surface, and the PDMS later peeled off. Using this type of system, immunoglobulins were patterned onto substrates with submicron resolution and used in enzyme-linked immunosorbent assay (ELISA)-type assays [16].

Microfluidics allows manipulation of cellular co-cultures in ways not possible with macroscopic techniques. Toner et al. have selectively delivered different cell suspensions in laminar flows to tissue culture substrates. Another technique uses microchannels saturated with protein solutions that adsorbs onto the surface exposed to the microflow. After microchannels are removed, only the proteins remain and cells attach selectively to these regions. Micropatterned cocultures have been created using patterns of collagen or fibronectin (Figure 4.2) [22].

Whitesides et al. have used three-dimensional (3D) microfluidic systems to pattern proteins and cells onto a surface. 3D microfluidic stamps are fabricated using a two-step photolithographic process and sealing two PDMS slabs together. This PDMS structure is then used as a conduit to deliver proteins (bovine serum albumin and fibronectin fluorescently labeled) and cells (bovine capillary endothelial cells (BCEs) and human bladder cancer cells (ECVs) shown in Figure 4.3). This technique allows for the creation of biologically relevant patterns on surfaces with only one additional step of pattern fabrication, alignment, and sealing [10].

Jeon et al. have developed a microfabricated neuronal culture device. The system is created from PDMS and consists of two compartments separated by a physical barrier

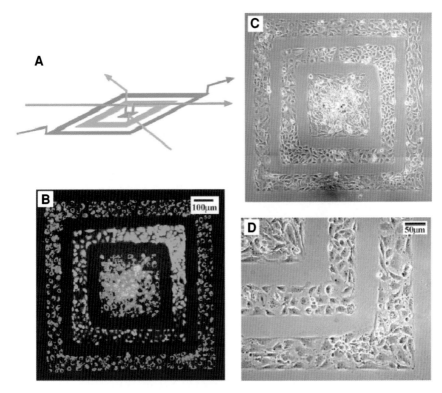

FIGURE 4.3. A 3D PDMS stamp shown in (A) is used to deposit two cell types onto a tissue culture dish in a concentric square pattern. Fluorescence (B) and phase-contrast (C and D) pictures show ECVs labeled in green CMFDA and BCEs labeled in a red DiI-conjugated acetylated low density lipoprotein. The cells are cultured with the stamp in place for 24 hours to grow and spread to confluency. The pictures are taken immediately after removing the stamp. (D) is an expanded view of the lower right corner in (C) [10]. (Reprinted with permission from Chiu, Jeon et al. 2000, Copyright 2000 National Academy of Sciences, USA).

with embedded micron-sized grooves. Neuronal cells are cultured into one compartment and after several days, neurites extend through the grooves and into the second compartment (Figure 4.4). They are also able to micropattern neurites on the surface through microcontact printing to direct neuronal attachment and orientation of the neuronal outgrowth [84].

4.4.2. Multiple Laminar Streams

In the previous section, each microfluidic channel transported one type of molecule or cell. The low Reynolds number flow characteristics in microfluidic channels, however, permits two or more streams of cells and biomolecules to flow next to each other inside of a single microfluidic channel without turbulent mixing. The only mixing at the interface between different streams takes place through diffusion. Researchers have exploited laminar flow in microfluidic networks to pattern proteins, cells, and planar lipid bilayers on substrates with micrometer-scaled resolution [58]. Multiple laminar streams are also

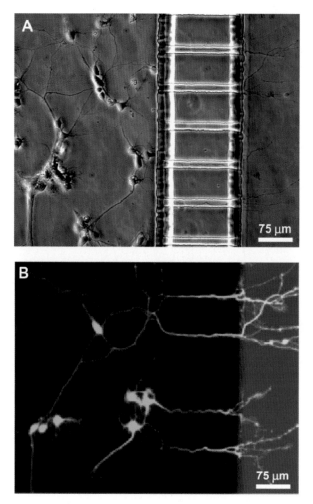

FIGURE 4.4. Neuronal cell culture inside microfabricated device. Calcein AM and Texas Red dextran are added to the neuritic chamber 1 hour before taking the pictures. (A) Phase micrograph after 4 days in culture of neurons extending to neuritic chamber (on the right). (B) Epifluorescence micrograph of same region with cells stained in green calcein AM, and the neuritic chamber in red dextran [84]. (Reprinted with permission from Taylor, Rhee et al. 2003. Copyright 2003 American Chemical Society)

useful in generating patterns or interfaces and gradients composed of (1) different adhesive regions, (2) different cell types, and (3) different solutions [79].

Multiple laminar streams can also generate and maintain gradients of chemicals. Cells naturally migrate in gradients of soluble molecules called chemoattractants in a process known as chemotaxis. In order to further examine this dynamic behavior, Jeon, Toner, Whitesides et al. have developed a technology that generates a stable, soluble chemoattractant gradient using a device fabricated by soft lithography. The device consists of a network of microfluidic channels with a gradient-generating portion and an observation

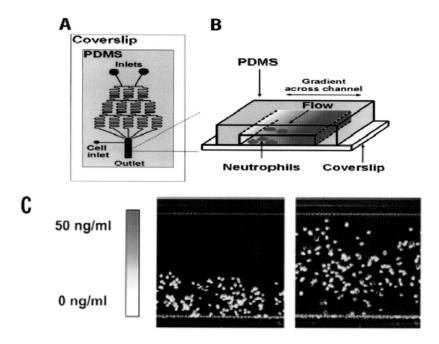

FIGURE 4.5. Schematic diagram of the microfluidic gradient generator. (A) Top view of device with gradient generating and observation regions. (B) 3D view of observation region where cells are exposed to chemoattractant gradients. (C) Micrograph of cells at the beginning of the experiment (0 min, left panel) deposited at the bottom of the field of view, and at the end of the experiment (90 min, right panel). Migration is subjected to a linear increase in IL-8 (0–50 ng/ml). Bar, 200 μm [40]. (Reproduced by permission from Jeon, Baskaran et al. 2002. Copyright 2002 Nature Publishing Group www.nature.com)

portion. (Figure 4.5A) The gradient-generating portion consists of a pyramidal branched array of channels that split, combine, and mix fluid streams. The channels recombine in the main channel to form a well-defined concentration gradient that spanned the width of the channel. (Figure 4.5B).

A gradient of interleukin-8 (IL-8) was formed in the device and the migration of neutrophils was observed. Figure 4.5C shows the results of one experiment where cells are initially positioned at the side of lower IL-8 concentration and subjected to a linear increase in IL-8 (1–50 ng/ml). The cells move towards the higher concentration. The ability to generate and maintain stable linear gradients allowed straightforward determination of chemotaxis coefficients demonstrating the power of microfluidics for quantitative cell biology. Comparison of cell migration in stable cliff-shaped and hill-shaped gradient profiles also allowed observation of complex migratory behaviors of the neutrophils previously unappreciated [40]. Whitesides et al. have also created substrate-bound gradients of proteins using laminar flows in microchannels. Linear gradients are created through streams that flow through long serpentine channels where mixing occurs. All streams converge into one channel where the gradient is established. Gradients of laminin and bovine serum albumin are demonstrated and neuronal axon orientation studied [18].

4.4.3. PARTCELL

Cellular transduction—the process of translating extracellular stimuli from the environment into responses such as growth, migration, or differentiation—occurs through subcellular components, including the cytoskeleton and various scaffolding proteins. These structural elements are distributed heterogeneously, but not randomly, throughout the cell with micron scale spatial variations. It has been proposed that structural order of these components is necessary for the propagation and processing of cellular signals. In order to study the role of such microscale organization, a method to perturb or manipulate specific domains within a cell with high molecular sensitivity and micron-scale spatial accuracy is needed [7]. Microfluidics can provide a solution. A technique, called PARTCELL (partial treatment of cells using laminar flows), positions the interface between multiple laminar flows over a single cell to deliver reagents with subcellular spatial selectivity.

In a typical set up, a microsystem is made from PDMS embossed with microchannel features sealed onto a glass surface. Streams containing different liquids are made to flow into three inlets and combine in a main channel to form parallel streams. (Figure 4.6a) The width of the streams and the position of the interface are adjusted by altering the relative amounts of fluid flowing in from the inlets. This setup allows for the patterned delivery of chemicals, colloids, and other reagents to a portion of a cell. (Figure 4.6b) Advantages to PARTCELL include: the facile generation of laminar flows which makes the procedure experimentally simple, the ability to pattern over delicate structures such as the cell surfaces, and the use of PDMS which makes fabrication easy and cell culture and analysis convenient. PARTCELL can be applied to a variety of studies including ones of cell dynamics, cell polarity, spatially regulated signaling, drug screening, and other cellular phenomena.

When cell permeable reagents are delivered to subcellular regions of surface-attached cells, a stable chemical concentration gradient is established and maintained within the cell. Such subcellular "positioning" of rapidly diffusing small molecules to a part of a

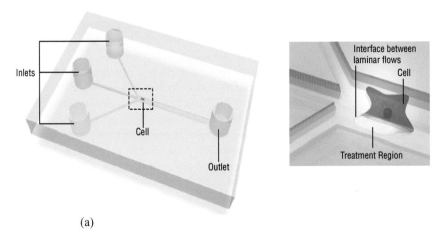

(a)

FIGURE 4.6. Schematic drawing of PARTCELL setup. (a) Three inlets combine to form parallel streams in main channel. (b) Close up view of cell placed at the interface between laminar streams.

cell interior is counter intuitive considering the small size of cells. PARTCELL makes this possible because the continuous fluid flows impose rapid influx of reagent to one cell region and efflux from another region [82]. This has been demonstrated in a living bovine capillary endothelial (BCE) cell by labeling two different mitochondria subpopulations within the cell with a green and red fluorescent dye respectively. In a similar procedure actin filaments were disrupted in selective regions of the actin microstructure by treatment with a membrane-permeable molecule that binds to actin monomers [83].

PARTCELL can also be used to manipulate the cell-substrate interactions. For example, trypsin can be delivered selectively to a portion of a cell resulting in selective trypsin-mediated detachment of the exposed region of the cell from the surface [82]. The method may be useful in obtaining insights into cell migration processes which often involve various matrix proteinases. PARTCELL also allows for the localized delivery of chemicals to only a portion of the cell surface. For example, flow of fluorescently labeled, acetylated, low-density lipoprotein (Ac-LDL) to only a portion of the surface led to spatially selective endocytosis of Ac-LDL [82].

PARTCELL uses microfluidics to perform subcellular manipulations that are difficult to perform otherwise. Limitations to PARTCELL include the need to seed the cells inside the channels and only being able to deliver the subcellular treatment in parallel laminar flows (as opposed to delivery to a point). Bradke et al. have developed a related technique for subcellular treatment and demonstrated the use of two pipettes to partially treat a neurite outgrowth with a small-molecule drug [6]. This method has the advantage of not requiring seeding of cells inside microfluidic channels, however, this method requires precise handling of micromanipulators and the open fluidic system needed makes the fluid flow more susceptible to disturbances and variations.

The ability to stimulate parts of single cells has led to discovery of new states and functions of cells not observed in traditional *in vitro* cultures. Bastiaens et al. first demonstrated that local stimulation of the epidermal growth factor receptor (EGFR) using epidermal growth factor (EGF) attached to microbeads leads to global phosphorylation of EGFR in the plasma membrane of MCF7 breast adenocarcinoma cells, in a ligand independent manner. Although microbeads served as a convenient tool for subcellular EGF delivery, the positioning of the microbeads on single cells was random in these studies. Furthermore, immobilization of EGF molecules on a comparatively huge microsphere alters cellular dynamics such as endocytosis of EGF. To overcome these limitations, Miyawaki et al used PARTCELL to perform subcellular stimulation of COS cells with a laminar flow stream of rhodamine-labelled EGF [56]. The PARTCELL experiments, which used special fluorescent indicators to visualize activation of Ras family G proteins [56] and tyrosine phosphorylation [45], were able to clarify the role of EGFR endocytosis as well as receptor density on the ligand independent lateral propagation (LILP) of EGF signals (Figure 4.7). Whereas many believed that LILP was a universal phenomena in cells, the PARTCELL experiments were able to show that endocytosis of activated EGFR could suppress LILP. Furthermore, it was shown that whether LILP occurs or not in a cell depends more generally on the amount of EGF stimulation, and receptor density [66]. Together with results from subsequent studies on the effect of tyrosine phosphatase inhibition on LILP and a mathematical analysis of EGF signaling [60], a picture is now emerging of how a reaction network of positive and negative regulators determines when and if LILP will occur within a cell (Tisher & Bastien, 2003). Although the physiological roles of LILP are still speculative, it may play a role in

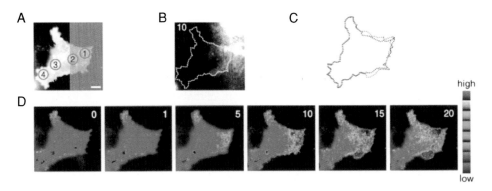

FIGURE 4.7. Local stimulation of normal COS cells with Rhodamine-EGF to visualize Ras activation. (A) Fluorescence image before stimulation. Violet region is exposed to rhodamine-EGF. Four ROIs are assigned across the laminar flows. (B) Rhodamine-EGF fluorescence after 10 minutes of stimulation. Cell is outlined with a dotted line. (C) Superimposed outlines of cell at 0 (black) and 20 (red) minutes. (D) Dual emission ratio images presented in intensity-modified display (IMD) mode. Right hand corner shows time in minutes after stimulation [66]. (Reproduced by permission from Sawano et al. 2002. Copyright 2002 Cell Press)

invasion and proliferation of cancer cells as EGFR is overexpressed in many carcinomas [66].

4.4.4. Microscale Integrated Sperm Sorter (MISS)

Laminar flows, which eliminates turbulent mixing, and the small dimensions of microfluidic systems, which allows rapid diffusion of small particles, has been used to separate particles based on their size. For example, Yager et al. have developed an H-filter in which two different liquid streams—one stream containing a mixture of different size particles such as cells and proteins, and another stream just composed of liquid—are introduced into a common channel. The two fluid streams flow parallel to one another, the large particles following the original streamline, while small particles diffuse rapidly between the streams and separate from the original solution. At the end of the main channel is a splitting junction which separates different regions of the main channel for downstream processing and analysis [30].

Such microfluidic interfaces can also be used for function-based cell sorting. Cho et al. designed a microfluidic system called the microscale integrated sperm sorter (MISS) to isolate motile sperm from nonmotile sperm and cellular debris [11]. This device also takes advantage of laminar flows in microfluidic channels, in a manner similar to H-filters, but applies the concept to sorting of cells. Motile sperm are isolated based on their ability to cross streamlines in a laminar fluid stream. The device can be used in clinical settings where small amounts of sperm need to be sorted. It can be used to select the most viable sperm for in vitro fertilization procedures and as disposable, at-home diagnostic tests for male infertility.

This simple, disposable device is made from poly(dimethylsiloxane) (PDMS) that is cast onto a silicon wafer with the desired microchannel features. After curing, the PDMS is peeled off and plasma oxidized to irreversibly seal it onto a glass surface. The design

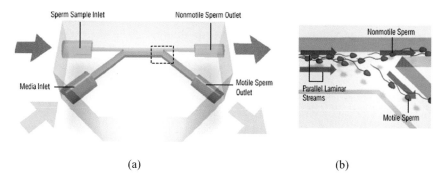

(a)   (b)

FIGURE 4.8. Schematic drawing of sperm sorting device. (a) Top view of device with two inlets and two outlet reservoirs. (b) Close up view of outlet channels where motile sperm are separated from nonmotile sperm.

contains two inlets (one for the sperm sample and one for the media) and two outlets (one for nonmotile sperm and one for motile sperm) shown in Figure 4.8a. Nonmotile sperm and debris flow along their initial streamlines and exit out one outlet, while the motile sperm swim into a parallel stream and exit out a separate outlet. (Figure 4.8b) The fluid flow is driven by a passively driven pumping system which uses horizontally oriented reservoirs for gravity-driven pumping. Gravity, surface tension, and channel resistance contribute to give a steady flow rate of sperm.

MISS is a self-contained, readily fabricated, functional microdevice that can manipulate living cells using microfluidics, avoiding the need for electronics or external power sources. It can be used in clinical settings and in the field of assisted reproductive technology. The device isolates sperm based on their motility and can be extended to the sorting of any other microorganisms that can swim efficiently. Other potential future uses include that for the preliminary screening of male infertility, as an indicator of toxicity, multiple connections or parallel integration of multiple MISS systems, and sorting on more subtle differences in motility [11].

4.4.5. Air-Sheath Flow Cytometry

Flow cytometry is a powerful tool to rapidly measure physicochemical characteristics of cells in liquid suspension using fluorescent probes that bind to specific cell-associated molecules (e.g. fluorochrome-labeled antibodies, nucleic acid probes, cell function probes, green fluorescent proteins). As labeled cells injected into the center of a fast stream of sheath liquid flow in single-file past a focused laser beam, they generate light scattering and fluorescence emission that are directed to detectors and converted into electric pulses. The measured amount of scattered light and fluorescence is used to enumerate cells at high rates ($\sim$25,000 cells/sec) and is correlated with the properties of cells such as shape, size [71]. The high-speed analytical capabilities of conventional flow cytometers have gained widespread use in routine laboratories and also, there have been many efforts to implement such functionalities into miniaturized bioanalytical systems. As a core component of flow cytometry, rapid delivery of cells suspended in narrow liquid streams is crucial for reliable analyses and has been an important theme in developing microfluidic systems that can be integrated with miniaturized optics to enable high-speed detection and enumeration of particulate samples [24, 54, 55, 74]. Microfabrication technologies promise opportunities to create new designs of flow geometries for transporting cells with microscale resolutions, however, large volumes of sheath liquid required to hydrodynamically focus streams of cells (250–1000 ml of sheath liquid for 1 ml of biological sample) remain as an issue in further reducing size requirements of miniaturized flow cytometry. Also, the use of conventional materials in microfabrication such as silicon or glass increases the cost and complexity of fluidic module fabrication.

In order to address the limiting issues associated with size and cost requirements of miniaturized flow cytometry, Huh et al. developed a new type of flow cytometer that uses a stable and disposable air-liquid two-phase microfluidic system to produce high-speed liquid streams of cells focused by air-sheath flows [36]. The main novelty of the system lies in the use of ambient air as a sheath fluid and the engineering of surface chemistries of disposable polymeric microchannels to stably maintain high-speed air-liquid two-phase flow configurations. As shown in Figure 4.3, a liquid sample containing cells is injected into a hydrophobic PDMS microfluidic channel and focused by air sheath flows driven by vacuum suction at the outlet to form multiple laminar streams of air and liquid. Experimental studies show that such two-phase microfluidic systems display unique characteristics useful to control fluid flows for better flow cytometry: (i) the spatial distribution of moving liquid and air is heavily affected by surface chemistries of channel walls. In hydrophobic channels, liquid streams are repelled from hydrophobic channel surfaces, which prevent lateral spreading of liquid sample and assist in sustaining continuous liquid streams without instabilities (Figure 4.3). Hydrophilic channel surfaces cause focused liquid streams to atomize into small droplets and quickly spread to form a lining of liquid along the sidewalls of microchannels (Figure 4.3). (ii) In concert with air and liquid flow rates, channel geometries play an important role in maintaining the stability of focused liquid streams in hydrophobic channels. For a fixed driving force for air flows, as the flow rate of liquid decreases, focused liquid streams become narrower and eventually, break up into small droplets due to two-phase flow instabilities. Hydrophobic channels with smaller heights delay break-up of liquid streams and permits formation of narrower continuous liquid streams at lower liquid flow rates [36, 37].

The high-speed air-liquid two-phase microfluidic systems are capable of stably transporting particulate samples in liquid streams and performing simple and volume-efficient basic flow cytometry to detect and count the number of fluorescent particles and cells (C_2C_{12} myoblasts) at moderate rates (∼100 samples/sec) using a photo-multiplier tube (PMT) mounted on an epi-fluorescent microscope [36]. Furthermore, air-liquid two-phase configuration in microscale opens new opportunities to use surface tension as an effective way to manipulate liquid streams. Such fluid flow switching capability is potentially useful for realizing miniaturized flow cytometry with particle/cell sorting capabilities. Huh et al. demonstrated manipulation of flow patterns of microscale air-liquid two-phase streams and precise positioning of liquid streams using a rapid and reversible change of surface energies by an electric potential—electrowetting [35]. As outlined in Fig. 4.9, an electric potential is applied to electrodes embedded in a channel floor coated with hydrophobic dielectric materials, and initiates surface energy changes to alter surface wettability in selective areas and to prompt liquid streams to quickly reposition to hydrophilic regions. This positional control of liquid streams is reversible, spatially selective, and operational at very high frequencies. If combined with optics that can feed out electric signals upon detection of cells to change surface energies and thus, the positions of liquid streams, this mechanism could

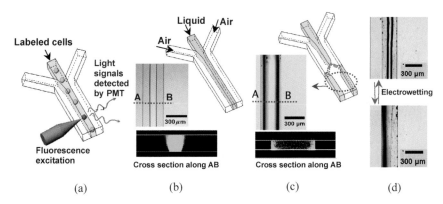

(a)   (b)   (c)   (d)

FIGURE 4.9. (a) Detection of fluorescently labeled cells transported by air-liquid two-phase streams. Simple fluorescence-assisted detection/enumeration system is built on an epi-fluorescence microscope equipped with a photo-multiplier tube (PMT). (b) A stable liquid stream focused by air sheath flows in a hydrophobic PDMS channel. (c) Spreading of a liquid stream in a hydrophilic PDMS channel prepared by plasma oxidation. Cross-sectional images were taken by confocal microscopy using a fluorescein solution as a working liquid. (d) Electrowetting-assisted flow pattern change and repositioning of a liquid stream. An electric potential applied to the left half of channel floor causes the liquid stream focused along the center to wet over the energized hydrophilic channel surface and rapidly move to its right.

potentially serve as a novel method to sort cells, adding new functionalities to air-sheath flow cytometry.

4.4.6. Immunoassays

A crucial part of many cellular assays is the detection and quantification of chemicals in the cellular microenvironment. Particularly useful are immunoassays which take advantage of the specific interactions between antigens and antibodies to identify and quantify the type and amount of analytes. When appropriately designed, immunoassays, which often take advantage of antigen-antibody interactions at the solid-liquid interface, can provide specific, precise, accurate, and reproducible estimates of levels of proteins in solution. Micro- and nanotechnology provides opportunities for integration and automation of immunoassays with microfluidic cell culture and sorting.

Immunoassays are an important analytical method with extremely high selectivity and sensitivity. The so-called sandwich assay system, for example, in which both capture and detection antibodies are used is often required for specific and sensitive determination in assays of biological samples such as sera. Conventional immunoassays using the 96-well plate format, require a long assay time, use large quantities of expensive reagents, and involve troublesome liquid-handling procedures. A microchip-based system overcomes these limitations and offers advantages such as enhanced reaction efficiency, simplified procedures, reduced assay time, and orders-of-magnitude reduction in consumption of samples and reagents [63]. Microchannels with dimensions in the 10–100 micrometer range are fabricated in glass, fused-silica substrates, or more recently PDMS using photolithography and etching techniques from the semiconductor industry [16]. The use of microchannels minimizes the analysis duration due to the reduced diffusion distance between immunological

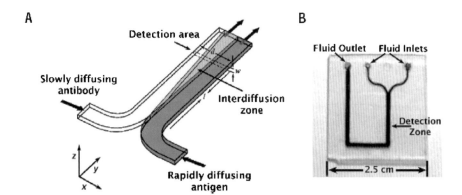

FIGURE 4.10. T-sensor device for diffusion immunoassay. (A) Schematic for diffusion assay with reagents pumped separately through inlets and meet at the junction. In the main channel the reagents mix only by diffusion. (B) Photograph of a microfluidic device used with the flow channel filled with dye to visualize [30]. (Reproduced by permission from Hatch 2004. Copyright 2004 Nature Publishing Group www.nature.com)

pairs. Nanoliter-scale fluid flow can be controlled using pumping methods such as electrokinetic pumping and hydrodynamic pressure [79].

Yager et al. have developed an immunoassay system in a T-sensor (Figure 4.10). Two solutions are pumped through separate inlets and combine into laminar streams that run parallel to each other in the main channel. Antigens in one stream are able to diffuse from one region into another and bind to antigen-specific antibodies. Based on the changes in the diffusive transport of the antigens after binding, the concentration of the antigen can be calculated [88]. Other examples of microchannel immunoassay systems are included in Table 4.1. Types are classified based on the type of interface that is involved- whether the antibodies are bound to channels, to microbeads, or are unbound. Technologies that are used include capillary electrophoresis (CE) based immunoassay systems. These systems involve separating the free antibody and antigen from the complex formed between the antigen and antibody using CE. Also commonly used are microbeads with adsorbed antibodies and flow-through systems.

It is often difficult to detect proteins in microchannels due to the small volumes (10 to 100 nL) involved and small amounts of analyte present [63]. To overcome this problem many have turned to amplifying the signal from each labeled molecule. Enzyme-linked immunosorbent assay (ELISA) or other immunosorbent assay systems that involve an antigen or antibody fixed onto a solid surface can be used to detect more analytes with a high sensitivity. In ELISA, the signal enhancement can be accomplished with liposome [61] or enzyme labels [19]. The most widely used enzymes in ELISA are alkaline phosphatase (ALP) or galactosidase (which catalyzes a hydrolysis reaction) and horseradish peroxidase or glucose oxidase (which catalyzes an electron transfer reaction) [63].

Future development efforts focus on improving the robustness of the microchip system and using new types of assays into the continuous-flow microchip format. New microchip designs that integrate both liquid-handling and compound dilution with assay functionality will allow us to eliminate many of the microplate preparation steps that are presently required for immunoassays. The goal towards higher density, smaller volume formats are limited fabrication processes, conventional liquid handling technology, and the difficulty in controlling evaporation in open systems as volumes continue to get smaller [67]. The

TABLE 4.1. Examples of microfluidic immunoassays.

Antibody-Antigen Interface	Technology	Setup	Surface	Antigen/Antibody	Detection	Ref.*
Liquid-liquid	T-sensor	Flow cell	Glass	Phenytoin/anti-phenytoin	Fluorescence	[31]
Bound to Channels	Microarray	Patterning by microfluidic networks	PDMS	Rabbit IgG/anti-rabbit IgG-FITC	Fluorescence	[3]
		Glass slides with Teflon dividers	Glass	Inflammatory cytokines induced by LPS or TNF-α/antibodies	Rolling circle amplification (RCS)	[69, 70]
	ELISA	Microchannels	PET microchannel	Mouse IgG/anti-mouse IgG	Electrochemical detection of D-Dimer alkaline phosphatase	[63]
Bound to beads	Bead-injection	Lab-on-valve Flow cell		Glutamic acid decarboxylase (GAD65)/Anti-GAD64	Absorbance of TMB substrate	[8]
	Packed bed	Packed bed of paramagnetic particles	Dynal Paramagnetic beads	-parathyroid hormone (PTH)/anti-PTH -IL-5/anti-IL-5	Fluorescence	[33]
	Two beads	Microfluidic device (Model PC-SC, Micralyne, Canada)		-TNF-α/anti-TNF-α -IL-6/anti-IL6	Confocal microscope fluorescence	[62]
	Luminex	Filter plat	Polystyrene beads		Fluorescently labeled anti-protein reporter antibodies	
Unbound	CE	Microchannels		Serum theophylline/Arginine-FITC	Laser induced fluorescence (LIF)	[86]
		Microchannels	Fused silica	Cortisol/anticortisol serum	Fluorescence	[44]
		Coupled to capillary reversed-phase LC	Clear Plexiglass	Glucagon/anti-glucagon	Laser induced fluorescence	[27]
		Microchip	Glass	Theophylline/anti-theophylline	Fluorescence	[9]

Other	ELISA	serial dilution network	PDMS/ polycarbonate membranes	Fluorescence	gp41, gp120/anti-gp41, anti-gp120	[41]

PDMS- Poly (dimethylsiloxane)
FITC- Fluorescein isothiocyanate
TNF- αTumor necrosis factor- α
PET- Polyethylene terephthalate
IL- interleukin
LC- liquid chromatography

IgG- Immunoglobulin G
LPS- lipopolysaccharide
ELISA- Enzyme linked immunosorbent assay
TMB- tetra-methyl benzidine
CE- capillary electrophoresis
gp- glycoprotein

*Bernard, A., Michel, B., Delamarche, E. (2001). "Micromosaic Immunoassays." *Analytical Chemistry* **73**: 8–12.
Carroll, A., Scampavia, L., Luo, D., Lernmark, A., Ruzicka, J. (2003). "Bead injection ELISA for the determination of antibodies implicated in type 1 diabetes mellitus." *Analyst* **128**: 1157–1162.
Chiem, N. and D. J. Harrison (1997). "Microchip-based capillary electrophoresis for immunoassays: Analysis of monoclonal antibodies and theophylline." *Analytical Chemistry* **69**(3): 373–378.
German, I. and R. T. Kennedy (2000). "Reversed-phase capillary liquid chromatography coupled on-line to capillary electrophoresis immunoassays." *Analytical Chemistry* **72**(21): 5365–5372.
Hatch, A., A. E. Kamholz, et al. (2001). "A rapid diffusion immunoassay in a T-sensor." *Nature Biotechnology* **19**(5): 461–465.
Hayes, M. A., N. A. Polson, et al. (2001). "Flow-based microimmunoassay." *Analytical Chemistry* **73**(24): 5896–5902.
Jiang, X. Y., J. M. K. Ng, et al. (2003). "A miniaturized, parallel, serially diluted immunoassay for analyzing multiple antigens." *Journal of the American Chemical Society* **125**(18): 5294–5295.
Kouty, L. B., D. Schmalzing, et al. (1996). "Microchip electrophoretic immunoassay for serum cortisol." *Analytical Chemistry* **68**(1): 18–22.
Roos, P., Skinner, CD. (2003). "A two bead immunoassay in a micro fluidic device using a flat laser intensity profile." *Analyst* **128**: 527–531.
Rossier, J., Girault, HH. (2001). "Enzyme linked immunosorbent assay on a microchip with electrochemical detection." *Lab on a chip* **1**: 153–157.
Schweitzer, B., S. Roberts, et al. (2002). "Multiplexed protein profiling on microarrays by rolling-circle amplification." *Nature Biotechnology* **20**(4): 359–365.
Schweitzer, B., Wiltshire, S., Lambert, J., O'Malley S., Kukanskis K, Zhu Z, Kingsmore SF, Lizardi PM, Ward DC. (2000). "Immunoassays with rolling circle DNA amplification: A versatile platform for ultrasensitive antigen detection." *Proc. Natl. Acad. Sci.* **97**(18): 10113–10119.
vonHeeren, F., E. Verpoorte, et al. (1996). "Micellar electrokinetic chromatography separations and analyses of biological samples on a cyclic planar microstructure." *Analytical Chemistry* **68**(13): 2044–2053.

combination of a cell-based bioassay system, nanoliter-scale liquid handling, integrated devices for compound dilution and assay functionality has the potential to create a truly advanced microfluidic bioassay system.

4.5. CONCLUSION

Interfaces can be found in numerous places in biology. At the microscale, important interfaces include those between cells, between different cells, between cells and their microenvironment, and between different chemicals in the body. Using traditional bioassay systems, it is difficult to recreate these biological boundaries with precise, microscale control. Microtechnology provides the tools that are necessary to overcome these limitations, providing a new type of interaction between the two disciplines of microtechnology and biology. Advanced bioassay systems, such as the ones described in this chapter take advantage of microfluidic interfaces to create microenvironments more suitable to studying fundamental questions in cell biology regarding the complex behavior of cells. Microfluidic interfaces also open new opportunities for immunoassays and cell sorting. In the future, further integration and development of these systems will enrich our understanding of physiological phenomena found in cell biology.

REFERENCES

[1] C.D. Bain. Motion of liquids on surfaces. *Chemphyschem*, 2(10):580–582, 2001.
[2] S.L.R. Barker, D. Ross et al. Control of flow direction in microfluidic devices with polyelectrolyte multilayers. *Analytical Chemistry*, 72(24):5925–5929, 2000.
[3] A. Bernard, B. Michel, and E. Delamarche. Micromosaic immunoassays. *Anal. Chem.*, 73(1):8–12, 2001.
[4] K. Bhadriraju and C.S. Chen. Engineering cellular microenvironments to cell-based drug testing improve. *Drug Discovery Today*, 7(11):612–620, 2002.
[5] R. Bharadwaj, J.G. Santiago et al. Design and optimization of on-chip capillary electrophoresis. *Electrophoresis*, 23(16):2729–2744, 2002.
[6] F. Bradke and C.G. Dotti. The role of local actin instability in axon formation. *Science*, 283(5409):1931–1934, 1999.
[7] R. Brock and T.M. Jovin. Heterogeneity of signal transduction at the subcellular level: microsphere-based focal EGF receptor activation and stimulation of Shc translocation. *J. Cell Sci.*, 114(13):2437–2447, 2001.
[8] A.D. Carroll, L. Scampavia, D. Luo, A. Lernmark, and J. Ruzicka. Bead injection ELISA for the determination of antibodies implicated in type 1 diabetes mellitus. *Analyst.*, 128(9):1157–1162, Sep. 2003.
[9] N.H. Chiem and D.J. Harrison. Monoclonal antibody binding affinity determined by microchip-based capillary electrophoresis. *Electrophoresis*, 19(16–17):3040–3044, Nov. 1998.
[10] D.T. Chiu, N.L. Jeon et al. Patterned Deposition of Cells and Proteins onto Surfaces by Using Three-dimensional Microfluidic Systems. *Proceedings of the National Academy of Sciences of the United States of America.* 97(6), pp. 2408–2413, 2000.
[11] B.S. Cho, T.G. Schuster et al. Passively driven integrated microfluidic system for separation of motile sperm. *Analytical Chemistry*, 75(7):1671–1675, 2003.
[12] S.K. Cho, H.J. Moon et al. Creating, transporting, cutting, and merging liquid droplets by electrowetting-based actuation for digital microfluidic circuits. *J. Microelectromech. Systems*, 12(1):70–80, 2003.
[13] T. Chovan and A. Guttman. Microfabricated devices in biotechnology and biochemical processing. *Trends in Biotechnology*, 20(3):116–122, 2002.
[14] A.A. Darhuber, J.P. Valentino et al. Thermocapillary actuation of droplets on chemically patterned surfaces by programmable microheater arrays. *J. Microelectromech. Systems* 12(6):873–879, 2003.
[15] J.M. Davis and S.M. Troian. Influence of attractive van der Waals interactions on the optimal excitations in thermocapillary-driven spreading. *Physical Review E*, 67(1), 2003.

[16] E. Delamarche, A. Bernard et al. Patterned delivery of immunoglobulins to surfaces using microfluidic networks. *Science*, 276(5313):779–781, 1997.

[17] E. Delamarche, A. Bernard, H. Schmidt, A. Bietsch, B. Michel, and H. Biebuyck. Microfluidic networks for chemical patterning of substrates: design and application to bioassays. *J. Am. chem. soc.*, 120:500–508, 1998.

[18] S.K.W. Dertinger, X.Y. Jiang et al. Gradients of Substrate-Bound Laminin Orient Axonal Specification of Neurons. *Proceedings of the National Academy of Sciences of the United States of America* 99(20): pp. 12542–12547, 2002.

[19] E. Diamandis and T.K. Christopoulos. Immunoassay. A. P. Inc. San Diego, 1996.

[20] H. Ding, K. Chakrabarty et al. Scheduling of microfluidic operations for reconfigurable two-dimensional electrowetting arrays. *Ieee Transactions on Computer-Aided Design of Integrated Circuits and Systems*, 20(12):1463–1468, 2001.

[21] D.C. Duffy, J.C. McDonald, O.J. Schueller, A., and G.M. Whitesides. Rapid prototyping of microfluidic systems in poly(dimethylsiloxane). *Anal. Chem.*, 70:4974-4984, 1998.

[22] A. Folch and M. Toner. Microengineering of cellular interactions. *Annual Review of Biomedical Engineering*, 2:227-+, 2000.

[23] A. Folch and M. Toner. Cellular micropatterns on biocompatible materials. *Biotechnology Progress*, 14(3):388–392, 1998.

[24] A.Y. Fu, C. Spence et al. A microfabricated fluorescence-activated cell sorter. *Nat. Biotechnol.*, 17(11):1109–1111, 1999.

[25] B.S. Gallardo, V.K. Gupta et al. ELectrochemical principles for active control of liquids on submillimeter scales. *Science*, 283(5398):57–60, 1999.

[26] B.D. Gates and G.M. Whitesides. Replication of vertical features smaller than 2 nm by soft lithography. *J. Am. Chem. Soc.*, 125(49):14986–14987, 2003.

[27] I. German and R.T. Kennedy. Reversed-phase capillary liquid chromatography coupled on-line to capillary electrophoresis immunoassays. *Anal Chem.*, 72(21):5365-5372, 1 Nov. 2000.

[28] J.T. Groves, S.G. Boxer et al. Electric Field-Induced Reorganization of Two-component Supported Bilayer Membranes. *Proceedings of the National Academy of Sciences of the United States of America* 94(25): pp. 13390–13395, 1997.

[29] J.T. Groves, N. Ulman et al. Micropatterning fluid lipid bilayers on solid supports. *Science*, 275(5300):651–653, 1997.

[30] A. Hatch, E. Garcia et al. Diffusion-based Analysis of Molecular Interactions in Microfluidic Devices. *Proceedings of the Ieee*, 92(1):126–139, 2004.

[31] A. Hatch, A.E. Kamholz et al. A rapid diffusion immunoassay in a T-sensor. *Nature Biotechnology*, 19(5):461–465, 2001.

[32] R.A. Hayes and B.J. Feenstra. Video-speed electronic paper based on electrowetting. *Nature*, 425(6956):383–385, 2003.

[33] M.A. Hayes, T.N. Polson, A.N. Phayre, and A.A. Garcia. Flow-based microimmunoassay. *Anal. Chem.*, 73(24):5896–5902, 15 Dec. 2001.

[34] H. Hisamoto, T. Saito, M. Tokeshi, A. Hibara, and T. Kitamori. Fast and high conversion phase-transfer synthesis exploiting the liquid-liquid interface formed in a microchannel chip. *Chemical Communications* 24:2662–2663, 2001.

[35] D. Huh, A.H. Tkaczyk et al. Reversible switching of high-speed air-liquid two-phase flows using electrowetting-assisted flow-pattern change. *J. Am. Chem. Soc.*, 125(48):14678–14679, 2003.

[36] D. Huh, Y.C. Tung et al. Use of air-liquid two-phase flow in hydrophobic microfluidic channels for disposable flow cytometers. *Biomedical Microdevices*, 4(2):141–149, 2002.

[37] D. Huh, H.H. Wei et al. Development of Stable and Tunable High-Speed Liquid Jets in Microscale for Miniaturized and Disposable Flow Cytometry. *Proceedings of 2nd IEEE-EMBS Special Topics Conference on Microtechnologies in Medicine & Biology*: pp. 449–452, 2002.

[38] R.F. Ismagilov, A.D. Stroock et al. Experimental and theoretical scaling laws for transverse diffusive broadening in two-phase laminar flows in microchannels. *App. Phys. Lett.*, 76(17):2376–2378, 2000.

[39] K.F. Jensen. Microchemical systems: Status, challenges, and opportunities. *Aiche J.*, 45(10):2051–2054, 1999.

[40] N.L. Jeon, H. Baskaran et al. Neutrophil chemotaxis in linear and complex gradients of interleukin-8 formed in a microfabricated device. *Nat. Biotechnol.*, 20(8):826–830, 2002.

[41] X. Jiang, J.M. Ng, A.D. Stroock, S.K. Dertinger, and G.M. Whitesides. A miniaturized, parallel, serially diluted immunoassay for analyzing multiple antigens. *J. Am. Chem. Soc.*, 125(18):5294–5295, 7 May 2003.

[42] T.J. Johnson, D. Ross et al. Rapid microfluidic mixing. *Anal. Chem.*, 74(1):45–51, 2002.

[43] P.J.A. Kenis, R.F. Ismagilov et al. Microfabrication inside capillaries using multiphase laminar flow patterning. *Science*, 285(5424):83–85, 1999.

[44] L.B. Koutny, D. Schmalzing, T.A. Taylor, and M. Fuchs. Microchip electrophoretic immunoassay for serum cortisol. *Anal. Chem.*, 68(1):18–22, 1 Jan 1996.

[45] K. Kurokawa, N. Mochizuki, Y. Ohba, H. Mizuno, A. Miyawaki, and M. Matsuda. A pair of fluorescent resonance energy transfer-based probes for tyrosine phosphorylation of the Crk11 adaptor protein in vivo. *J. Biol. Chem.*, 276:31305–31310, 2001.

[46] P. Lam, K.J. Wynne et al. Surface-tension-confined microfluidics. *Langmuir*, 18(3):948–951, 2002.

[47] J. Lee, H. Moon et al. Electrowetting and electrowetting-on-dielectric for microscale liquid handling. *Sensors and Actuators a-Physical* 95(2-3):259–268, 2002.

[48] P.C.H. Li and D.J. Harrison. Transport, manipulation, and reaction of biological cells on-chip using electrokinetic effects. *Anal. Chem.* 69(8):1564–1568, 1997.

[49] V. Linder, H.K. Wu et al. Rapid prototyping of 2D structures with feature sizes larger than 8 mu m. *Anal. Chem.* 75(10):2522–2527, 2003.

[50] J.C., W.D. Love, H.O. Jacobs, and G.M. Whitesides. Microscope projection photolithography for rapid prototyping of masters with micron-scale features for use in soft lithography. *Langmuir*, 17:6005–6012, 2001.

[51] M. Madou. Fundamentals of Microfabrication, 2002.

[52] J. McDonald, D.C. Duffy, J.R., D.T., C., H.W. Anderson, O.J. Schueller, G.M. Whitesides. Fabrication of microfluidic system in poly(dimethysiloxane). *Electrophoresis*, 21:27–40, 2000.

[53] J.C. McDonald and G.M. Whitesides. Poly(dimethylsiloxane) as a material for fabricating microfluidic devices. *Accounts of Chemical Research*, 35(7):491–499, 2002.

[54] R. Miyake, H. Ohki et al. Investigation of sheath flow chambers for flow cytometers—(Micro machined flow chamber with low pressure loss). *Jsme Int. J. Series B-Fluids and Thermal Eng.* 40(1):106–113, 1997.

[55] R. Miyake, H. Ohki et al. Flow cytometric analysis by using micro-machined flow chamber. *Jsme Int. J. Series B-Fluids and Thermal Eng.*, 43(2):219–224, 2000.

[56] N. Mochizuki, S. Yamashita, K. Kurokawa, Y. Ohba, T. Nagai, A. Miyawaki, and M. Matsuda. Spatiotemporal images of growth-factor-induced activation of Ras and Rap1. *Nature*, 411:1065–1068, 2001.

[57] P. Paik, V.K. Pamula et al. Electrowetting-based droplet mixers for microfluidic systems. *Lab on a Chip*, 3(1):28–33, 2003.

[58] A. Papra, A. Bernard et al. Microfluidic networks made of poly(dimethylsiloxane), Si, and Au coated with polyethylene glycol for patterning proteins onto surfaces. *Langmuir*, 17(13):4090–4095, 2001.

[59] M.G. Pollack, R.B. Fair et al. Electrowetting-based actuation of liquid droplets for microfluidic applications. *App. Phys. Lett.*, 77(11):1725–1726, 2000.

[60] A.R. Reynolds, C. Tischer, P.J. Verveer, O. Rocks, and P.I. Bastiaens. EGFR activation coupled to inhibition of tyrosine phosphatases causes lateral signal propagation. *Nat. Cell Biol.*, 5(5):447–453, May 2003.

[61] M. Roberts and R.A. Durst. Investigation of liposome based immunomigration sensor for the detection of polychlorinated-biphenyls. *Anal. Chem.*, 67:482–491, 1995.

[62] P. Roos and C.D. Skinner. A two bead immunoassay in a micro fluidic device using a flat laser intensity profile for illumination. *Analyst.*, 128(6):527–531, Jun 2003.

[63] J. Rossier and H.H. Girault. Enzyme linked immunosorbent assay on a microchip with electrochemical detection. *Lab on a chip*, 1:153–157, 2001.

[64] K. Sato, A. Hibara et al. Microchip-based chemical and biochemical analysis systems. *Advanced Drug Delivery Reviews*, 55(3):379–391, 2003.

[65] K. Sato, A. Hibara, M. Tokeshi, H. Hisamoto, and T. Kitamori. Integration of chemical and biochemical analysis systems into a glass microchip. *Analytical Sciences*, 19:15–22, 2003.

[66] A. Sawano, S. Takayama, M. Matsuda, and A. Miyawaki. Lateral propagation of EGF signaling after local stimulation is dependent on receptor density. *Developmental Cell*, 3:245–257, 2002.

[67] J.A. Schwartz, J.V. Vykoukal et al. Droplet-based chemistry on a programmable micro-chip. *Lab on a Chip*, 4(1):11–17, 2004.

[68] P.V. Schwartz. Meniscus force nanografting: Nanoscopic patterning of DNA. *Langmuir*, 17(19):5971–5977, 2001.

[69] B. Schweitzer, S. Wiltshire, J. Lambert, S. O'Malley, K. Kukanskis, Z. Zhu, and S.F. Kingsmore. Inaugural article: immunoassays with rolling circle DNA amplification: a versatile platform for ultrasensitive antigen detection. *Proc. Natl. Acad. Sci. USA*, 97(18):10113–10119, 29 Aug 2000.

[70] B. Schweitzer, S. Roberts, B. Grimwade, W. Shao, M. Wang, Q. Fu, Q. Shu, I. Laroche I, Z. Zhou, V.T. Tchernev, J. Christiansen, M. Velleca, S.F. Kingsmore, P.M. Lizardi, and D.C. Ward. Multiplexed protein profiling on microarrays by rolling-circle amplification. *Nat. Biotechnol.*, 20(4):359–365, Apr. 2002.

[71] H.M. Shapiro. Practical Flow Cytometry. New York, Wiley-Liss, 1995.

[72] J. Shim, T.F. Bersano-Begey, X. Zhu, A.H. Tkaczyk, J.J. Linderman, and S. Takayama. Micro- and Nano-technologies for studying cellular function. *Current Topics in Medicinal Chemistry*, 3:687–703, 2003.

[73] S.K. Sia and G.M. Whitesides. Microfluidic devices fabricated in poly(dimethylsiloxane) for biological studies. *Electrophoresis*, 24(21):3563–3576, 2003.

[74] D. Sobek, A.M. Young et al. A Microfabricated Flow Chamber for Optical Measurement in Fluids. *Proceedings of MEMS '93*: pp. 219–224, 1993.

[75] H.A. Stone and S. Kim. Microfluidics: Basic issues, applications, and hallenges. *Aiche J.*, 47(6):1250–1254, 2001.

[76] A.D. Stroock, S.K.W. Dertinger et al. Chaotic mixer for microchannels. *Science*, 295(5555):647–651, 2002.

[77] A.D. Stroock, M. Weck et al. Patterning electro-osmotic flow with patterned surface charge (vol. 84, pg. 3314, 2000). *Phys. Rev. Lett.*, 86(26):6050–6050, 2001.

[78] A.D. Stroock and G.M. Whitesides. Controlling flows in microchannels with patterned surface charge and topography. *Acc. Chem. Res.* 36(8):597–604, 2003.

[79] S.A. Sundberg. High-throughput and ultra-high-throughput screening: solution- and cell-based approaches. *Curr. Opin. Biotechnol.* 11(1):47–53, 2000.

[80] S. Takayama, J.C. McDonald et al. Patterning cells and their Environments using Multiple Laminar Fluid Flows in Capillary Networks. *Proceedings of the National Academy of Sciences of the United States of America* 96(10): pp. 5545–5548, 1999.

[81] S. Takayama, E. Ostuni et al. Laminar flows—Subcellular positioning of small molecules. *Nature*, 411(6841):1016–1016, 2001.

[82] S. Takayama, E. Ostuni et al. Selective chemical treatment of cellular microdomains using multiple laminar streams. *Chem. Biol.*, 10(2):123–130, 2003.

[83] S. Takayama, E. Ostuni et al. Topographical micropatterning of poly(dimethylsiloxane) using laminar flows of liquids in capillaries. *Advanc. Mater.* 13(8):570-+, 2001.

[84] A.M. Taylor, S.W. Rhee et al. Microfluidic multicompartment device for neuroscience research. *Langmuir*, 19(5):1551–1556, 2003.

[85] M. Unger, H.P. Chou, T. Thorsen, A. Scherer, and S.R. Quake. Monolithic microfabricated valves and pumps by multilayer soft lithography. *Science*, 288(5463):113–116, 2000.

[86] F. von Heeren, E. Verpoorte, A. Manz, and W. Thormann. Micellar electrokinetic chromatography separations and analyses of biological samples on a cyclic planar microstructure. *Anal. Chem.*, 68(13):2044–2053, 1 Jul. 1996.

[87] A. van Oudenaarden and S.G. Boxer. Brownian ratchets: Molecular separations in lipid bilayers supported on patterned arrays. *Science*, 285(5430):1046–1048, 1999.

[88] B.H. Weigl and P. Yager. Microfluidics—Microfluidic diffusion-based separation and detection. *Science*, 283(5400):346–347, 1999.

[89] G.M. Whitesides, E. Ostuni et al. Soft lithography in biology and biochemistry. *Annual Review of Biomedical Engineering*, 3:335–373, 2001.

[90] H. Wu, T.W. Odom, D.T. Chiu, and G.M. Whitesides. Fabrication of complex three-dimensional microchannel systems in PDMS. *J. Am. Chem. Soc.*, 125(2):554–559, 2003.

[91] T.L. Yang, E.E. Simanek et al. Creating addressable aqueous microcompartments above solid supported phospholipid bilayers using lithographically patterned poly(dimethylsiloxane) molds. *Anal. Chem.*, 72(11):2587–2589, 2000.

[92] B. Zhao, J.S. Moore et al. Surface-directed liquid flow inside microchannels. *Science*, 291(5506):1023–1026, 2001.

[93] B. Zhao, J.S. Moore et al. Principles of surface-directed liquid flow in microfluidic channels. *Anal. Chem.* 74(16):4259–4268, 2002.

[94] B. Zhao, N.O.L. Viernes et al. Control and applications of immiscible liquids in microchannels. *J. Am. Chem. Soc.*, 124(19):5284–5285, 2002.

5

Multi-phenotypic Cellular Arrays for Biosensing

Laura J. Itle[1,2], Won-Gun Koh[1], and Michael V. Pishko[3]

[1]*Department of Chemical Engineering, The Pennsylvania State University, University Park, PA 16802-4420*

[2]*The Huck Institute for the Life Sciences, The Pennsylvania State University, University Park, PA 16802-4420*

[3]*Departments of Chemical Engineering, Materials Science & Engineering, and Chemistry, The Pennsylvania State University, University Park, PA 16802-4420*

5.1. INTRODUCTION

The term "biosensor" is a broad based term referring to a sensor that uses a biological molecule as the sensing element. A "cell-based" biosensor, then, would utilize a prokaryotic or eukaryotic cell or cell line as the sensing agent (Figure 5.1). Single phenotype biosensors use one type prokaryotic or eukaryotic cell line as the sensing elements, which are contrasted to single-cell biosensors that may use, as the name implies, one cell as the sensing element. Though many methods have measured chemical toxins have utilized chemical, nucleic acid, and antibody approaches [1], sensing technology as a whole would be improved through the introduction of whole cell based biosensors. Whole cell-biosensors have several inherent advantages over DNA, RNA, and protein arrays. Foremost, a whole cell biosensor can offer functional information, i.e., information about the effect of a stimulus on a living system [2]. Functional information includes the effects of stimuli on cell health (toxicity) as well as cell function. Secondly, the use of cell-based biosensors eliminates the need for costly purification steps, such as the isolation and retrieval of RNA and DNA. A cell based biosensor also provides natural signal amplification of a response through cellular pathways and cellular cascades [3].

The use of cell-based biosensors has rapidly advanced in the last decade for environmental monitoring, sensing for chemical and biological warfare agents, and high-throughput

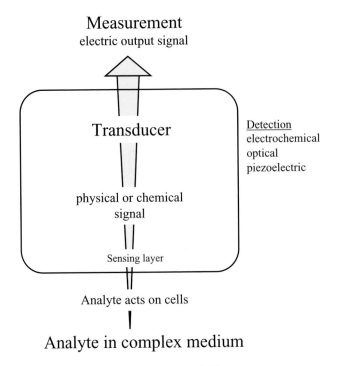

FIGURE 5.1. Schematic of a Biosensor.

drug screening. Of particular importance is the application of cell-based systems to drug discovery, because cells represent the ultimate target for pharmaceuticals. In this instance, a multi-phenotypic biosensor would allow systemic information to be obtained for a drug candidate in a one-pass system. For example, a drug candidate could be tested on hepatocytic and macrophagic cells simultaneously to gauge that candidates affect on liver function and immune response in a single assay step.

Additionally, with the increasing number of drug candidates available in the recent years, the demands for high-throughput drug screening system with lowered costs [4] have stimulated development of assay miniaturization [5–7]. To move towards assay miniaturization, significant efforts have focused on the fabrication of microarrays of living cells using a variety of cell patterning techniques. Cellular arrays are being coupled with fluorescence technology (either whole cell fluorescence or green-fluorescent protein transfected cells) and incorporated into microfluidic devices to optically detect the physiological changes of cells by drug candidates. As an alternative to measuring the optical properties of fluorescent cells, sensors can utilize the electrochemical properties of cells. Changes in the electrical properties of neural, cardiac, or pancreatic beta cells can be directly measured in response to changes in external environment [3].

In order to become a widely used sensing tool, multi-phenotypic biosensors must meet several technical criteria. First, the fabrication of cellular arrays must be relatively easy, inexpensive, and non-toxic. These fabrication methods should be conducive to storage of the biosensor and provide for easy maintenance and portability of cell lines. There must be a non-invasive method of ascertaining, recording, and displaying physiological changes in cells. In this chapter, we focus on the development of multi-phenotypic biosensors by addressing fabrication methods, methods of measuring physiological changes, current examples of multi-phenotypic systems, and future work in the development of multi-phenotypic biosensors.

5.2. FABRICATION OF MULTI-PHENOTYPIC ARRAYS

Cell patterning, the ability to control the positions of cells on a surface, is important for various biotechnologies, including the development of cell-based biosensors. Patterned arrays of single or multiple cell types in culture serve as a model system for exploration of fundamental cell biology, especially cell/cell, cell/surface, and cell/matrix interactions. In biosensor applications, accurate positioning of "sensing" cells is critical for monitoring the status of the cells. Patterns of cells on solid supports can be generated by: patterning adhesive molecules, integrating surface modification and microfabrication techniques, as in photolithography and soft lithography, and immobilizing cells in three-dimensional, patterned hydrogels.

5.2.1. Surface Patterning

Direct pattering of cells on surfaces can be accomplished using chemisorption or-physisorption of cells or cellular adhesion molecules, or more recently, through quasi-ordered patterning using electrophoretic deposition (EPD). Initial attempts to pattern neurons on solid substrates utilized chemically modified glass treated with silane monolayers to promote cellular adhesion and treated with a perfluorinated alkylsilane to prevent cellular adhesion [8–11]. An alternative method of selectively patterning one phenotype of cells is electrophoretic deposition (EPD). A method originally used for patterning fine particles, EPD uses either ac or dc fields to pattern yeast on an electrode surface [12]. The main advantage of this method is that the patterning is reversible; removal of the electric field disrupts the pattern. However, the direct patterning of cells to the surface does not necessarily allow the patterning of specific cells in specific regions.

5.2.2. Photolithography

Photolithography is the most commonly used method of cell patterning and can be divided into two main classes. In the first method, micropatterns are generated using light, photoresist, and a photomask. The photoresist is exposed to ultraviolet light through the photomask and developed, resulting in the desired patterns. Subsequently, materials such as cell-adhesion proteins are applied on the patterned substrates. Finally, the photoresist is stripped. Incubation with cell solution results in the desired pattern. Techniques involving the deposition of adhesion promoting materials include the formation of self-assembled monolayers (SAMs) of silanes [9], bioactive peptides [13], sulfated hyaluronic acids [14], and attachment factors [15].

Conversely, adhesive surfaces can be patterned with non-adhesive materials. Early methods involved simply patterning a glass substrate with photoresist [16, 17]. This method worked for primary cardiomyocytes, but proved unreliable for other cell lines because cells adhered to the photoresist. The patterning of a protein resistant material, poly(ethylene) oxide, proved to be ideal growing cells on localized regions of substrate [17]. Another method for patterning non-adhesive micropatterns involves the precise regional modification of surfaces via photochemical fixation of phenyl azido-derivatized polymers on substrates [18]. For the precise control of cell positioning, hydrophobic surfaces are micropatterned with a hydrophilic polymer so that cells adhere and grow only on the hydrophobic region, creating cellular patterns [19].

5.2.3. Soft Lithography

As an alternative to photolithography, Whitesides and colleagues have developed a set of patterning methods referred to as "soft lithography" [20]. This method utilizes a aoft elastomeric material for pattern transfer and modification. Currently, the most common soft lithographic techniques used for cell patterning are microcontact printing (μ-CP) and patterning with microfluidic networks.

Microcontact printing (μ-CP) method is based on the pattern transfer of the material of interest from the poly(dimethylsiloxane) (PDMS) stamp onto the substrate surfaces. A common utilization of of μ-CP involves the transfer of a self-assembled alkanethiol monolayer from PDMS stamp to a gold or silver thin film [20, 21]. The terminal position of an alkanethiol can possess a variety of functional groups, making it possible to control the properties of surface. For example, an alkanethiol terminated with oligo(ethylene glycol) or ethylene glycol can form SAMs that resist the adhesion of protein and cells. By patterning the formation of SAMs using μ-CP into regions that promote and resist adsorption of protein, attachment of cells to surfaces could be confined to rows 10–100 μm, or to islands whose size was the same as that of single cell. Thus far, this method has allowed for the patterning of cells, but not the selective patterning of multiple phenotypes.

The use of microfluidic networks, which can deliver solutions of interest to selected regions of a substrate, is a simple method for patterning several biological molecules simultaneously. Microfluidic channels can be fabricated by bringing PDMS into conformal contact with a photoresist master, curing the PDMS, and simply peeling the master away [22]. Microfluidic systems have distinctive properties as a result of their small dimensions. One notable feature is that the flow of liquid inside microchannels has a low Reynolds number and is laminar. When two or more streams of laminar flow are joined into a single stream, the combined streams flow parallel to each other without any turbulent mixing. This ability to generate and sustain parallel streams of different solutions without mixing gives microfluidic patterning an inherent advantage in the generation of multi-phenotypic arrays.

Biomolecules can be introduced at the entrance of the microfluidic channel; the liquid will be pulled into the channels via capillary action. Initially, immunoglobulins were patterned with submicron resolution on a variety of substrates [23, 24]. Microchannels were also used to produce patterns of cells on biocompatible substrates [25]. Here, a single cell type was placed into a microfluidic channel via capillary action, while the adjacent channels were filled with phosphate buffered saline solution (PBS). After this single cell line was in place, a secondary cell line was introduced into the channels previously containing PBS. In this way, multiple cell lines were patterned directly to cell adherent surfaces using microfluidic channels.

5.2.4. Poly(ethylene) Glycol Hydrogels

A final method for fabricating ordered arrays of cells involves the encapsulation and immobilization of cells in hydrogels. Cell encapsulation has been used to help cells achieve a more *in vivo* like state under culture conditions. Encapsulation materials range from natural materials (i.e., alginate, chitosan, and collagen) to totally synthetic materials (i.e., poly(ethylene) oxide (PEO), poly(ethylene) glycol (PEG), hydrolyzed poly(vinyl) alcohol, or poly(propylene) fumarate). These natural and synthetic polymer chains can form highly hydrated polymeric materials termed hydrogels.

An method of encapsulation ideal for biosensing applications is the use of poly(ethylene) glycol (PEG). PEG is non-degradable, hydrophilic polymer that has been widely employed as a biomaterial to obtain biocompatibility because of its remarkable non-adhesive nature towards protein and cells [26–28]. Typically, PEG has been incorporated onto biomaterial surfaces via surface grafting [29], plasma polymerization [30], or physical adsorption of PEG-containing block copolymers [17, 31]. Initial research focused on using PEG or oligoethylene glycol (OEG) as a surface modification technique to control protein or cell adhesion. For example, poly(ethylene) glycol patterns have been used to generate high density cellular arrays on glass, by defining specific regions of glass substrata for cell attachment [19, 32].

Different molecular weight of PEGs can be easily converted into acrylates such as PEG diacrylate (PEG-DA), and polymerization of acrylated PEG in the presence of light and photoinitiator yields a highly cross-linked hydrogel network [33]. These networks can be easily patterned using photolithography [34]. PEG hydrogels have a high equilibrium water content, and it was demonstrated that microporosity of PEG hydrogels could be easily altered [33, 35, 36]. The aqueous environment of PEG hydrogels is not only appropriate for the encapsulation of various biomolecules such as proteins, nucleic acids, and even whole cells, but allows for rapid transport of small molecules, including nutrients and analytes of interest, to the encapsulated cell or groups of cells. Furthermore, PEG hydrogels have been shown to be both biocompatible and non-fouling in complex environments [37]. It is also possible to modify PEG hydrogels with extracellular matrix adhesion peptides such as the RGD integrin binding domain, which enhances cell spreading [38–42].

Because of these characteristics, PEG hydrogels have been evaluated for multiple *in vivo* uses, including implanted glucose sensors [43, 44], drug delivery devices [45, 46], and tissue engineering [47–49]. Multiple mammalian cell lines have been encapsulated within PEG hydrogels, including murine fibroblasts, macrophages, and hepatocytes [42, 50], rat osteoblasts [51], embryonic rat cortical neurons and astrocytes [52], human tracheal scar fibroblasts [53], and human hepatoma (HepG2) cells [54]. Additionally, the transparent nature of PEG hydrogels also makes them suitable for various optical detection schemes when they are used in biosensing applications.

5.3. DETECTION METHODS FOR CELL BASED SENSORS

After selecting a method of patterning arrays of cells for sensing mechanisms, a method of measuring physiological changes within the cells needs to be selected. Two main methods for the detection of physiological changes of cells are employed in biosensor application: electrical and optical. These methods include the use of microelectronics, naturally luminescent proteins, green fluorescent proteins (GFPs), and intracellular fluorescent probes. In this section, we highlight the applications of these detection methods in the fabrication of whole cell sensors and their potential application to multi-phenotypic sensors.

5.3.1. Microelectronics

Modern drug development often lies in the generation of products that interact with neurons and cause the generation of electrical singles. Consequently, nerve cells used in biosensing applications are often immobilized on a micro-machined silicon support, which allows for the integration of microelectronic devices, including microelectrodes and miniature field-effect transistors (FETs) [55]. The use of microelectrodes and miniature FETs is predicated on the knowledge that drug-receptor binding results in electron flow along the cleft between a neuronal membrane and an electrically active substrate [55].

The generation of sensors that utilize the measurement of electrical singles is dependent on the integration of neuronal cells with silicon chips. In the development of the neuron-FET system, neurons were mounted on thin insulating layers of a "gate" oxide on silicon [56]. When a positive bias voltage is applied to the surface, mobile positive electrons accumulate near the surface such that a change in the action potential of the neuron lifts the voltage gate, causes a shift in electron density and a change in the voltage profile [56–58].

As an alternative to using FETs, neuronal cells can be grown on electrode arrays with analog amplifiers measuring any changes in voltage [59]. This method was demonstrated to be used for the determination of small concentrations of strychnine [59]. Neuron based biosensors relying on the measurement of electrical signals have been shown to be useful in measuring multiple pharmacological agents, including the excitors acethylcholine and glutamate, and the inhibitors, glycine and GABA [60]. The use of cells in conjunction with electrical interfaces is not limited to neurons. Cell adhesion, spreading, mortality, as well as the detection of analytes, including glucose, can be measured by monitoring changes in electrode impedance [61–64].

5.3.2. Fluorescent Markers For Gene Expression and Protein Up-regulation

In order to use a bioluminescent reporting system in a whole cell biosensor, the analyte of interest should cause a change in cellular gene expression resulting in the production of the bioluminescent protein. If we simply wish to detect the presence of analyte, the emission of any light is a sufficient outcome. If we are interested in the concentration of the analyte in the system, the sensor should be able to be calibrated such that the emitted light is proportional to the concentration of the analyte.

The luciferase gene has been introduced into a number of prokaryotic and eukaryotic organisms to monitor specific gene expression via the emission of light [65–68]. This system has been derived from a number of organisms (Figure 5.2A), but the most commonly used

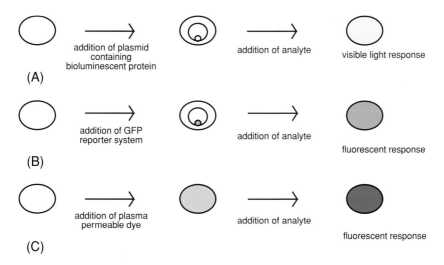

FIGURE 5.2. Optical Detection Methods (A) Bioluminescent Proteins, (B) Green Fluorescent Protein (C) Whole Cell Fluorescence.

system involves the promotor-less *lux* genes from *Vibrio fischeri*. The *lux* gene can be introduced either with a plasmid based system [69, 70] or a transposon delivery system [71]. The use of bioluminescent bacteria for the sensing of environmental toxins is varied. Researchers have used bioluminescent whole cell sensors to detect toxins such as mitomycin C [72], organic contaminants such as benzene [71], toluene [71], xylene [71], naphthalene [70, 73], 3,5-dicholorphenol [74], metal ions such as copper [74], lead [74], and nutrients such as iron [69] and phosphates [75, 76].

For applications other than environmental sensing, i.e., high throuput drug screening, prokaryotic cells and simple eukaryotic cells would not yield as pertinent information as the use of mammalian cells. Bioluminescent systems have been introduced into several mammalian systems, including monkeys [77, 78] and rabbits [79]. Unfortunately, bioluminescent protein expression generally requires the addition of a cofactor or substrate for luminescence [80]. The necessity of a cofactor or substrate for maximal activity is a distinct disadvantage for a self-contained biosensor. In contrast to the bioluminescent gene system involving luciferase, light emission from green fluorescent protein (GFP) does not require any additional cofactors or substrates from its native species *Aequorea victoria victoria* [81], which allows for real time detection in cells (Figure 5.2B). The isolation and sequencing of GFP jellyfish [82] allowed GFPs to be readily incorporated into a variety of cell systems. Subsequently, GFP was incorporated in prokaryotic systems (*E. coli*) and eukaryotic systems with green fluorescence visible for at least 10 minutes when illuminated with 450 to 490 nm light [81]. GFPs have been introduced into mammalian cell lines, such as HeLa cells and 3T3 fibroblast cells, using transient transfection [83].

GFPs can be used in conjunction with multi-phenotypic biosensors to measure the up-regulation of gene expression [81, 84], intracellular protein trafficking [85], and to selectively target specific organelles [86]. For example, GFP has been coupled with TlpA gene of *Salmonella* with uses coiled-coil proteins to sense real-time temperature changes

[87]. Other fluorescent reporter genes have also been developed, including blue and red fluorescent proteins.

Clearly, the possible uses of GFPs as reporters for sensing are just beginning to be explored. Unfortunately, as with the *lux* system, the use of a GFP reporting system relies on the ability of researchers to know which specific gene or protein to monitor and then to generate the relevant transfected cell line. While the specificity generated by this method could be advantageous for screening drugs whose ultimate goal is the up-regulation of a specific protein, it is not necessarily applicable for the determination of cell viability or changes in intracellular environment.

5.3.3. Intracellular Fluorescent Probes for Small Molecules

The development of a variety of fluorescent dyes for the sensing of intracellular compounds has led to another class of whole cell biosensor. These biosensors measure the effect of an agent on the intracellular chemistry, rather than looking at the up-regulation in specific gene expression (Figure 5.2C). Some of these fluorescent dyes look solely at cell viability, while others look at specific intracellular conditions (i.e., pH) or the generation of signaling biomolecules (i.e., nitric oxide).

When assessing the toxicity of a drug candidate or the presence of a deadly chemical or biological warfare agent, a whole cell sensor need only be monitored for total cell viability. A variety of optically active methods for ascertaining cell viability have been employed. Ethidium monoazide (EMA) has been used to measure the percent of viable cells using flow cytometry [88]. However, EMA is excluded by viable cells, making it difficult to use in sensing applications. Fluorescein diacetate and ethidium bromide used in combination allows for the visualization of live and dead cells simultaneously, with live cells appearing green and membrane damaged cells appearing red [89, 90]. A host of other esterase based probes have developed because of their ability to diffuse through the cell membrane and be cleaved to form a fluorescently active, stable compound. The most commonly used of these ester based products is calcein AM which is converted intracellularly to its optically active form, calcein [91, 92]. Calcein AM is usually used in conjunction with an ethidium homodimer to stain for non-viable cells.

Physiological changes in intracellular pH can be indicators of a variety of including changes in multi-drug resistance [93, 94] and apoptosis [95–97]. The optical measurement of pH changes intracellularly relies on fluorophores that show measurable deviations in emission with slight changes in pH. Intracellular cytosolic pH ranges from 6.8 to 7.4, with slight deviations of a fraction of a pH unit indicating a change in homeostasis. Three classes of fluorophores are commonly used to measure intracellular pH: fluorescein, seminaphthofluorescein (SNALF), and seminaphtorhodafluor-1 (SNARF) derivatives. Because fluorescein tends to leak from cell membranes, the carboxy derivative of fluorescein and its cell permeable ester is more commonly used. 98–100 Fluorescein also has the disadvantage of having a pK_a of ~6.5, which is lower than the physiological range. Unlike fluorescein, SNALF and SNARF have pK_a values in a more physiological range. SNALF and SNARF have a dual-emission and dual-excitation spectra [101], making them useful for optically based sensing [102, 103]. Cell permeable membrane analogs, such as seminaphtorhodafluor-1-acetoxymethylester (SNARF-1-AM) are also used.

Nitric oxide is a small, quickly dissipating biomolecule. Nitric oxide has a variety of *in vivo* functions, including activity as a blood pressure regulator, neurotransmitter, and indicator of general cell health [104, 105]. The detection of nitric oxide is traditionally accomplished via the detection of nitrate and nitrite in cell media using the Greiss reagent [106]. While this method does produce an optically active product, it is not a viable method for real-time sensing of nitric oxide production. The real-time sensing of intracellular nitric oxide can be accomplished by using a fluorescent dye, 4,5-diaminofluorescein (DAF-FM), and its cell permeable form, 4,5-diaminofluorescein diacetate (DAF-FM diacetate) [107, 108]. In the presence of nitric oxide, DAF-FM converts to a fluorescently active benzotriazole derivative [107, 108]. DAF-FM has been used for the real-time monitoring of the effect of shear stress on endothelial cells [109] and the effects of methylpyridinium ion (MPP(+ l)) on the expression of the NO-dependent increase in Bax protein in neuroblastoma cells [110]. Both these applications highlight the potential for DAF-FM to be incorporated into whole cell biosensors for the real-time monitoring of intracellular nitric oxide as the product of an applied agent.

Any method of signal transduction in a multi-phenotypic array would have to allow for differentiating in the signal emanating for each cell line. For example, an array used for pharmaceutical screening would need to be able to determine not only that a drug candidate caused a change in electrical signal, but distinguish whether or not that change in signal was because of a response from a neuron, cardiac cells, or both. Optical methods of detection allow for the detection scheme to be controlled at the cellular level, where as microelectronics require integration of cells during fabrication steps. Using bioluminescent and fluorescent proteins as reporters allow for detection allow for increased specificity of the sensor, but require a great deal of cellular manipulation before being incorporated into a sensor. Whole cell stains, though not giving the specificity of a protein-based expression system, are easy to use and monitor. For example, the cytotoxicity of an agent can be readily ascertained by simply monitoring a change in cell color. Optical methods can also be used in conjunction within any of the cell patterning techniques outlined in section 2, while the use of microelectronics would require modifications of the patterning techniques.

5.4. CURRENT EXAMPLES OF MULTI-PHENOTYPIC ARRAYS

The ideal multi-phenotypic array would incorporate multiple cell types that could be monitored for "organ" specific responses. For example, a hybrid array of liver cells, kidney cells, and fat cells could be used to assess the effect of a drug candidate on metabolic and excretory systems. A single test sensor would be able to provide a wide range of information about an agents' activity on the whole body, rather than being limited to traditional *in vitro* methods that require the testing each cell type in succession, or having to move prematurely to more expensive animal testing. This "cell culture analogue" method for integrating pharmacokinetic testing and *in vitro* culture [111] extends to the development of multi-phenotypic biosensors.

The development of cellular arrays containing multiple phenotypes has been achieved in a variety of methods. One method of generating multi-phenotype arrays is the direct patterning of mammalian cell clusters each containing a different fluorescently visible cDNA [112, 113]. As discussed earlier, the laminar flow properties of microfluidic

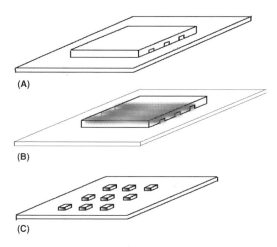

FIGURE 5.3. Patterning hydrogel microarrays using microfluidic channels. (A) A PDMS mold is reversibly fixed to a support material. (B) Cell-containing polymer precursor solutions are used to fill microfluidic channels. (C) UV-exposure, removal of the mask, and developing with water results in an array of cell containing microstructures.

channels have been used to selectively pattern *E. coli* and chicken erythrocytes onto glass substrates [25].

The same laminar flow principle was used to simultaneously generate a single array of PEG-hydrogels with each array element containing either murine hepatocytes, macrophages, or fibroblasts [42]. Combining the encapsulation procedure with the patterning procedure allows for simultaneous patterning of multiple cell lines while retaining all the advantages of cell encapsulation. In brief, cells were suspended in a PEG-diacrylate precursor solution containing photoinitiator. This cell containing solution was placed into microfluidic channels (Figure 5.3B). By shining ultraviolet light through a photomask, exposed regions of polymer precursor solutions were polymerized. Removal of the PDMS mold and rinsing away of any unreacted precursor solution resulted in an array of hydrogel microstructures (Figure 5.3C). Figure 5.4 shows a single array of hepatocytes, fibroblasts, and macrophages.

5.5. FUTURE WORK

The development of commercially feasible multi-phenotypic biosensor is contingent on the integration of cells with fabrication and measurable signals. These requirements are coupled with the need to be able to affordably produce the sensors in large quantities, the ability to preserve and store the sensors for long periods of time without appreciable losses in cell viability, and the need for the sensors to operate under all conditions. Cell based biosensors must also meet the rigorous demands placed on all sensors: selectivity, sensitivity, specificity, stability, reliability, and rapidity.

The patterning method used to generate a multi-phenotypic array must limit the use of cytotoxic reagents, be able to accommodate multiple cell lines without major losses

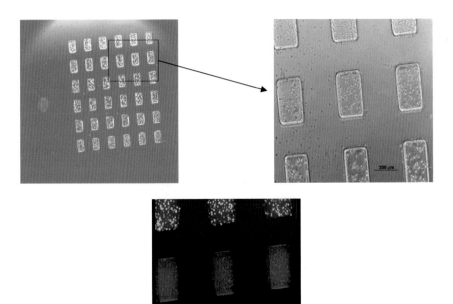

FIGURE 5.4. A hydrogel array containing murine fibroblasts (green), macrophages (red), and hepatocytes (blue).

of cell viability, maintain total isolation of each cell phenotype, provide for the adequate mass transfer of dissolved gases and nutrients, and be easy enough to allow for mass production. A further challenge is that the fabrication technique must be amenable to long term preservation, either via refrigeration or cryopreservation.

The choice of either electrical or optical methods in conjunction with cell-based biosensors is often dependent on the range of analytes to be detected. Ideally, within a multiphenotypic biosensor, one would be able to incorporate not only a variety of cell lines but also incorporate multiple sensing techniques. For example, changes in voltage might be measurable using the FET method and with a voltage sensitive fluorphore [114]. Regardless of the method chosen, some sort of portable device that measures either electrical signal or optical intensity must accompany the biosensor. A hypothetical device such as this would likely have to provide appropriate environmental conditions for the cellular array.

Multi-phenotypic arrays should also be incorporated into sample delivery devices such as microfludic channels. Microfluidic channels are useful in the development of contained biosensors because of the small volume of sample needed. The integration of life cells within microfluidic channels has been accomplished using single cell lines, including immobilized *E. coli* [115] and rat mucosal mast cells [116]. A schematic of a potential multi-phenotypic microfluidic device is shown in Figure 5.5.

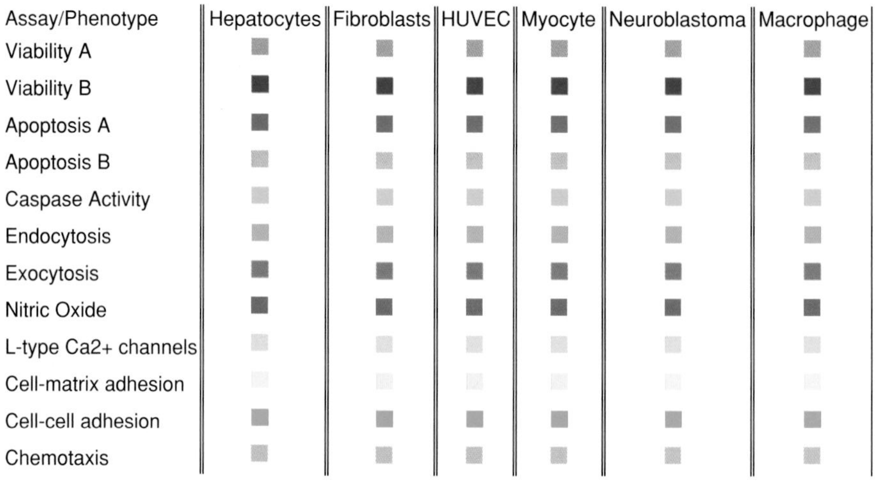

Assay/Phenotype	Hepatocytes	Fibroblasts	HUVEC	Myocyte	Neuroblastoma	Macrophage
Viability A	■	■	■	■	■	■
Viability B	■	■	■	■	■	■
Apoptosis A	■	■	■	■	■	■
Apoptosis B	■	■	■	■	■	■
Caspase Activity	■	■	■	■	■	■
Endocytosis	■	■	■	■	■	■
Exocytosis	■	■	■	■	■	■
Nitric Oxide	■	■	■	■	■	■
L-type Ca2+ channels	■	■	■	■	■	■
Cell-matrix adhesion	■	■	■	■	■	■
Cell-cell adhesion	■	■	■	■	■	■
Chemotaxis	■	■	■	■	■	■

FIGURE 5.5. Schematic of a potential multi-phenotypic biosensor. Six different cell lines are isolated in hydrogel arrays in six different microfluidic channels. Several cellular assays can be performed using one array.

Multi-phenotypic arrays offer the advantage of providing more information about cellular responses than existing arrays and lend themselves to miniaturization, which could lead to high degrees of multiplexing. Multi-phenotypic cell based biosensors show promise in the area of high-throughput drug screening, the detection of biochemical warfare agents, and the detection of environmental toxins. While the use of nucleic acid and protein arrays is well established, the uses of multi-phenotypic cellular arrays are in the infancy of their development, creating an exciting and dynamic area of research.

REFERENCES

[1] B.M. Paddle. *Biosens. Bioelectron.*, 11:1079–1113, 1996.
[2] L. Bousse. *Sens. Actu. B*, 34:270–275, 1996.
[3] C. Ziegler. *Fresenius J. Anal. Chem.*, 366:552–559, 2000.
[4] J. Hodgson. *Nat. Biotechnol.*, 19:722–726, 2001.
[5] J. Knight. *Nature*, 418:474–475, 2002.
[6] T. Chovan and A. Guttman. *Trends Biotechnol.*, 20:116–122, 2002.
[7] C. Ziegler and W. Gopel. *Curr. Opin. Chem. Biol.* 2:585–591, 1998.
[8] M. Matsuzawa, R.S. Potember, D.A. Stenger, V. Krauthamer, *J. Neurosci. Methods.* 50:253–260, 1993.
[9] B.J. Spargo, M.A. Testoff, T.B. Nielsen, D.A. Stenger, J.J. Hickman, and A.S. Rudolph. *Proc. Natl. Acad. Sci. U.S.A.*, 91:11070–11074, 1994.
[10] M. Matsuzawa, P. Liesi, and W. Knoll. *J. Neurosci. Methods*, 69:189–196, 1996.
[11] W. Ma, Q.Y. Liu, D. Jung, P. Manos, J.J. Pancrazio, A.E. Schaffner, J.L. Barker, and D.A. Stenger. *Brain Res. Dev. Brain Res.*, 111:231–243, 1998.
[12] V. Brisson and R.D. Tilton. *Biotechnol. Bioeng.*, 77:290–295, 2002.
[13] J.F. Clemence, J.P. Ranieri, P. Aebischer, and H. Sigrist. *Bioconjug. Chem.*, 6:411–417, 1995.
[14] G. Chen, Y. Ito, Y. Imanishi, A. Magnani, S. Lamponi, and R. Barbucci. *Bioconjug. Chem.*, 8:730–734, 1997.
[15] S. Rohr, R. Fluckiger-Labrada, and J.P. Kucera. *Pflugers Arch.*, 446:125–132, 2003.
[16] S. Rohr, D.M. Scholly, and A.G. Kleber. *Circ. Res.*, 68:114–130, 1991.

[17] V.A. Liu, W.E. Jastromb, and S.N. Bhatia. *J. Biomed. Mater. Res.*, 60:126–134, 2002.

[18] T.A.M. Sugwara. *T. Macromolecules*, 27:7809–7814, 1994.

[19] M.L. Amirpour, P. Ghosh, W.M. Lackowski, R.M. Crooks, and M.V. Pishko. *Anal. Chem.*, 73:1560–1566, 2001.

[20] R. Singhvi, A. Kumar, G.P. Lopez, G.N. Stephanopoulos, D.I. Wang, G.M. Whitesides, and D.E. Ingber. *Science*, 264:696–698, 1994.

[21] M. Mrksich, L.E. Dike, J. Tien, D.E. Ingber, and G.M. Whitesides. *Exp. Cell. Res.*, 235:305–313, 1997.

[22] D. Duffy, J.C. MacDonald, O.J. Schueller, and G.M. Whitesides. *Anal. Chem.*, 70:4974–4984, 1998.

[23] E. Delamarche, A. Bernard, H. Schmid, B. Michel, and H. Biebuyck. *Science*, 276:779–781, 1997.

[24] E. Delamarche, A. Bernard, H. Schmid, A. Bietsch, B. Michel, and H. Biebuyck. *J. Am. Chem. Soc.*, 120:500–508, 1998.

[25] S. Takayama, J. McDonald, Cooper, Ostuni, Emanuele, Liang, N. Michael, Kenis, J.A. Paul, Ismagilov, F. Rustem, Whitesides, and M. George. *Proceedings of the National Academy of Sciences of the United States of America*, 96:5545–5548, 1999.

[26] C.S. Pale-Grosdemange, E.S., K.L. Prime, and G.M. Whitesides. *J. Am. Chem. Soc.*, 113:12–20, 1991.

[27] E. Ostuni, L. Yan, and G.M. Whitesides. *Coll. Surf. B: Biointerfac.*, 15:3–30, 1999.

[28] N. Huang, R. Michel, J. Voros, M. Textor, R. Hofer, A. Rossi, D.L. Elber, J.A. Hubbell, and N.D. Spencer. *Langmuir*, 17:489–498, 2001.

[29] N.A. Alcantar, E.S. Aydil, and J.N. Israelachvili. *J. Biomed. Mater. Res.*, 51:343–351, 2000.

[30] K.R. Kamath, M.J. Danilich, R.E. Marchant, and K. Park. *J. Biomater. Sci. Polym. Ed.*, 7:977–988, 1996.

[31] A. Park, B. Wu, and L.G. Griffith. *J. Biomater. Sci. Polym. Ed.*, 9:89–110, 1998.

[32] A. Revzin, R.G. Tompkins, and M. Toner. *Langmuir*, 19:9855–9862, 2003.

[33] G.M. Cruise, D.S. Scharp, and J.A. Hubbell. *Biomaterials*, 19:1287–1294, 1998.

[34] A. Revzin, R.J. Russell, V.K. Yadavalli, W.G. Koh, C. Deister, D.D. Hile, M.B. Mellott, and M.V. Pishko. *Langmuir*, 17:5440–5447, 2001.

[35] M.B. Mellott, K. Searcy, and M.V. Pishko. *Biomaterials*, 22:929–941, 2001.

[36] R. Russell, A.C. Axel, K.L. Shields, and M.V. Pishko. *Polymer*, 42:4893–4901, 2001.

[37] N. Wisniewski and M. Reichert. *Coll. Surf. B Biointerfac.*, 18:197–219, 2000.

[38] S.P. Massia and J.A. Hubbell. *Anal. Biochem.*, 187:292–301, 1990.

[39] S.P. Massia and J.A. Hubbell. *Ann. N Y Acad. Sci.*, 589:261–270, 1990.

[40] S.P. Massia and J.A. Hubbell. *J. Biomed. Mater. Res.*, 25:223–242, 1991.

[41] B.K. Mann, A.T. Tsai, T. Scott-Burden, and J.L. West. *Biomaterials*, 20:2281–2286, 1999.

[42] W.G. Koh, L.J. Itle, and M.V. Pishko. *Anal. Chem.*, 75:5783–5789, 2003.

[43] C.A. Quinn, R.E. Connor, and A. Heller. *Biomaterials*, 18:1665–1670, 1997.

[44] K. Podual, F.J. Doyle, 3rd; and N.A. Peppas. *Biomaterials*, 21:1439–1450, 2000.

[45] R.A. Scott and N.A. Peppas. *Biomaterials*, 20:1371–1380, 1999.

[46] N.A. Peppas, K.B. Keys, M. Torres-Lugo, and A.M. Lowman. *J. Control. Rel.*, 62:81–87, 1999.

[47] D.K. Han, K.D. Park, J.A. Hubbell, and Y.H. Kim. *J. Biomater. Sci. Polym. Ed.*, 9:667–680, 1998.

[48] J.S. Temenoff, K.A. Athanasiou, R.G. LeBaron, and A.G. Mikos. *J. Biomed. Mater. Res.*, 59:429–437, 2002.

[49] P.J. Martens, S.J. Bryant, and K.S. Anseth. *Biomacromolecules*, 4:283–292, 2003.

[50] W.G. Koh, A. Revzin, and M.V. Pishko. *Langmuir*, 18:2459–2462, 2002.

[51] J.A. Burdick and K.S. Anseth. *Biomaterials*, 23:4315–4323, 2002.

[52] S.M. O'Connor, J.D. Andreadis, K.M. Shaffer, W. Ma, J.J. Pancrazio, and D.A. Stenger. *Biosens. Bioelectron.*, 14:871–881, 2000.

[53] X. Zheng Shu, Y. Liu, F.S. Palumbo, Y. Luo, and G.D. Prestwich. *Biomaterials*, 25:1339–1348, 2004.

[54] V. Liu and S.N. Bhatia. *Biomed. Microdev.*, 4:257–266, 2002.

[55] M. Keusgen. *Naturwissenschaften*, 89:433–444, 2002.

[56] P. Fromherz, A. Offenhausser, T. Vetter, and J. Weis. *Science*, 252:1290–1293, 1991.

[57] R. Weis, B. Muller, and P. Fromherz. *Phys. Rev. Lett.*, 76:327–330, 1996.

[58] G. Zeck and P. Fromherz. *Proc. Natl. Acad. Sci. U.S.A.*, 98:10457–10462, 2001.

[59] A. Harsch, C. Ziegler, and W. Gopel. *Biosens. Bioelectron.*, 12:827–835, 1997.

[60] G.W. Gross, B.K. Rhoades, H.M. Azzazy, and M.C. Wu. *Biosens. Bioelectron.*, 10:553–567, 1995.

[61] J.H. Luong, M. Habibi-Rezaei, J. Meghrous, C. Xiao, K.B. Male, and A. Kamen. *Anal. Chem.*, 73:1844–1848, 2001.

[62] C. Xiao, B. Lachance, G. Sunahara, and J.H. Luong. *Anal. Chem.*, 74:1333–1339, 2002.

[63] C. Xiao, B. Lachance, G. Sunahara, and J.H. Luong. *Anal. Chem.*, 74:5748–5753, 2002.

[64] C. Tlili, K. Reybier, A. Geloen, L. Ponsonnet, C. Martelet, H.B. Ouada, M. Lagarde, and N. Jaffrezic-Renault. *Anal. Chem.*, 75:3008–3012, 2003.

[65] J. Engebrecht, M. Simon, and M. Silverman. *Science*, 227:1345–1347, 1985.

[66] G. Kirchner, J.L. Roberts, G.D. Gustafson, and T.D. Ingolia. *Gene*, 81:349–354, 1989.

[67] J. King, P.M. DiGrazia, B. Applegate, R. Burlage, J. Sanseverino, P. Dunbar, F. Larimer, and G.S. Dayler. *Science*, 249:778–780, 1990.

[68] F. Marincs and D.W. White. *Appl. Environ. Microbiol.*, 60:3862–3863. 1994.

[69] K.A. Durham, D. Porta, M.R. Twiss, R.M. McKay, and G.S. Bullerjahn. *FEMS Microbiol. Lett.*, 209:215–221, 2002.

[70] A. Heitzer, K. Malachowsky, J.E. Thonnard, P.R. Bienkowski, D.C. White, and G.S. Sayler. *Appl. Environ. Microbiol.*, 60:1487–1494, 1994.

[71] B.M. Applegate, S.R. Kehrmeyer, and G.S. Sayler. *Appl. Environ. Microbiol.*, 64:2730–2735, 1998.

[72] M.B. Gu, G.C. Gil, and J.H. Kim. *Biosens. Bioelectron.*, 14:355–361, 1999.

[73] R.S. Burlage, G.S. Sayler, and F. Larimer. *J. Bacteriol.*, 172:4749–4757, 1990.

[74] C. Lagido, J. Pettitt, A.J. Porter, G.I. Paton, and L.A. Glover. *FEBS Lett.*, 493:36–39, 2001.

[75] P.P. Schreiter, O. Gillor, A. Post, S. Belkin, R.D. Schmid, and T.T. Bachmann. *Biosens. Bioelectron.*, 16:811–818, 2001.

[76] M.A. Dollard and P.J. Billard. *Microbiol. Methods*, 55:221–229, 2003.

[77] J.R. de Wet, K.V. Wood, M. DeLuca, D.R. Helinski, and S. Subramani. *Mol. Cell. Biol.*, 7:725–737, 1987.

[78] G.A. Keller, S. Gould, M. Deluca, and S. Subramani. *Proc. Natl. Acad. Sci. U.S.A.*, 84:3264–3268, 1987.

[79] U. Deuschle, R. Pepperkok, F.B. Wang, T.J. Giordano, W.T. McAllister, W. Ansorge, and H. Bujard. *Proc. Natl. Acad. Sci. U.S.A.*, 86:5400–5404, 1989.

[80] S.J. Rosochacki and M. Matejczyk. *Acta. Microbiol. Pol.*, 51:205–216, 2002.

[81] M. Chalfie, Y. Tu, G. Euskirchen, W.W. Ward, and D.C. Prasher. *Science*, 263:802–805, 1994.

[82] D.C. Prasher, V.K. Eckenrode, W.W. Ward, F.G. Prendergast and M.J. Cormier. *Gene*, 111:229–233, 1992.

[83] J. Pines. *Trends Genet*, 11:326–327, 1995.

[84] E. Yeh, K. Gustafson, and G.L. Boulianne. *Proc. Natl. Acad. Sci. U.S.A.*, 92:7036–7040, 1995.

[85] M. Girotti and G. Banting. *J. Cell. Sci.*, 109 (Pt 12):2915–2926, 1996.

[86] R. Rizzuto, M. Brini, P. Pizzo, M. Murgia, and T. Pozzan. *Curr. Biol.*, 5:635–642, 1995.

[87] R.R. Naik, S.M. Kirkpatrick, and M.O. Stone. *Biosens. Bioelectron.*, 16:1051–1057, 2001.

[88] M.C. Riedy, K.A. Muirhead, C.P. Jensen, and C.C. Stewart. *Cytometry*, 12:133–139, 1991.

[89] T.J. Nikolai, M.V. Peshwa, S. Goetghebeur, W.S. Hu. *Cytotechnology*, 5:141–146, 1991.

[90] B.C. Patel, J.M. Courtney, J.H. Evans, and J.P. Paul. *Biomaterials*, 12:722–726, 1991.

[91] L.S. De Clerck, C.H. Bridts, A.M. Mertens, M.M. Moens, and W.J. Stevens. *J. Immunol. Methods*, 172:115–124, 1994.

[92] N.G. Papadopoulos, G.V. Dedoussis, G. Spanakos, A.D. Gritzapis, C.N. Baxevanis, and M. Papamichail. *J. Immunol. Methods*, 177:101–111, 1994.

[93] S. Simon, D. Roy, and M. Schindler. *Proc. Natl. Acad. Sci. U.S.A.*, 91:1128–1132, 1994.

[94] M.M. Hoffman, L.Y. Wei, and P.D. Roepe. *J. Gen. Physiol.*, 108:295–313, 1996.

[95] D. Perez-Sala, D. Collado-Escobar, and F. Mollinedo. *J. Biol. Chem.* 270:6235–6242, 1995.

[96] G.W. Meisenholder, S.J. Martin, D.R. Green, J. Nordberg, B.M. Babior, and R.A. Gottlieb. *J. Biol. Chem.*, 271:16260–16262, 1996.

[97] R.A. Gottlieb, J. Nordberg, E. Skowronski, and B.M. Babior. *Proc. Natl. Acad. Sci. U.S.A.*, 93:654–658, 1996.

[98] Y. Maeda, K. Tanaka, Y. Koga, X.Y. Zhang, M. Sasaki, G. Kimura, and K. Nomoto. *J. Immunol. Methods*, 157:117–123, 1993.

[99] M.L. Graber, T.E. Dixon, D. Coachman, K. Herring, A. Ruenes, T. Gardner, and E. Pastoriza-Munoz. *Am. J. Physiol.*, 250:F159–F168, 1986.

[100] P. Breeuwer, J.C. de Reu, J.L. Drocourt, F.M. Rombouts, and T. Abee. *App. Environ. Microbiol.*, 63:178–185, 1997.

[101] S. Bassnett, L. Reinisch, and D.C. Beebe. *Am. J. Physiol.*, 258:C171–C178, 1990.

[102] H. Szmacinski and J.R. Lakowicz. *Anal. Chem.*, 65:1668–1674, 1993.

[103] Z. Xu, A. Rollins, R. Alcala, and R.E. Marchant. *J. Biomed. Mater. Res.*, 39:9–15, 1998.

[104] J.S. Beckman and W.H. Koppenol. *Am. J. Physiol.*, 271:C1424–C1437, 1996.

[105] J. MacMicking, Q.W. Xie, and C. Nathan. *Annu. Rev. Immunol.*, 15:323–350, 1997.

[106] L.C. Green, D.A. Wagner, J. Glogowski, P.L. Skipper, J.S. Wishnok, and S.R. Tannenbaum. *Anal. Biochem.*, 126:131–138, 1982.

[107] H. Kojima, Y. Urano, K. Kikuchi, T. Higuchi, Y. Hirata, and T. Nagano. *Angew. Chem. Int. Ed. Engl.*, 38:3209–3212, 1999.

[108] Y. Itoh, F.H. Ma, H. Hoshi, M. Oka, K. Noda, Y. Ukai, H. Kojima, T. Nagano, and N. Toda. *Anal. Biochem.*, 287:203–209, 2000.

[109] J.F. Ye, X.X. Zheng, and L.X. Xu. *Sheng Wu Hua Xue Yu Sheng Wu Wu Li Xue Bao (Shanghai)*, 35:296–300, 2003.

[110] J. Dennis and J.P. Bennett Jr. *J. Neurosci. Res.*, 72:76–88, 2003.

[111] T.H. Park and M.L. Shuler. *Biotechnol. Prog.*, 19:243–253, 2003.

[112] J. Ziauddin and D.M. Sabatini. *Nature*, 411:107–110, 2001.

[113] R.Z. Wu, S.N. Bailey, and D.M. Sabatini. *Trends Cell. Biol.*, 12:485–488, 2002.

[114] D. Braun and P. Fromherz. *Phys. Rev. Lett.*, 86:2905–2908, 2001.

[115] J. Heo, K.J. Thomas, G.H. Seong, R.M. Crooks. *Anal. Chem.*, 75:22–26, 2003.

[116] Y. Matsubara, Y. Murakami, M. Kobayashi, Y. Morita, and E. Tamiya. *Biosens. Bioelectron.*, 19:741–747, 2004.

6

MEMS and Neurosurgery

Shuvo Roy[1], Lisa A. Ferrara[2], Aaron J. Fleischman[1],
and Edward C. Benzel[3]

[1]*Co-Director, BioMEMS Laboratory, Department of Biomedical Engineering, The Cleveland Clinic Foundation, 9500 Euclid Avenue, ND20, Cleveland, Ohio 44195*
[2]*Director, Spine Research Laboratory, The Cleveland Clinic Foundation, 9500 Euclid Avenue, S80, Cleveland, Ohio 44195*
[3]*Chairman, Spine Institute, Vice-Chairman, Department of Neurosurgery, The Cleveland Clinic Foundation, 9500 Euclid Avenue, S80, Cleveland, Ohio 44195*

PART I: BACKGROUND

6.1. WHAT IS NEUROSURGERY?

Neurosurgery is the branch of medicine that concerns itself with the diagnosis and surgical treatment of disorders affecting the nervous system, both centrally and peripherally. The central nervous system consists of the brain and spinal cord (Figure 6.1). It is not capable of full regeneration after injury, which is in striking contrast to the peripheral nervous system. The brain and spinal cord are the higher processing centers that regulate and control the peripheral nervous system. The latter is directly responsible for movement, speech, and action. Thus, it is the neurosurgeon's charge to restore and preserve these functions. The neurosurgeon surgically tackles such entities as head trauma, brain injuries, spinal cord injuries, degenerative spine disease, aneurysms, tumors, and congenital malformations of the brain, skull, and spine.

6.2. HISTORY OF NEUROSURGERY

The art of neurosurgery dates back to the Neolithic time (late stone age) with evidence of brain surgery found in the unearthed remains from this era [4, 15, 21, 27, 32]. Many ancient

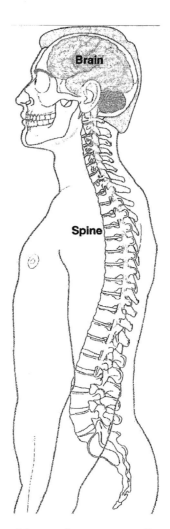

FIGURE 6.1. Schematic illustration of the central nervous system. Neurosurgery refers to the operative and subsequent non-operative care of disorders that affect the central, peripheral and autonomic nervous system as well as the management of pain. Neurosurgeons also treat disorders of the vascular and skeletal components supporting the central nervous system.

cultures practiced the art of trephination (alternately spelled trepanation), which involves drilling holes in the head, presumably to release "evil spirits" or "bad humors". Figure 6.2 presents a photograph of a trephined skull that was recovered from an archeological site in South America. Although the resulting health effects of trephination are difficult to assess today, archaeological evidence of bone healing in trephined skulls suggests that ancient patients not only survived the operation, but also went on to live for a long while afterwards [4, 27, 30]. Early papyrus writings from Egypt demonstrate evidence of brain surgery in Africa as early as 3000 BC [8, 9, 14]. Pre-historic evidence of brain surgery existed in the Pre-Incan civilization in Peru, as early as 2000 BC [4]. The early surgical tools were made

FIGURE 6.2. Photograph of a prehistoric human skull from Peru showing evidence of trephination (arrow). *(Courtesy: Museum of Man, San Diego, CA)*

from volcanic rock or ancient metals (Figure 6.3) [21]. Archeological evidence indicates brain surgery was used for the treatment of organic diseases, osteomyelitis, and head injuries [4, 8, 27, 32].

The early Greeks used trephination in the treatment of head injuries [21, 23, 31]. Hippocrates (460–370 BC) documented many of their early medical accomplishments. He was also the first to describe numerous neurologic conditions, which most likely developed from battlefield injuries, as well as the use of a trephine for brain contusions. He was aware of the underlying dural sheath covering the brain and recommended such techniques as irrigation during surgery to reduce overheating and injury to intracranial structures.

Aulus Cornelius Celsus was the brain surgeon for ancient Rome during the first century AD. Celsus advanced neurosurgery from the point where Hippocrates left off [31]. Unlike Hippocrates, Celsus performed surgery on depressed skull fractures. He demonstrated that injury to the lower spine could cause paralysis of the leg as well as incontinence (9). Celsus was also responsible for describing the symptoms of brain injury in great detail [21].

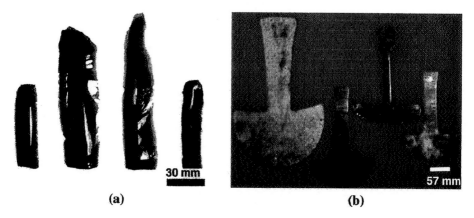

(a)   **(b)**

FIGURE 6.3. Photographs of instruments for ancient brain surgery showing: (a) obsidian blades; and (b) copper and bronze tumi knives. *(Courtesy: Museum of Man, San Diego, CA)*

Paul of Aegina (625–690 AD) was the last great Byzantine physician. His most notable writings, *The Seven Books of Paul of Aegina,* contained discussions on the treatment of head injuries and the use of a trephine [31]. He also categorized skull fractures into groups: fissure, incision, expression, depression, arched fracture, and in infants, dent. Paul also endorsed surgery for spine fractures [9]. He classified skin incisions and was the first to postulate that hemorrhage may cause hydrocephalus (increased cerebrospinal fluid volume within the brain) [18].

The practice of neurological surgery and Arabic medicine was introduced to medieval Europe around 1050 AD by Constantinus Africanus [15, 31]. Roger of Salerno was the first to write on the topic of surgery in Italy. He studied in Baghdad where he was influenced by Arabic medicine. His work on neurosurgery significantly influenced medieval medical practices. His writings, *Practica Chirurgiae,* discuss dural tear assessment and cerebrospinal fluid leakage in depressed skull fractures by having the patients hold their breath (Valsalva maneuver) while watching for leaks and air bubbles [21]. Roger was also a pioneer in the techniques for managing nerve injury, and, in particular, he emphasized the rejoining of severed nerves [9].

The seventeenth century exhibited rapid growth in science and medicine. Scientists like Isaac Newton, Francis Bacon, William Harvey, and Robert Boyle made significant advances in physics, experimental design, knowledge of the circulatory system, and physiologic chemistry. Education regarding scientific thought and open public forums were used to transmit information and to discuss the ideas developed in scientific societies. One of the most notable scientists of this era, Thomas Willis (1621–1675), published an accurate anatomic study of the brain [31]. In fact, the circle of Willis bears his name. Willis also introduced the concept of the "neurology", or the doctrine of neurons, but the term did not enter general medical usage until the eighteenth century.

The nineteenth century was the age of anesthesia, antisepsis and aseptic technique, and cerebral localization [21, 31, 33]. These innovations were great advances for the establishing neurosurgery as an independent field. Anesthesia provided freedom from pain during surgery. Antisepsis and aseptic techniques were used to minimize the risk of infection by creating sterile fields surrounding the surgical site. Cerebral localization techniques were used as diagnostic and decision making tools [33]. It was cerebral localization, the concept that the brain is divided into segments that correspond to particular functions, which aided in the diagnosis of brain lesions or injury. Paul Broca (1824–1880) localized the speech to a particular region of the brain, Broca's area [35]. He based his discovery on the early work of Ernest Auburtin (1825–1893) who had a patient with a frontal lobe defect [31]. Auburtin abolished speech by applying pressure to the frontal lobe, which returned when the pressure was released.

Finally, it was Harvey Cushing (1869–1939), the father of American Neurosurgery, who ushered in the era of modern neurosurgery [3, 10, 19]. Cushing studied at Johns Hopkins under William Halsted (1852–1922), a premier general surgeon, where he excelled in meticulous surgical technique. Cushing contributed extensively to the field of neurosurgery. Among his many innovations are the introduction of the practice of recording the vitals signs during anesthesia (devised while still a medical student), and the use of silver clips to control bleeding [19]. He was an early proponent of the then-new technique of x-raying [3, 19]. His most memorable work was that on pituitary surgery, which was published as a monograph in 1912 and formed the basis for further studies on the pituitary gland and classification of brain tumors [31].

6.3. CONVENTIONAL NEUROSURGICAL TREATMENTS

Today, neurosurgery is considered a medical discipline that provides the operative and non-operative care of disorders that affect the central, peripheral and autonomic nervous system (including their supportive structures and vascular supply), and the operative and non-operative management of pain. It encompasses disorders of the brain, skull, pituitary gland, spinal cord, cranial and spinal nerves, and the autonomic nervous system. Some common neurosurgical strategies that are currently used to treat disorders of the brain, nerves, and spine are described below.

6.3.1. Hydrocephalus

Hydrocephalus is a condition where excess cerebrospinal fluid (CSF) results in enlargement of ventricles of the brain due to conditions such as cerebral atrophy and/or failure of development of the brain [31]. The cerebrospinal fluid (CSF), which is normally clear and colorless, bathes the brain and spinal cord and protects them from injury [35]. This fluid normally flows through one ventricle to the next, and is kept over the surface of the brain and down the spinal cord before being finally absorbed into the blood stream. When the circulation or absorption of CSF is blocked, or if there is excess production, the volume of the brain becomes excessively larger then normal. This increased volume manifests as enlarged heads in babies since their skulls are not completely fused (Figure 6.4).

There are several types of hydrocephalus. An obstructive type of hydrocephalus where the CSF is under increased pressure is termed *tension hydrocephalus* [35]. The terms *communicating* and *non-communicating hydrocephalus* differentiate between a complete or incomplete obstruction of the aqueduct of the brain. Communicating hydrocephalus refers to communication between the ventricles and the spinal subarachnoid space within the brain, whereas, non-communicating is the contrary. Chronic hydrocephalus is a similar condition that has been present for months or years in an individual. Causes of hydrocephalus may

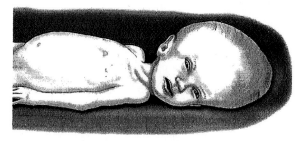

FIGURE 6.4. Graphical illustration showing the effect of hydrocephalus in infants. The increased volume of cerebrospinal fluid (CSF) within the brain causes head enlargement.

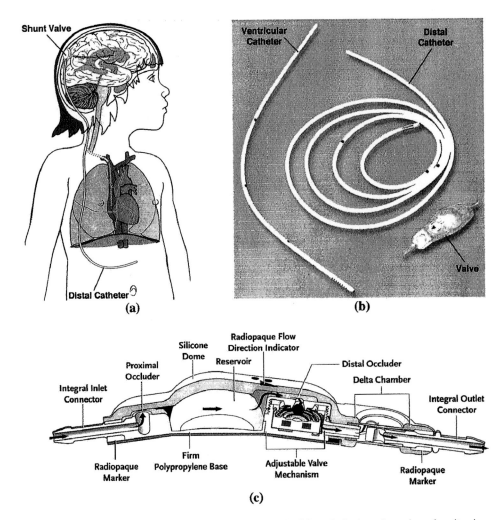

FIGURE 6.5. Treatment of hydrocephalus. Excess CSF is drained from the brain to the peritoneal cavity via a surgically implanted shunt (a). The shunt (b) comprises a short catheter inserted into the ventricle, a one-way valve underneath the skin, and a longer catheter that typically drains into the abdominal cavity. Recent advances in shunt design include the development of valves with features to prevent overdrainage and allow for telemetric adjustment (c). *(Courtesy: Medtronic Inc)*

include tumors, cysts, trauma, subarachnoid hemorrhages, infection, subdural hematomas, and Paget's disease [28, 35]. The result of hydrocephalus is the accumulation of cerebrospinal fluid within the brain and increased intracranial pressure (ICP).

The goal of hydrocephalus treatment is to minimize or prevent brain damage by improving CSF flow. Surgical interventions are the primary treatment for hydrocephalus. This includes direct removal of the obstruction, if possible. However, the most common conventional approach involves the surgical placement of a CSF shunt within the brain to bypass the obstructed area (Figure 6.5) [6, 34]. The CSF shunt consists of three primary components:

(1) a shorter catheter (~2 mm diameter) that is usually inserted into the ventricle through a small hole in the skull; (2) a valve to regulate CSF flow; and (3) a longer catheter to drain excess CSF. When a critical ICP value is exceeded, the valve opens and allows excess CSF to flow from the ventricles through the shorter catheter and drain into the peritoneal cavity through the longer catheter. Alternate sites for drainage include the pleural space and the heart. Drainage of CSF ceases when the ICP falls below the critical threshold and closes the valve.

There are numerous surgical approaches for shunt placement to treat hydrocephalus; however, the success rate of shunting is approximately 70%, with a complication rate as high 20% [28, 31, 36]. Complications such as overdraining the shunt when the ICP is too low will leave the patient with headaches and possibly subdural hematomas (bleeding below the dura of the brain). Insufficient draining is another complication that may occur if the ventricles fail to shrink after shunt placement. Shunt malfunction and infection further complicate the treatment of hydrocephalus [1, 5, 28]. For instance, there is a heightened risk of shunt blockage in the first few months following placement. The proximal end of the catheter can become obstructed by the chloroid plexus, which lines the ventricles, or blood that may be generated during catheter placement. Infections are primarily caused by bacteria (staphylococci) that are derived from the patient's skin flora during surgery [1, 17]. The bacteria colonize the inner surfaces of the catheters to create biofilms, which are difficult to treat without shunt removal. The use of anti-infective-catheters may minimize the complications resulting from shunt infection. Typically, the shunt material is impregnated or coated with antimicrobial substances such as rifampin and sparfloxin, which will kill the colonizing bacteria [17].

6.3.2. Brain Tumors

Brain tumors are traditionally classified by their cell types. For example, gliomas (neuroglial tumors) arise from cells derived from the primitive neuroepithelium, while meningiomas arise from arachnoid cap cells [36]. Symptoms of a brain tumor can include headaches (resulting from compression and increased ICP), seizures, cognitive or personality changes, eye movement abnormalities, nausea or vomiting, speech disturbances, and memory loss. Primary malignant brain tumors grow rapidly and destroy normal brain tissue, yet seldom metastasize (spread to other tissues in the body). In contrast, metastatic secondary brain tumors (cancer that spreads from other parts of the body to the brain) occur at some point in 10–15% of persons with cancer and are the most common type of brain tumor [26].

Chemotherapeutic agents, radiation treatment, and surgical excision are conventional methods to diminish or remove the tumors to regain normal neurological function [16]. Unfortunately, invasiveness or encroachment of the tumor upon healthy brain tissue is the deterrent to cure for most brain tumors. This behavior is especially true of gliomas, which are the most common form of primary tumors and exhibit a propensity to infiltrate healthy tissue rapidly while increasing malignant potential [36]. Surgery is the chief form of treatment for brain tumors such as meningiomas, which lie within the membranes covering the brain, or in parts of the brain that can be removed without damaging critical neurological functions. Because a tumor will recur if any tumor cells are left behind, the neurosurgeon's goal is to remove the entire tumor whenever possible. Radiation therapy and chemotherapy, in

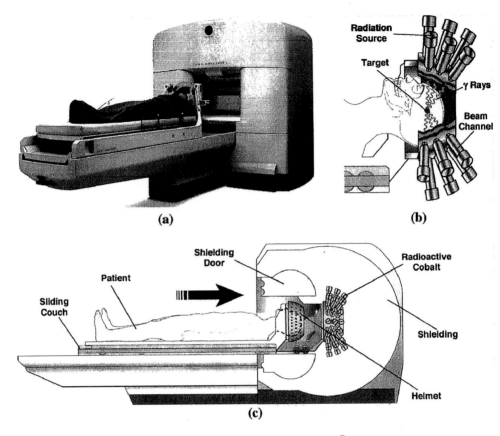

FIGURE 6.6. Sterotactic radiosurgery of brain tumors. In the Gamma Knife® unit (a), the supine patient's head is covered by a helmet (b), which is mounted on a cradle (c) that slides into a shielded portal. The tumor site is irradiated with gamma rays from multiple ^{60}Co sources that are located around the helmet. *(Courtesy: Elekta AB)*

general, are used as secondary or adjuvant treatment for tumors that cannot be cured by surgery alone.

Radiation therapy is mainly used after surgery for tumors that cannot be removed completely, as well as for cases in which surgery would involve too great a risk to the patient. Conventional radiation therapy (also termed teletherapy) directs external beams of x-rays from a linear accelerator (LINAC) or gamma rays from a ^{60}Co radioisotope towards the tumor from different directions sequentially to achieve an additive effect [31]. Typical treatment regimens deliver a daily dose of 1.8–2.0 Gy (Gray) to achieve a total dose of 50–60 Gy over 5–7 weeks [36]. An alternate approach is brachytherapy, which refers to a continuous irradiation of the tumor via radioactive seeds implanted into the tumor [31].

Stereotactic radiosurgery is a non-invasive therapeutic alternative for treating brain disorders. Both Gamma Knife® (Figure 6.6) and XKnife™ are types of tools for stereotactic radiosurgery. In radiosurgery, healthy tissue is spared while the tumor is targeted with either "pencil-thin" beams of gamma rays (Gamma Knife®) simultaneously, or a series

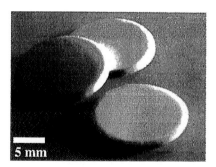

FIGURE 6.7. Photograph of GLIADEL® wafers for the chemotherapy of malignant brain tumors. These poly-meric wafers are comprised of a biodegradable copolymer and a chemotherapeutic agent (carmustine). The wafers are implanted into the brain cavity after tumor resection surgery to kill any residual tumor cells that might be present. *(Courtesy: Guilford Pharmaceuticals Inc)*

of intersecting coronal arc beams of x-rays (XKnife™). With stereotactic radiosurgery techniques, a higher dose of radiation is delivered to the specific site (or tumor) of 15–20 Gy and is usually given in one day.

Chemotherapy works to destroy tumor cells with drugs that may be given either alone or in combination with other treatments. A key problem with chemotherapy has been the difficulty in delivering sufficient amounts of drug directly to the tumor while sparing normal brain cells [11, 31]. Another problem is the blood-brain barrier mechanism that normally serves to keep harmful substances out of the brain—unfortunately, this same blood-brain barrier can also work to keep potentially helpful drugs out of the brain. Although chemother-apy is usually given by mouth or injected in the vein, some new techniques of intratumoral chemotherapy use either small pumps or biodegradable wafers (Figure 6.7) to place the drug inside the tumor [11, 12].

6.3.3. Parkinson Disease

This disease was originally described by James Parkinson in 1817 and was characterized by involuntary tremulous motion with decreased muscle strength [35]. Clinical symptoms consist of slowness of voluntary movement, resting tremor, stooped posture, axial instability, rigidity, and festinating gait. The aging Parkinson patient develops rigid spine and limbs contributing to a shuffled gait. The characteristic tremor of a hand, or a few fingers, is often an initial symptom of the disease.

Parkinson's disease is a progressive neurological disorder that results from degeneration of neurons in a region of the brain that controls movement [7, 29, 37]. This degeneration creates a shortage of the brain signaling chemical (neurotransmitter) known as dopamine, which allows smooth, coordinated function of the body's muscles and movement. Without dopamine, the nerve cells cannot properly transmit messages, thereby resulting in the loss of muscle function (Figure 6.8). Unfortunately, there is no known treatment that will halt or reverse the neural degeneration of this disease [35]. Instead, pharmaceutical agents or surgical options can be utilized to minimize the effects of deleterious symptoms on the Parkinson patients. A common medication is Levodopa (also known as L-dopa), which is a

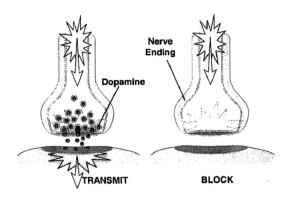

FIGURE 6.8. Schematic illustrations showing the role of dopamine in Parkinson's disease. In a normal brain, dopamine acts as a neurotransmitter between nerve endings. In Parkinson patients, the neurons that produce dopamine become degenerated, which, in turn, triggers a chain reaction of abnormal signaling that results in impaired mobility.

metabolic precursor of dopamine. Its active ingredient, L-dihydroxyphenylalanine, is able to cross the blood brain barrier in to the basal ganglia, where it is decarboxylated to form dopamine, replacing the missing neurotransmitter.

A surgical means for the treatment of Parkinson's disease is that of ablative surgical therapy, involving the stereotactic placement of lesions [37]. Typically, a surgeon uses a heat probe to destroy a small region of brain tissue that is abnormally active in either the globus pallidus, ventrolateral thalamus, or subthalamic nucleus, contralateral to the affected side of the body. Optimal results are generally obtained in younger patients where unilateral tremors or rigidity are present. An alternate approach to lesions is transplantation of cells at the target site to replace neurons for dopamine production [7, 20, 37]. However, challenges with respect to source of cells and immunosuppression must still be overcome for transplantation to become a reality.

Deep brain stimulation (DBS) is a recent neurosurgical advance in the treatment of Parkinson's disease that could soon replace traditional ablative procedures [25, 37]. DBS uses a surgically implanted, battery-operated medical device called a neurostimulator, which is similar to a heart pacemaker and approximately the size of a stopwatch. It delivers electrical stimulation to targeted areas in the brain that control movement, blocking the abnormal nerve signals that cause tremor and other symptoms. The surgical procedure involves placing a thin metal electrode ($\sim$1 mm diameter) into targeted brain region, while the battery and control electronics are housed in a titanium case implanted under the skin of the chest (Figure 6.9). All parts of the stimulator system are internal; there are no wires coming out through the skin. To improve control of symptoms, the stimulator parameters can be adjusted by the surgeon after surgery using wireless telemetry module.

6.3.4. Degenerative Disease of the Spine

The human spine was not "designed to last" 80 or more years. As the population ages, spine degeneration looms as an ever-increasing health care problem. Degeneration of the spine refers to breakdown of the normal architecture of the various components of the spine

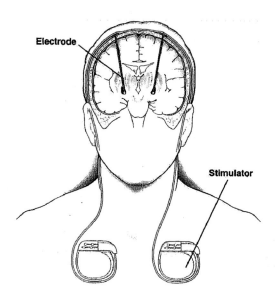

FIGURE 6.9. Graphical illustration showing the deep brain stimulation (DBS) system for the treatment of Parkinson's disease. A pacemaker-like device is implanted under the skin in the chest and connected to electrodes inserted into the brain. Programmed electrical stimulation is delivered to the brain to block abnormal nerve signals that cause tremors, which are characteristic of Parkinson patients.

(vertebral body, facet joint, and disc) due to the natural aging process. For example, discs bulge, joints enlarge, ligaments increase in size, and osteophytes (calcified spurs) form [31]. Degeneration results in mechanical instability of the spine, which can ultimately manifest as clinical symptoms including pain and spine deformity. Typically, in mid-life (~40 years), disc degeneration and hemiation become prevalent. Disc hemiation can cause spinal cord and/or nerve root compression, which, in turn, can result in spinal cord dysfunction and paralysis, pain in the arm or leg (e.g., sciatica), or neck and back pain. Beyond 60 or 70 years, the spine degeneration process manifests as spondylosis with indications of stenosis and compression fractures. Lumbar stenosis symptoms can be particularly quite disabling, ranging from myelopathy (spinal cord dysfunction with an accompanying difficulty with walking and with hand functions) and neurogenic claudication (difficulties with walking with the onset of disabling leg and back pain).

Surgery can be used to treat each of the aforementioned degenerative processes if non-surgical approaches (physical therapy and medication) fail [2]. Decompression surgery can be performed from the front or back of the spine to restore or increase size of the spinal canal and relieve nerve root compression. Stabilization surgery may be performed to minimize pain and correct deformities. For patients with disc herniation, fusion surgery is performed to immobilize adjacent vertebra and eliminate mechanical instability (undesirable motion) of the spine. In a typical spine fusion surgery, the degenerated disc is removed and the resulting intervertebral space is packed with bone graft. The bone graft stimulates the formation of new bone between the vertebral end plates until there is a rigid union (no motion) between the vertebral segments. The fusion region must be supported during the healing process (6–12 months) after surgery using spinal instrumentation such as rods, plates, cages, wires,

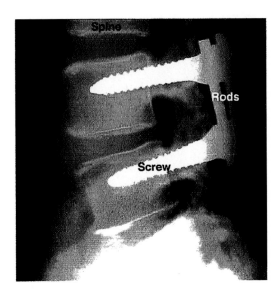

FIGURE 6.10. X-ray image showing spinal instrumentation implanted during fusion surgery to treat degenerative disc disease. The rods and screws support and immobilize the fusion segment while new bone forms in the intervertebral space to create a rigid union between the vertebral endplates.

hooks, and screws (Figure 6.10). The spinal instrumentation initially functions as the load-bearing element that immobilizes the fusion segment during the early unstable stages of bone grafting and healing. However, the success of a fusion across a destabilized spinal segment is often difficult for the surgeon to assess. Conventional methods to monitor healing of the spine after surgery are based on a combination of patient history and imaging tools such as x-rays, magnetic resonance imaging (MRI), and computed tomography (CT). Unfortunately, these "snapshot" techniques cannot provide accurate real-time information on the fusion status or the possibility of spinal instrumentation failure, which, in turn, can lead to unnecessary and costly revision surgeries.

An alternative to spine fusion may be the replacement of a degenerated disc with an artificial construct [13]. The underlying premise is that the replacement disc would act as a "shock absorber" and maintain the natural mechanics of the spine. Unlike fusion, the use of artificial discs would still preserve motion between vertebral segments. Various artificial disc designs are currently in clinical trials to evaluate their safety and efficacy.

6.4. EVOLUTION OF NEUROSURGERY

Neurosurgery has evolved from its earliest roots in the practice of trephination to a sophisticated intellectual and physical exercise in an exquisitely complex three-dimensional (3D) environment of the nervous system [21]. The scope of neurosurgery has largely been defined by technological progress to enhance the operative environment as well as the development of new tools for patient diagnosis and surgical therapy. For example, advanced imaging technology such as magnetic resonance imaging (MRI) and computerized tomography (CT) is transitioning from preoperative neurodiagnostics to the intraoperative neurosurgical environment. In addition, endoscopic procedures on the spine using high-resolution video cameras to visualize surgical anatomy are increasing for discectomy, tumor resection, and deformity correction [24]. Indeed, the continued evolution of neurosurgery will be driven principally by concurrent advances in science and technology. Scientific advances will provide greater insight into the molecular mechanisms of disease, which, in turn, will modify the strategy for neurosurgical treatment. Concurrent technological progress will enhance the drive to decrease patient morbidity and reduce overall healthcare costs. Principal technological directions would likely include the introduction of sophisticated robotics and integration of sensors for physiological monitoring [22]. Ultimately, however, the successful incorporation of scientific advances and technological progress will require that neurosurgeons not only identify significant unmet clinical needs, but also be willing to partner with scientists and engineers to co-develop appropriate solutions.

PART II: APPLICATIONS

6.5. MEMS FOR NEUROSURGERY

There is tremendous interest in the development of microelectromechanical systems (MEMS) for medical applications. In the most general sense, MEMS refers to miniature components that are fabricated using a combination of techniques that were originally borrowed from the microelectronics fabrication industry, and then modified for the production of microstructures, micromachines, or microsystems such as sensors and actuators [71–73]. The MEMS fabrication techniques are often collectively and synonymously termed microfabrication and/or micromachining. Currently, there are numerous research, development, and commercialization efforts underway to create high-performance clinical devices by exploiting the potential for size miniaturization and integration with microelectronics afforded by microfabrication and micromachining techniques [45–47, 62, 69, 71].

A variety of neurosurgical procedures—cranial and spinal—is amenable to the application of MEMS technology [38, 40, 71]. Cranial surgery involves surgery on the skull and its contents, the brain. Such surgery may be as simple as a trephination-like operation that is utilized to evacuate superficial blood clots, such as subdural hematomas, to craniotomies that are designed to approach deep seated aneurysms, tumors, and a variety of other pathologies. Spine surgery involves surgery on the vertebra, discs, and spinal cord. Such surgery may range from discectomy, where a herniated disc is partially or completely removed, to fusion that is intended to stabilize adjacent motion segments.

Brain surgery may be broadly classified into three fundamental categories: vascular, tumor, and trauma. Cranial vascular procedures involve surgery for aneurysms, arterial venous malformations, and a variety of hemorrhages. The application of current MEMS devices to this arena will be limited due to size and biocompatibility constraints. However, the assessment of pressures within blood vessels and vascular malformations, as well as the management and obliteration of pathological structures, could ultimately benefit from advances in MEMS technology such as miniature pressure sensors and smart surgical tools.

Brain tumor surgery may most certainly be enhanced by MEMS technology in the near future. The knowledge of intracranial pressure following surgical procedures and within surgical cavities will likely drive the insertion of MEMS devices. This motivation may also be applied to vascular and trauma surgery as well. The knowledge of intracranial pressure following the surgery for a malignant brain tumor may be critical in the days, weeks, and even months following a surgical procedure. Trauma surgery is often associated with the monitoring of intracranial pressure. Following the performance of a craniotomy for the evacuation of an intracranial hemorrhage due to trauma, the placement of a MEMS device for the monitoring of pressure could revolutionize the management of head injury patients. Hydrocephalus could similarly be monitored to provide the surgeon a knowledge of abnormal pathology or shunt failure [54].

Spine surgery may also be grouped into three primary categories: degenerative disease, tumor, and trauma. The utilization of MEMS technology for the surgical management of degenerative disease may become commonplace. The loads applied to spinal implants may be monitored via strain gages and pressure sensors [38, 46, 56, 71]. Intradiscal pressure and pressure within the bone parenchyma may also be monitored with minimally invasive or surgically applied MEMS devices. Bone healing and disc regeneration, as well as spinal cord regeneration, can also be enhanced by MEMS technology.

6.6. OBSTACLES TO NEUROSURGICAL EMPLOYMENT OF MEMS

There are specific scientific, technical, and economic challenges that must be over-come before MEMS can become successfully transitioned into neurosurgery. Many of the challenges arise from the hostile environment of the human body [52, 72]. The corrosive effects of body fluids can disrupt and injure or destroy implanted MEMS devices. The implantation of foreign substances, such as silicon, initiates the body's elaborate defense mechanisms that begin with inflammation and end with fibrotic encapsulation. Therefore, the biocompatibility of implantable MEMS devices is a major concern. Recent advances in evaluation of biocompatibility of MEMS materials of construction are described later in this section.

Another challenge relates to implantable devices that must communicate information to the neurosurgeon [71]. For example, implanted sensors must transmit information from the implant to external monitoring instruments. Wires inserted through the skin can be used to connect the external instruments to the implanted sensors. However, this approach reduces patient mobility and there is an increased likelihood of infection. Consequently, a wireless telemetry system would be highly desirable.

The development cost of MEMS devices can be prohibitive on the "front end." The significant expense is due in part to the capital investment that is required for the cleanroom-based equipment and processes required for microfabrication [71]. However, once the technology is developed, it can become very cost effective on the "business end". Consequently, prototyping of MEMS devices is often accomplished using shared or multi-user facilities and processes. This attraction of this approach is that the high fabri-cation costs are distributed over multiple users, thereby making it economically feasible for cyclical design iterations. Examples of multi-user services include MUMPs (MEMSCAP Inc., Durham, NC) and SUMMiT (Sandia National Laboratories, Albuquerque, NM). The primary downside of prototyping via multi-user services is the design of the MEMS device can be limited by process and equipment constraints.

6.6.1. Biocompatibility Assessment

The deployment of implantable MEMS devices based on silicon and related micro-electronics materials has generally relied on protective packaging approaches to isolate the MEMS component from the hostile body environment [52, 72]. For example, MEMS pressure sensors are encapsulated with biocompatible silicone gels to isolate the piezore-sistive sensor from body fluids. The protective packaging approaches can result in two primary drawbacks: attenuation of signal/stimulus that must be communicated between the physiological environment and device; and, increased size that detracts from the benefits of miniaturization particularly when working in constrained spaces or at the cellular level. Unfortunately, there is a paucity of data on the biocompatibility of MEMS materials of construction. Often, previous efforts on biocompatibility testing of MEMS materials were usually limited to specific applications, and often, relied on non-standard [42, 44]. For example, the pre-testing procedures employed in many studies did not correspond to acceptable sterilization protocols for a clinical device. Although such efforts are valuable in shedding insight into device performance *in vivo* within a research context, the regulatory scrutiny associated with medical devices will require biocompatibility evaluation of MEMS devices based on standardized tests.

Materials commonly used in the fabrication and packaging of standard MEMS devices were recently evaluated for cytotoxicity using the ISO 10993 biocompatibility testing standards [15]. The material set comprised of: silicon (Si, 500 μm-thick), silicon dioxide (SiO_2, 0.5 μm-thick), silicon nitride (Si_3N_4, 0.2 μm-thick), polycrystalline silicon (polysil-icon, 0.5 μm-thick), silicon carbide (SiC, 0.5 μm-thick), titanium (Ti, 0.5 μm-thick), and SU-8 (50 μm-thick) (Table 6.1). Except for the silicon substrates, all the other materials were formed by standard thin film deposition or coating techniques onto 100-mm diameter, (100)-oriented silicon wafers, which were procured from commercial suppliers. Half of the wafers was packaged in a sealed sterilizable pouch and exposed to 2X-gamma sterilization of 2.5 Mrad per exposure. The remaining half of the wafers was autoclaved utilizing steam sterilization using a 25 min wrap, 10 min dry algorithm at 121°C. Random wafers were

TABLE 6.1. *In vitro* cytotoxicity of MEMS materials.

Material	Reactivity (0–4)
Positive Control (tin stabilized polyvinylchloride)	4
Negative Control (high density polyethylene)	0
Si (monocrystalline silicon)	0
SiO_2 (silicon dioxide)	0
Polysilicon (polycrystalline silicon)	0
Si_3N_4 (silicon nitride)	0
SiC (monocrystalline 3C silicon carbide)	0
Ti (sputtered titanium)	0
SU-8 (epoxy photoresist)	0

subjected to scanning electron microscopy (SEM) pre- and post-sterilization to look for evidence of sterilization-induced damage.

Cytotoxicity tests were performed following the ISO 10993-5 standard: "Test for Cytotoxicity—In Vitro Methods". A single extract of the test article was prepared using single strength Minimum Essential Medium (1X MEM) supplemented with 5% serum and 2% antibiotics. Each test extract was then placed onto three separate confluent monolayers of L-929 mouse fibroblast cells which had been propagated in 5% CO_2. Two milliliters of the test extract, the reagent control, the negative control (high density polyethylene), and the positive control (tin stabilized polyvinylchloride) were placed into each of the three 10 cm^2 test wells and were incubated at 37°C in the presence of 5% CO_2 for 48 hours. The test well contents were then examined microscopically (100X) to determine any change in cell morphology and the percent lysis. Test well contents were also examined for confluence of the monolayer, and color as an indicator of resulting pH. Results were scored on a scale of 0–4, where 0 represented the best case—no adverse reaction whatsoever—and 4 represented the worst case—complete cell lysis. A score of 2 or below is considered acceptable for many implantable applications.

SEM evaluation revealed no discernable damage induced by either radiation or steam sterilization. For the cytotoxicity testing, all MEMS materials were graded 0, which is as good as the negative. The data from this evaluation suggests that MEMS materials are suitable candidates for the development of implantable medical devices. However, it should be noted further testing will still be required to validate MEMS devices for specific neurosurgical applications [46, 52].

6.7. OPPORTUNITIES

There are many challenges yet to be overcome. This implies that the clinical use of MEMS for the management of neurological disorders is not yet on the immediate horizon. Nonetheless, the capabilities of MEMS technology hold great promise to revolutionize the discipline of neurosurgery. The routine employment of MEMS technology in the neurosurgical arena obviously hinges on the biocompatibility and the wireless telemetry issues previously mentioned. Nevertheless, the opportunities are abundant and the potential for future technological advances and applications are nearly endless. Examples of shorterterm opportunities that are under investigation in academic, corporate, and government laboratories are described next.

6.7.1. Intracranial Pressure Monitoring

Intracranial pressure monitoring is currently achieved by the placement of a percutaneous wire or tube into the substance of the brain (parenchyma) or the ventricular system. This obligatorily is associated with a moderate to significant risk of infection and a relative urgency to remove the device. Therefore, these devices are often removed prematurely due to fears of catastrophic infection complications. If a non-intrusive monitoring and telemetric device was placed into the brain (e.g., to monitor intracranial pressure), the advantage for the management of head injury patients (on a subacute basis) and brain tumor patients (on a

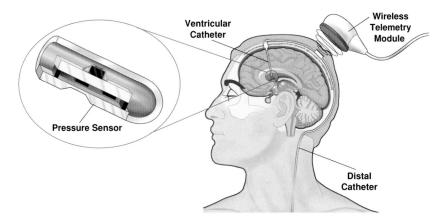

FIGURE 6.11. Schematic depiction of a concept for an intracranial pressure (ICP) measurement system. A MEMS pressure sensor is located at the tip of the ventricular catheter. The pressure information is transmitted via wireless telemetry from an implanted coil to an external telemetry module.

subacute and long term basis) would be substantial. Decisions regarding treatment could be readily made and catastrophic neurologic deterioration could be prevented or aborted.

Figure 6.11 presents a schematic overview of a proposed *Programmable Automatic Shunt System* for the treatment of hydrocephalus [51]. The design of the shunt system incorporates, in a single implanted unit, MEMS sensors for the measurement of ICP and cerebrospinal fluid (CSF) flow, an adjustable valve for CSF drainage, and microelectronics for signal processing and telemetry [58]. The implant transmits the pressure and flow information via the implanted coil antenna to the neurosurgeon, who, in turn, issues commands to adjust the CSF drainage valve. The system could potentially be modified to work based on internal feedback without neurosurgeon input. This latter strategy embraces and, in fact defines, the concept of a smart system—or in the specific case discussed here, a "smart shunt." With such a smart shunt, for example, an elevation of intracranial pressure would signal a modification of outflow resistance so that CSF flow would be increased. Recent efforts in the development of a smart shunt system include the investigations into microvalves and packaging strategies [41, 59].

The employment of intracranial pressure monitoring techniques may even be applied outside the immediate medical environment (e.g., hospital or clinic). For example, the employment of MEMS-derived pressure sensing technology with modern telemetric capabilities might permit a parent of a child with hydrocephalus to monitor intracranial pressure. Such a system could alert the parent to an impending catastrophe related to increased cerebrospinal fluid pressure, well in advance of clinical manifestations. A system in which a "red light—yellow light—green light" concept is employed could be used to alert a parent via an external sensing device that is simply positioned close to the child's head with telemetric and MEMS technology. Then, an evaluation of the child's intracranial pressure could be made and grossly presented to the parent in the form of a "green light, yellow light, or red light", thus providing an early warning system. The neurosurgeon could then act accordingly following a call from the parents. Finally, the surgeon should have, in his/her office, a monitoring unit that would provide precise pressure data and waveforms. This approach would facilitate further analysis of the patient's status, thus making the clinical decision-making process more accurate and safer.

6.7.2. Neural Prostheses

Neural prostheses include devices that facilitate the manipulation of the nervous system. They may be used to augment or diminish (e.g., ablate) neurological function. They are becoming increasingly important in neurosurgical applications due to the myriad of recent advances in the functional neurosurgery and surgical epilepsy arenas. With such devices as implanted electrodes for the modulation of undesirable spontaneous motion in Parkinson's patient, feedback to the surgeon and to the device itself (smart system) may be of extreme clinical significance. For example, rather than providing a fixed electrical current pattern to a specific region of the brain to compensate for a pathological movement, a "smart prosthesis" could record electrical activity from the surrounding brain and modify electrode stimulation, based on the input from the surrounding brain. This type of input could perhaps even help the surgeon direct the initial placement of the electrode. Furthermore, integration of locomotive actuators onto the electrodes could enable the creation of a "smart neural prosthesis" that would propel itself to a more optimal recording and stimulation location.

Figure 6.12 presents examples of microelectrode arrays that have been fabricated using MEMS technology to link the nervous system to microelectronics circuitry [50, 66, 68]. These neural probes were designed for multichannel sensing and stimulation in the cerebral cortex. The micromachined silicon substrate supports an array of metallic film conductors that are insulated above and below by dielectric films. Openings in the upper layer along the probe shank defined stimulating or recording sites, which are inlayed with gold or iridium oxide for interfacing with brain tissue. At the rear of the probe, integrated circuits provided

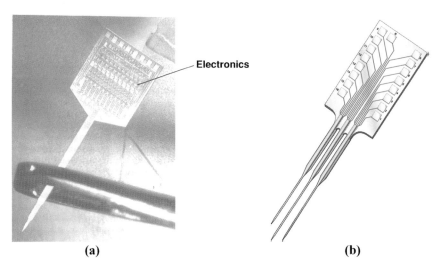

(a)   **(b)**

FIGURE 6.12. Examples of neural microprobes fabricated by MEMS technology. (a) Photograph showing integrated electronics and microprobe shank, which is passed through the eye of a needle. (b) Schematic depiction of a multi-shank probe.

FIGURE 6.13. SEM image of an electroplated microprobe for deep brain stimulation (DBS) applications. This probe design exhibits smooth geometrical taper while maintaining sharp tips necessary for ready penetration into brain tissue. *(Courtesy: UCLA)*

signal amplification, filtering, and a multiplexed interface to the output leads, which are connected to external circuitry through a percutaneous plug. The applicability of this neural prosthesis system could be further enhanced if the electrical connection through the wires was substituted by wireless telemetry.

Recent investigations into the use of MEMS technology for the development of neural probes have focused on addressing biocompatibility for long-term performance, integration of fluidic ports for drug delivery, and novel fabrication strategies to achieve shank flexibility and three-dimensional geometry [49, 64]. Flexibility is desirable to minimize tissue damage that can be caused by relative micro-motion between the implanted shank and the brain. However, the shank must be sufficiently stiff to penetrate into the brain without buckling. One feasible strategy to meet the structural criteria appears to be a combination shank comprising a polymeric material, such as polyimide or benzocyclobutene (BCB), and silicon [57, 75]. Other considerations for penetration of neural probes into the brain are requirements for sharp tips and smooth geometrical taper. A MEMS probe that meets these criteria is shown in Figure 6.13 [65]. The probe design could incorporate multiple electrodes while promising to maintain the taper and length characteristics of the conventional micro-wires.

6.7.3. Drug Delivery Systems

Drugs and other substances, that require precise instillation techniques, could be delivered by MEMS systems. This is particularly so for agents that require micro dosages (e.g., bone morphogenic proteins, some chemotherapeutic agents, or dopamine and related agents). Such a device could be placed at the time of a surgical fusion procedure to stimulate bone fusion, within a tumor bed to precisely apply chemotherapeutic agents, or into the deep brain substance to treat movement disorders. A controlled and precise drug release could then be achieved.

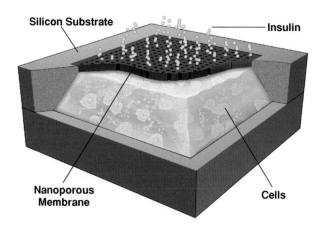

FIGURE 6.14. Schematic cut-away depiction of a MEMS biocapsule for transplantation of pancreatic islet cells. The nanometer-sized pores in the membrane allow for nutrient and insulin transport but ensure an immunoprivileged environment for the islet cells.

Figure 6.14 presents a schematic depiction of a micromachined silicon immunoisolating biocapsule that was developed to house pancreatic islet cell transplants for insulin therapy [42]. This MEMS device consists of a cell chamber and filter membrane with 20 nm pores. The pore size is large enough to allow for insulin and oxygen exchange, but small enough to impede the passage of viruses and immune molecules. The microfabricated biocapsule could also be used in the neurosurgical arena to place neuronal xenografts into the brain. The transplanted neurons could be used to supplement electrical stimulation or lesion surgery for neurologically impaired patients. The biocapsule could also be used to create drug delivery systems in which the release of the drug is controlled by its diffusion across the micromachined membrane, while the membrane protects the patient and the xenograft by creating an immunoprivileged environment within the capsules.

6.7.4. Smart Surgical Instruments and Minimally Invasive Surgery

The capability of MEMS could enable a set of next-generation surgical tools. For example, the incorporation of pressure sensors, strain gauges, or biochemical sensors into surgical instruments could enable smart instruments that would be able to distinguish between different types of tissue, both with regard to tissue density and biochemical makeup. Furthermore, the incorporation of MEMS sensors and actuators could allow for precision cutting or local manipulation of tissue with unprecedented control, thereby minimizing nondesirable tissue damage. A possible example of such smart tools is illustrated in Figure 6.15 [55]. The employment of sensors and actuators that modify, limit, and amplify selected surgeon-generated motion could significantly enhance surgeon precision, accuracy, and speed. This can be accomplished by the employment of systems that, among other things, filter out non-desirable repetitive motion (e.g., tremors), while enhancing desirable motion (e.g., precision) and even amplifying other motions (e.g., extension of an instrument, such as

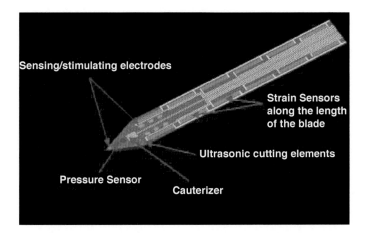

FIGURE 6.15. Schematic illustration of a smart tool for minimally invasive surgery. The sensors would enable the
neurosurgeon to distinguish between different types of tissue, while the ultrasonic cutting elements and cauterizer
would enable precision cuts with minimal non-specific damage. *(Courtesy: Verimetra Inc)*

a micro-dissection tool, under enhanced control). The availability of such smart instruments
should also enhance the capabilities of minimally invasive procedures in neurosurgery.

The use of piercing probes and cutting blades for neurosurgical applications ranges
from needles for micro-syringes to scalpels for surgery. These pointed and edged tools are
typically mass-produced with metals (e.g., stainless steel) and glass, or are individually
hand-ground from harder materials, such as diamond and ceramics. The mass-produced
components may be inexpensive. However, they may become blunt and degrade relatively
quickly over time. In contrast, diamond and ceramic-based tools exhibit better performance,
but at relatively higher cost. The use of MEMS technology for the fabrication of silicon
probes and blades combine the performance characteristics of the hard-material tools with
the cost advantages of mass production. Specifically, the miniaturization and batch fabrication characteristics of MEMS technology enable the fabrication of arrays of microprobes
and microblades simultaneously on a single substrate. In contrast to the grinding processes
used to sharpen metallic tools, the etching processes used to create microfabricated devices could provide smoother surfaces and sharper edges, as shown in Figure 6.16 [45].
Consequently, meticulous dissection of fine membranous structures, such as the arachnoid
membrane, could be accomplished with minimal effort and with significant precision. In
addition, the wear and strength characteristics of silicon devices can be further enhanced
through the use of thin film coatings of silicon nitride, which can be uniformly deposited
on the shaped points and edges.

All of the aforementioned surgeon-enhancing strategies are of significant relevance
to the minimally invasive surgery arena. The enhancement of precision and safety (e.g.,
the minimization of tremor) could significantly advance the field, which is currently
severely limited by the surgeon's physical limitations. Furthermore, MEMS and related
microelectronics can create an environment for the development of enhanced workstations,

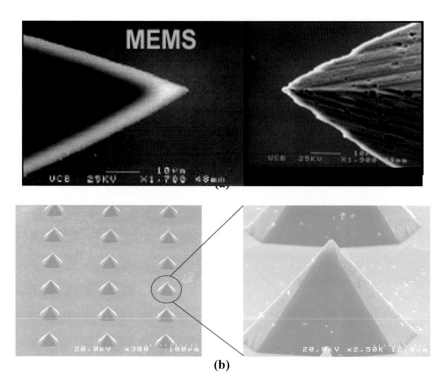

FIGURE 6.16. SEM images of MEMS surgical sharps. (a) Sharper and smoother cutting edges are provided by the etching processes used in the fabrication of MEMS devices. (b) Array of piercing microprobes. *(Courtesy: Verimetra Inc)*

fly-by-wire technologies, and electronically enhanced tools that revolutionize the minimally invasive neurosurgical environment, which at the current time, is relatively stagnant.

6.7.5. In Vivo *Spine Biomechanics*

The application of MEMS to the spine arena is centered about its ability to monitor pressure, strains, and loads. Pressure within bone or within a cage, as well as strains and loads applied to interbody, ventral, and dorsal implants, are clearly of value to the researcher and the practicing surgeon. The telemetric transmission of this information to the treating physician could provide information that is indicative of fusion status, implant competence, and implant failure. A characteristic pattern of fusion and non-union most certainly exists on a clinical situation-specific basis as suggested in Figure 6.17. Such patterns are obviously speculative at this point. However, they are probably consistent within specific clinical applications. Furthermore, they most likely can differentiate fusion from non-fusion (pseudoarthrosis) status. Current methods for fusion assessment involve radiographic imaging techniques and the subjective assessment of patient pain. Conventional radiographic techniques, (CT, MRI, x-ray) provide only a snapshot in time of the condition [39]. Artifacts from the spinal instrumentation often disrupt the image and skew the clinical impression. In fact, the imaging of threaded interbody fusion constructs and other metallic implants often

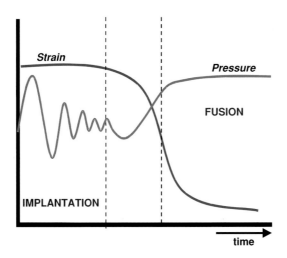

FIGURE 6.17. Graph showing possible relative variations in strains on spinal implants and pressure within bone graft versus time. The dashed lines depict the phases of fusion maturation.

cannot accurately display the bone-metal interface where the fusion is initiated [39]. Finally, patient derived information is highly subjective and a clinically accurate and complete history is not usually obtained.

The assessment of pressures, loads, and strains on an on-line basis via telemetric techniques permits the spine surgeon to assess biomechanical parameters *in vivo*. This forms the basis for the *in vivo* biomechanics laboratory.

Each patient with such monitoring capabilities inserted essentially is, in fact, such a laboratory. The *in vivo* biomechanics laboratory concept may be of much greater significance to the future of spine surgery than is intuitively apparent at the outset. Figure 6.18 presents the concept for a proposed fusion monitoring system based on implantable wireless

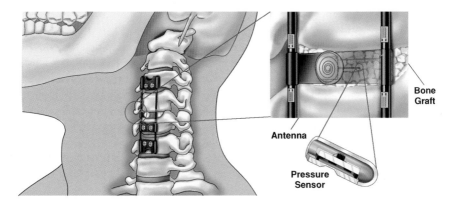

FIGURE 6.18. Schematic depiction of a concept for an implantable fusion assessment system using MEMS pressure sensors and strain gauges. The strain gauges monitor biomechanical loading of the cervical plate while the pressure sensor is used to assess pressure fluctuations within the bone graft. The measured pressure and load information is transmitted to an external receiver via wireless telemetry.

MEMS sensors that are mounted on spinal implants and within bone [38, 71]. The miniature size, integration with electronics, and manufacturability of MEMS sensors suggest that the corresponding fusion monitoring system could be readily transitioned into routine clinical practice for spine patients undergoing vertebral stabilization surgery. The combination of a MEMS sensor and a telemetric information transmission system can provide the surgeon with invaluable, otherwise unobtainable, data that is accurate, online, and real-time. This data is derived from a system (the combination of the patient and a MEMS-based implant) that requires no assumptions to generate the information. It only measures actual and realized parameters such as loads, strains, and pressures. Conversely, with traditional biomechanical laboratory testing or finite element modeling (FEM) studies, multiple assumptions are made. These include assumptions about paraspinous muscle activity and their contribution to stability, bone density, the applicability of the test performed to the clinical situation at hand, etc. If, in a given laboratory experiment, 10 such assumptions were made, and each was associated with a 20% chance of error (both are conservative estimates), a significant error would be introduced in the final analysis. This error, in fact, is portrayed by the predicted accuracy of such an experiment. If each assumption is associated with an 80% accuracy (20% chance of error), 10 assumptions would drop the accumulative accuracy to $0.8^{10} = 0.107 = 10.7\%$ accuracy [39].

In vivo biomechanical testing makes no assumptions. If the device or system is accurate, it simply provides *in vivo*, real-time information. If this can be routinely and readily achieved, the traditional biomechanics laboratory and finite element modeling strategies for spine applications will essentially become obsolete.

The implications to the treating physicians, as well as the scientist (e.g., engineer) and implant manufacturer regarding the strains and loads applied to a device, as well as the pressure within cages, is clearly of clinical, research, and marketing value. What such information permits is the *in vivo* monitoring of biomechanical parameters; hence the term "*in vivo* biomechanical testing" and the development of a new entity—the "*in vivo* biomechanics laboratory".

6.7.6. Neural Regeneration

The miniaturization and electronics integration capabilities of MEMS technology may also be applicable to the development of interfaces for neural regeneration [43, 53]. Peripheral nerve injuries often recover since collagen and other structurally stabilizing substances are present in the peripheral nerve. This is not so in the central nervous system. Central nervous system regeneration may be facilitated by MEMS technology in four ways: [38] employment of hydraulic and micro-hydraulic principles to MEMS applications (microfluidics) via the employment of very small tubes to facilitate neural growth or the pumping of fluids; [39] facilitation of neural growth by the application of electrical potentials; [40] application of growth stimulation agents, such as growth factors, via MEMS driven drug delivery systems; and [41] utilization of silicon substrates for the attainment of spatially directed neuronal regeneration.

Figure 6.19 presents a proposed neuronal regeneration interface using MEMS technology [53]. The interface consists of a perforated silicon substrate that is designed to spatially constrain and direct the growth of regenerating nerve fibers from opposite directions. This design also includes metal electrodes that are patterned on the silicon substrate. These

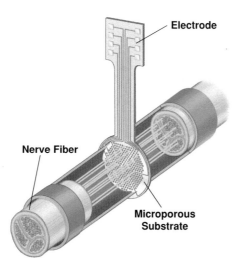

FIGURE 6.19. Schematic illustration of a MEMS-based neural regeneration interface. The perforated substrate spatially constrains the growing ends of the nerve fibers in close proximity. The electrodes are incorporated for recording and stimulation.

electrodes are incorporated in order to record axon signals and stimulate growth during the regeneration process.

Recent investigations into the use of MEMS technology for neural regeneration have focused on the development of hybrid neural interfaces, which incorporate a biological construct to enhance the integration of the synthetic implant into a host's nervous system [48]. The biological constructs are typically either cells that outgrow from the interface into the host or neurophillic factors that coerce the host to grow into the interface [61, 63, 70, 74]. Figure 6.20 presents SEM images of a sieve electrode that incorporates proteins and nerve guide tubes to encourage axons to grow through the pores [63]. Nine recording

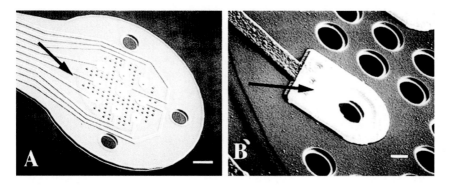

FIGURE 6.20. SEM images of a sieve electrode to encourage nerve cells to grow through the pores. (a) There are silicon leads (arrow) and nine iridium-coated active sites, which are 5–8 μm in diameter. (b) Close up of an 8 μm-diameter active site showing the iridium coating (arrow).

sites are integrated via polysilicon leads into a silicon ribbon cable. The silicon leads are insulated from the external environment and internal diaphragm with dielectric layers.

6.8. PROSPECTS FOR MEMS IN NEUROSURGERY

The future of MEMS applications to the neurosurgical arena is immense. In fact, even the clinical employment of smart systems, the application of MEMS technology to the molecular biology arena, and the modification of cell growth via MEMS technology is imminently within reach. MEMS and related technologies can be utilized to enhance bone fusion, improve recovery following spinal cord and peripheral nerve injury, and for the management of patients with brain tumors and traumatic brain injury. The incorporation of MEMS devices into surgical robotics and navigation will enhance minimally invasive neurosurgery. Placement of sensors into operative tools for structural and physiological monitoring will minimize unnecessary damage to tissues while providing critical biological information. Ultimately, advances in the development of hybrid neural interfaces coupled with increased understanding at the cell/molecular biology level could lead to intriguing prospects for bionic integration [60]. What is readily achievable today is the monitoring of clinical parameters via minimally invasive means on a long-term basis [71].

The integration of MEMS technology into the medical milieu and the "neurosurgery milieu" faces many obstacles. Neurosurgeons must become more aware of the potential and limitations of MEMS technology. This effort will require educational initiatives that will bring together developers of MEMS technology with users in neurosurgery. For example, new training programs directed towards medical residents and research fellows will need to be established. In addition, comprehensive review articles and didactic textbooks must be written. The successful application of MEMS to the neurosurgical arena requires the merging of fields and disciplines, the merging of technologies and the collective gathering of expertise. Collegiality, communication, and significant investments will be required. Risks must be taken. However, the benefits derived will far outweigh the risk and the time and money invested.

ACKNOWLEDGEMENTS

We would like to thank David Schumick of the Department of Medical Illustrations at the Cleveland Clinic Foundation for his generous assistance with the original graphics that are extensively used throughout this article. We are also grateful to Dr. Michele Migliulo of Verimetra Inc., Pittsburgh, PA, and Dr. Jack W. Judy of the University of California, Los Angeles, for their generosity in providing illustrations and SEM images. In addition the following organizations are credited for our courtesy use of their graphics: Museum of Man (San Diego, CA), Medtronic Inc. (Minneapolis, MN), Elekta AB (Stockholm, Sweden), Guilford Pharmaceuticals Inc. (Baltimore, MD).

REFERENCES

[1] R. Bayston, W. Ashraf, and C. Bhundia. Mode of action of an antimicrobial biomaterial for use in hydro-cephalus shunts. *J. Antimicrob. Chemother.*, 53:778–782, 2004.

[2] E.C. Benzel. *Biomechanics of Spine Stabilization.* Rolling Meadows, IL, American Association of Neurological Surgeons, 2001.

[3] P.M. Black. Harvey cushing at the Peter Brigham hospital. *Neurosurgery,* 45:990–1001, 1999.

[4] D. Campillo. Neurosurgical pathology in prehistory. *Acta Neurochirurg.,* 70:275–290, 1984.

[5] J.M. Drake, J.R.W. Kestle, and S. Tuli. CSF shunts 50 years on—past, present and future. *Child's Ner. Syst.,* 16:800–804, 2000.

[6] J.M. Drake and C. Sainte-Rose. *The Shunt Book.* Cambridge, MA, Blackwell Science, 1995.

[7] R. Drucker-Colin and L. Verdugo-Diaz. Cell transplantation for Parkinson's disease: present status. *Cell. Mol. Neurobiol.,* 24:301–316, 2004.

[8] S. El Gindi. Neurosurgery in Egypt: past, present, and future—from pyramids to radiosurgery. *Neurosurgery,* 51:795–796, 2002.

[9] S.T. Goodrich. History of spine surgery in the ancient and medieval worlds. *Neurosurg. Focus,* 16:E2, 2004.

[10] S.H. Greenblatt. The crucial decade: modem neurosurgery's definitive development in Harvey Cushing's early research and practice, 1900 to 1910. *J. Neurosurg.,* 87:964–971, 1997.

[11] D.R. Groothuis. The blood-brain and blood-tumor barriers: a review of strategies for increasing drug delivery. *Neuro-Oncology,* 2:45–59, 2000.

[12] C. Guerin, A. Olivi, J.D. Weingart, H.C. Lawson, and H. Brem. Recent advances in brain tumor therapy: local intracerebral drug delivery by polymers. *Investigat. New Drugs,* 22:27–37, 2004.

[13] R.D. Guyer and D.D. Ohnmeiss. Intervertebral disc prostheses. *Spine,* 28:S15–S23, 2003.

[14] L.F. Haas. Papyrus of Ebers and Smith. *J. Neurol., Neurosurg., Psychiat.,* 67:578, 1999.

[15] B. Karger, S. Hubert, and B. Brinkmann. Arrow wounds: major stimulus in the history of surgery. *World J. Surg.,* 25:1550–1555, 2001.

[16] A.H. Kaye and E.R. Laws, Jr. Churchill Livingstone, London, New York, 2001.

[17] W. Kohen, C. Kolbenschlag, S. Teske-Keiser, and B. Jansen. Development of a long-lasting ventricular catheter impregnated with a combination of antibiotics. *Biomaterials,* 24:4865–4869, 2003.

[18] J.G. Lascaratos, I.G. Panourias, and D.E. Sakas. Hydrocephalus according to Byzantine writers. *Neurosurgery,* 55:214–220, 2004.

[19] E.R. Laws, Jr. Neurosurgery's man of the century: Harvey Cushing—the man and his legacy. *Neurosurgery,* 45:977–982, 1999.

[20] O. Lindvall, Z. Kokaia, and A. Martinez-Serrano. Stem cell therapy for human neurodegenerative disorders-how to make it work. *Nat. Med.,* 10:S42–S50, 2004.

[21] C.Y. Liu and M.L.J. Apuzzo. The genesis of neurosurgery and the evolution of the neurosurgical operative environment: part I-prehistory to 2003. *Neurosurgery,* 52:3–19, 2003.

[22] C.Y. Liu, M. Spicer, and M.L.J. Apuzzo. The genesis of neurosurgery and the evolution of the neurosurgical operative environment: part II-concepts for future development, 2003 and beyond. *Neurosurgery,* 52:20–35, 2003.

[23] G. Martin. Was Hippocrates a beginner at trepanning and where did he learn? *J Clin. Neurosci.,* 7:500–502, 2000.

[24] D.E. McDonnell. History of spinal surgery: one neurosurgeon's perspective. *Neurosurg. Focus,* 16:El, 2004.

[25] E.B. Montgomery. Two advancements in the management of Parkinson disease. *Cleveland Clin. J. Med.,* 69:639–643, 2002.

[26] R.A. Morantz and J.W. Walsh. Marcel Dekker, New York, 1994.

[27] J. Piek, G. Lidke, T. Terberger, U. von Smekal, and M.R. Gaab. Stone age surgery in Meckelburg-Vorpommem: a systematic study. *Neurosurgery,* 45:147–151, 1999.

[28] I.K. Pople. Hydrocephalus and shunts: what the neurologist should know. *J. Neurol. Neurosurg. Psych.,* 73:i17–i22, 2002.

[29] D. Purves, G.J. Augustine, D. Fitzpatrick, L.C. Katz, A.-S. LaMantia, J.O. McNamara, and S.M. Williams. Sinauer Associates, Sunderland, MA, 2001.

[30] G.E. Rawlings, 3rd, and E. Rossitch, Jr. The history of trephination in Africa with a discussion of its current status and continuing practice. *Surg. Neurol.,* 41:507–513, 1994.

[31] S.S. Rengachary and R.H. Wilkins. Mosby, St. Louis, 1994.

[32] M. Rutkow. Trephination: how did they do it? *Archives of Surgery,* 135:1119, 2000.

[33] J.L. Stone Paul Broca and the first craniotomy based on cerebral localization. *J. Neurosurg.* 75:154–159, 1991.

[34] C. Toporek and K. Robinson. *Hydrocephalus: A Guide for Patients, Families, and Friends.* Beijing; Sebastopol, O'Reilly, 1999.

[35] M. Victor and A.H. Ropper McGraw-Hill, New York, 2001.

[36] L.W. Way and G.M. Doherty. McGraw-Hill, New York, 2003.

[37] T.A. Zesiewicz and R.A. Hauser Neurosurgery for Parkinson's disease. *Seminar. Neurol.*, 21:91–101, 2001.

[38] E. Benzel, L. Ferrara, S. Roy, and A. Fleischman. Micromachines in spine surgery. *Spine*, 29:601–606, 2004.

[39] E.C. Benzel. *Biomechanics of Spine Stabilization*. Rolling Meadows, IL, AANS Publications, 2001.

[40] E.C.Benzel, L.A. Ferrara, S. Roy, and A.J. Fleischman. Biomaterials and implantable devices: discoveries in the spine surgery arena. *Clini. Neurosurg.*, 49:209–225, 2002.

[41] S. Chung, J.K. Kim, K.C. Wang, D.-C. Han, and J.-K. Chang. Development of MEMS-based cerebrospinal fluid shunt system. *Biomed. Microdev.*, 5:311–321, 2003.

[42] T.A. Desai, W.H. Chu, J.K. Tu, G.M. Beattie, A. Hayek, and M. Ferrari. Microfabricated immunoisolating biocapsules.*Biotechno. Bioeng.*, 57:118–120, 1998.

[43] D.J. Edell. A peripheral nerve information transducer for amputees: long-term multichannel recordings from rabbit peripheral nerves. *IEEE Trans. Biomed. Eng.*, 33:203–214, 1986.

[44] D.J. Edell, V.V. Toi, V.M. McNeil, and L.D. Clark. Factors influencing the biocompatibility of insertable silicon microshafts in cerebral cortex. *IEEE Trans. Biomed. Eng.*, 39:635–643, 1992.

[45] L.A. Ferrara, A.J. Fleischman, E.C. Benzel, and S. Roy. Silicon dermabrasion tools for skin resurfacing applications. *Med. Eng. Phys.*, 25:483–490, 2003.

[46] L.A. Ferrara, A.J. Fleischman, D. Togawa, T.W. Bauer, E.C. Benzel, and S. Roy. An in vivo biocompatibility assessment of MEMS materials for spinal fusion monitoring. *Biomed. Microdev.*, 5:297–302, 2003.

[47] A.J. Fleischman. Miniature high frequency focused ultrasonic transducers for minimally invasive imaging procedures. *Sens. Actu. A: Phys.*, 103:76–82, 2003.

[48] W.M. Grill, J.W. McDonald, P.H. Peckham, W. Heetderks, J. Kocsis, and M. Weinrich. At the interface: convergence of neural regeneration and neural prostheses for restoration of function. *J. Rehabilitat. Res. Develop.*, 38:633–639, 2001.

[49] J.W. Judy. *International Conference of IEEE Engineering in Medicine and Biology Society.* Cancun, MEXICO, 2003.

[50] D.T. Kewley, M.D. Hills, D.A. Borkholder, I.E. Opris, N.I. Maluf, C.W. Storment, J.M. Bower, and G.T.A. Kovacs. Plasmaetched neural probes. *Sen. Actu.*, A58:27–35, 1997.

[51] W.H. Ko, C.W. Meyrick, and H.L Rekate. Cerebrospinal fluid control system. *Proceedings of the IEEE*, 76: 1226–1235, 1988.

[52] G. Kotzar, M. Freas, P. Abel, A. Fleischman, S. Roy, C. Zorman, J.M. Moran, and J. Melzak. Evaluation of MEMS materials of construction for implantable medical devices. *Biomaterials*, 23:2737, 2002.

[53] G.T.A. Kovacs, C.W. Storment, and J.M. Rosen. Regeneration microelectrode array for peripheral nerve recording and stimulation. *IEEE Trans. Biomed. Eng.*, 39:893–902, 1992.

[54] J.S. Kroin, R.J. McCarthy, L. Stylos, K. Miesel, A.D. Ivankovich, and R.D. Penn. Long-term testing of an intracranial pressure monitoring device. *J. Neurosurg.*, 93:852–858, 2000.

[55] K.S. Lebouitz and M. Migliuolo. Verimetra, Inc., USA, 2002.

[56] E.H. Ledet, B.L. Sachs, J.B. Brunski, C.E. Gatto, and P.S. Donzelli. Real-time in vivo loading in the lumbar spine: part 1. Interbody implant: load cell design and preliminary results. *Spine*, 25:2595–2600, 2000.

[57] K.-K. Lee, J. He, A. Singh, S. Massia, G. Ehteshami, B. Kim, and G. Raupp. Polyimide-based intracortical neural implant with improved structural stiffness. *J. Micromech. Microeng.*, 14:32–37, 2004.

[58] A.M. Leung, W.H. Ko, T.M. Spear, and J.A. Bettice. Intracranial pressure telemetry system using semicustom integrated circuits. *IEEE Trans. Biomed. Eng.*, 33:386–395, 1986.

[59] L.-A. Liew and V.M. Bright. *2000 IEEE-EMBS International Special Topic Conference on Microtechnologies in Medicine and Biology*, Lyon, FRANCE, 2000.

[60] C.Y. Liu, M. Spicer, and M.L.J. Apuzzo. The genesis of neurosurgery and the evolution of the neurosurgical operative environment: part II-concepts for future development, 2003 and beyond. *Neurosurgery*, 52:20–35, 2003.

[61] M.P. Maher, J. Pine, J. Wright, and Y.-C. Tai. The neurochip: a newmultielectrode device for stimulating and recording from cultured neurons. *J. Neurosci. Methods*, 87:45–56, 1999.

[62] D.V. McAllister, M.G. Allen, and M.R. Prausnitz. Microfabricated microneedles for gene and drug delivery. *Ann. Rev. Biomed. Eng.*, 2:289–313, 2000.

[63] A.F. Mensinger, D.J. Anderson, C.J. Buchko, M.A. Johnson, D.C. Martin, P.A. Tresco, R.B. Silver, and S.M. Highstein. Chronic recording of regenerating VIIIth nerve axons with a sieve electrode. *J. Neurophys.*, 83:611–615, 2000.

[64] P.S. Motta and J.W. Judy. In *IEEE-EMBS Special Topic Conference on Microtechnologies in Medicine and Biology*. Madision, WI, pp. 251–254, 2002.

[65] P.S. Motta and J.W. Judy. *Solid-State Sensor and Actuator Workshop*. Hilton Head, SC, 2002.

[66] K. Najafi. Solid-state microsensors for cortical nerve recordings. *IEEE Eng. Med. Biol.*, 13:375–387, 1994.

[67] K. Najafi and J. Hetke. Strength characterization of silicon microprobes in neurophysiological tissues. *IEEE Trans. Biomed. Eng.*, 37:474–481, 1990.

[68] K. Najafi, K.D. Wise, and T. Mochizuki. A high-yield IC-compatible multichannel recording array. *IEEE Trans. Elect. Dev.*, 32:1206–1211, 1985.

[69] D.L. Polla, A.G. Erdman, W.P. Robbins, D.T. Markus, J. Diaz-Diaz, R. Rizq, Y. Nam, H.T. Brickner, A. Wang, and P. Krulevitch. Microdevices in medicine. *Annu. Rev. Biomed. Eng.*, 2:551–576, 2000.

[70] P.J. Rousche, D.S. Pellin, D.P. Pivin, J.C. Williams, R.J. Vetter, and D.R. Kipke. Flexible polyimide based intracortical electrode arrays with bioactive capability. *IEEE Trans. Biomed. Eng.*, 48, 2001.

[71] S. Roy, L.A. Ferrara, A.J. Fleischman, and E.C. Benzel. Microelectromechanical systems and neurosurgery: a new era in a new millennium. *Neurosurgery*, 49:779–797; (discussion) 797–778, 2001.

[72] S. Roy and A.J. Fleischman. Cytotoxicity evaluation for microsystems materials using human cells. *Sens. and Mat.*, 15:335–340, 2003.

[73] S. Roy and M. Mehregany., Introduction to MEMS. In H. Helvajian (ed.), *Microengineering Aerospace Systems*, The Aerospace Press, El Segundo, CA, pp. 1–28, 1999.

[74] B. Schlosshauer, T. Brinker, H.-W. Muller, J.-U. Meyer. Towards micro electrode implants: in vitro guidance of rat spinal cord neurites through polyimide sieves by Schwann cells. *Brain Res.*, 903:237–241, 2001.

[75] A. Singh, K. Lee, J. He, G. Ehteshami, S. Massia, and G. Raupp. *International Conference of IEEE Engineering in Medicine and Biology Society* Cancun, MEXICO, pp. 3364–3367, 2003.

II

Drug Delivery

7

Vascular Zip Codes and Nanoparticle Targeting

Erkki Ruoslahti

The Burnham Institute, Cancer Research Center, 10901 North Torrey Pines Road, La Jolla, CA 92037

The use of nanoparticles in medicine will require sensors that can serve as guidance molecules for targeting the particles to appropriate sites in the body. Another set of sensors will be needed to allow the particles to respond to features at the target, such as inflammation, cell death, etc. Compounds that target the vascular endothelium provide one answer to the guidance and sensing problems. The endothelium of blood vessels is readily accessible from the blood stream, and the vessels in different tissues carry unique molecular signatures. Pathological lesions also put their signature on the vasculature; in tumors, both blood and lymphatic vessels differ from normal vessels. Peptides and antibodies that recognize vascular signatures have been shown to be useful in directing therapeutic agents to targets such as tumors. The targeting can enhance the efficacy of the therapy while reducing side effects. Combining the targeting technology with nanoparticles can take us a step closer to truly smart nanodevices.

7.1. INTRODUCTION

The concept of a "magic bullet" as a targeted treatment of disease has been around since Paul Ehrlich realized the unique specificity of antibodies. There are two ways of designing a targeted therapy: a drug or other therapeutic moiety can be so specific that it will affect only the intended target, or it can be physically targeted to the appropriate site in the body. Physical targeting could potentially make it possible to design "smart" nanoparticles that seek out the site of disease. The targeting would concentrate the particles at that site for diagnostic or therapeutic purposes. Targeted nanoparticles have been designed and some of them are about to enter clinical trials [31, 41]. The target recognition event could also serve as a sensor that would trigger a change in the particle. This change could then be detectable for diagnostic purposes, or could trigger the release of a drug or other treatments. The result would be a smart nanodevice of a kind that does not exist yet.

Vascular addresses can serve in targeting of diagnostic and therapeutic agents, including nanoparticles and nanodevices. The vasculature is an attractive tissue for targeting purposes because it is accessible through the blood stream and because extensive heterogeneity exists in endothelial cells that form the inner lining of blood vessels and lymphatics. Individual tissues express their own endothelial markers, and the vessels in pathological lesions, such as tumors, differ from the vessels in normal tissues [37]. In this chapter, I review some of the recent developments in vascular addresses and discuss the use of probes that recognize such addresses in nanoparticle targeting.

7.2. IN VIVO PHAGE DISPLAY IN VASCULAR ANALYSIS

We use bacteriophage, a virus that infects common bacteria, as a tool in probing the vasculature for specific changes. The phage can be engineered to display a vast collection (library) of peptides (more than 10^9 of them) on its surface. By injecting such a library into the blood stream of a live mouse, we can select peptides that make the phage home to a given target [4, 7, 22, 33]. We have also recently expanded the tissue targeting to include the screening of libraries of small molecular weight chemicals [7]. When screening phage libraries, we intravenously inject the library into a mouse, rescue the phage from the target tissue, and repeat the process several times to derive a phage pool that selectively homes to the target tissue. More recently, we have included an *ex vivo* step to generate a phage pool enriched in sequences capable of recognizing the target tissue [22]. The homing peptides and their binding molecules (receptors) discovered in this manner have revealed a wealth of heterogeneity in the blood and lymphatic vessels (Fig. 7.1). These results are discussed next.

7.3. TISSUE-SPECIFIC ZIP CODES IN BLOOD VESSELS

Our phage studies, as well as phage and antibody studies by others, have revealed an unexpected extent of tissue-specific molecular heterogeneity in the vasculature of various normal tissues. Tissue-specific vascular homing peptides have been obtained for at least 15 different tissues [3, 13, 17, 32, 39]. Importantly, we have been successful with each individual normal tissue we have chosen for targeting so far, suggesting that every tissue puts its own signature on the vasculature. The list includes both major organs, such as the brain, lungs, heart, and kidneys, and minor ones such as the prostate [3, 17, 39]. With our current technology, we obtain phage that home to the target tissue with a selectivity that can reach several hundred fold [22, 42]. Only a few of the molecules that the tissue-specific vascular homing peptides or antibodies recognize are known. Proteases are strongly represented among them: dipeptidyl peptidase IV [25] and membrane dipeptidase [36] are markers of lung vasculature; and another peptidase, aminopeptidase P, is selectively expressed in breast gland vasculature [17]. However, other types of molecules can also serve as tissue-specific endothelial markers [42].

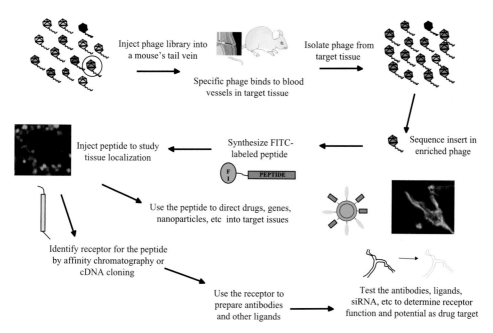

FIGURE 7.1. **Schematic representation of the protocol for *in vivo* phage screening and homing peptide characterization**. A library containing as many as several billion peptides and expressed on the surface of bacteriophage is intravenously injected into a mouse. A few minutes later, the target tissue is removed, and a phage pool enriched in phage carrying homing peptides for the target tissue is rescued from it by infecting bacteria. A highly enriched pool is obtained by repeating the process with the first pool. Sequencing of the DNA encoding the peptides displayed by the phage reveals the sequences of the enriched peptides. The peptides can be used to deliver labeling molecules (micrograph on the left; [35]), drugs, or nanoparticles to a target tissue. A quantum dot coated with a peptide (for homing) and polyethylene glycol (for solubility) is schematically depicted [2]. The micrograph on the right shows green tumor-homing quantum dots that have accumulated in tumor vessels (labeled with a red dye). The homing peptides can also be used to identify the receptors for the peptides. Some of the receptors are functionally important in the target vasculature, and compounds that affect their function may destroy the target vasculature [29].

7.4. SPECIAL FEATURES OF VESSELS IN DISEASE

It is well known that inflammation causes changes in the endothelium of adjacent vessels. The expression of adhesion molecules that bind leukocytes from blood is increased, and the production of cytokines that attract and activate leukocytes is turned on (Springer, 1994). Among other pathologies, the molecular features that characterize the vasculature of regenerating tissues and tumors have been extensively studied. In both cases, new vessels sprout from existing ones in a process termed angiogenesis, and the growing vessels are different from normal vessels. Further stimulating the growth of angiogenic vessels, or targeting them for destruction are important therapeutic goals; the former to facilitate regeneration of damaged tissue, and the latter to starve a tumor.

The markers of angiogenesis include increased expression of receptors for vascular endothelial growth factor (VEGF) receptors [1] and integrins [15, 27]. Indeed, one of the peptides identified by *in vivo* screening of phage libraries for tumor homing recognizes αvβ3 and αvβ5 [4], two of the integrins that are selectively expressed in angiogenic vessels. Other proteins selectively expressed in angiogenic vessels include aminopeptidase N; its involvement in angiogenesis was established by *in vivo* phage display [34]. The form of aminopeptidase N expressed in tumor vasculature may be distinct from those expressed by normal cells [10]. A similar situation exists with a fibronectin splicing variant that is selectively expressed in angiogenic blood vessels, whereas fibronectin, in general, is ubiquitous [20].

We recently used one of our tumor-homing peptides to show that angiogenic endothelial cells, unlike resting endothelial cells, express nucleolin at the cell surface [8]. Antibodies prepared against nucleolin also specifically recognize blood vessels both in tumors and in a model of wound healing angiogenesis. Various tumor cells also express cell surface nucleolin. Thus, peptides, antibodies, and nucleic acid probes that bind to nucleolin can specifically target both tumor blood vessels and tumor cells. Nucleolin can serve as a receptor for tumor-targeting quantum dots [2].

Detailed analysis of the specificity of the tumor-homing peptides we have selected from phage libraries shows that some of the peptides recognize vascular markers that are common to all tumors, but some only bind to the vessels in a limited range of tumors. The former recognize angiogenesis-related proteins. The molecules recognized by the peptides reactive with a restricted range of tumors are not known, but it is likely that the tumor tissue sends the endothelium the signals necessary for the expression of specific molecules. We are currently attempting to identify tumor-homing peptides that would be specific for the blood vessels or lymphatics of a given tumor type, such as prostate cancer or breast cancer.

We have recently discovered an additional degree of vascular diversity in tumors. A collaborative study with Dr. Douglas Hanahan's laboratory using transgenic mouse tumor models showed that some homing peptides were capable of distinguishing between the vessels of pre-malignant and malignant lesions (while not recognizing the normal vasculature). This finding indicates that the molecular features evolve as tumorigenesis progresses [22, 26]. Interestingly, the islet tumor set of peptides did not recognize the various stages of the skin tumors, and vice versa. These studies reveal an unexpected degree of diversity in the vasculature of tumors and premalignant lesions. As *in vivo* phage display and similar analyses are extended beyond tumor vasculature, it is likely that vascular markers for lesions in other diseases will be identified. Initial studies on atherosclerotic plaques have already shown promise in this regard [24].

The angiogenesis-specific proteins revealed by phage display are likely to be functionally important in the angiogenesis process. One of the first tumor-homing peptides we found was a peptide containing the RGD (arginine-glycine-aspartic acid) sequence, which we had many years ago shown to be central to the recognition of extracellular matrix proteins by integrins [38]. This particular peptide homes to tumor vasculature because it is specific for the αvβ3 and αvβ5 integrins, which are up-regulated in angiogenic vessels [37]. Blocking the activity of these integrins is known to inhibit angiogenesis [16]. Aminopeptidase N, identified as an angiogenesis marker with another phage peptide, is also functionally important in angiogenesis [34]. A third example is provided by cell surface-expressed nucleolin. Anti-nucleolin antibodies (S. Christian, M. Akerman, J. Pilch, and E. Ruoslahti, in preparation) inhibit angiogenesis, and nucleolin-binding aptamers suppress tumor cell proliferation [5]. Finally, a peptide that recognizes tumor endothelial cells in tumor lymphatics and also binds to tumor cells [28] inhibits tumor growth when systemically administered to tumor-bearing mice [29]. These results underscore the point that the molecular markers of tumor vasculature are also likely to be functionally important for the maintenance of the vasculature. Moreover, the lymphatic homing peptide provided the first demonstration that tumor lymphatics are molecularly different from normal lymphatics. The availability of peptides that home to tumor lymphatics, along with peptides that home to tumor blood vessels, may allow a two-pronged attack on tumors.

7.5. DELIVERY OF DIAGNOSTIC AND THERAPEUTIC AGENTS TO VASCULAR TARGETS

Peptides that home to specific vascular sites, such a tumors, can carry a payload to that site. When the payload is a drug, the delivery will concentrate the drug at the target site, and increased efficacy and lesser side effects should result. We have used the peptides that specifically bind to the $\alpha v \beta 3$ and $\alpha v \beta 5$ integrins (RGD motif peptide) or to aminopeptidase N (NGR motif peptide) to target doxorubicin in this manner [4]. Another laboratory has used these same peptides to deliver tumor necrosis factor α into tumors [9, 10].

A particularly illuminating example of homing peptide targeting is our use of the RGD and NGR peptides to target an anti-bacterial peptide to the vasculature at various disease sites. The payload in this case is an amphipathic peptide that disrupts bacterial and mito-chondrial membranes and induces apoptosis if internalized by mammalian cells. Systemic treatment of tumor-bearing mice with conjugates in which this pro-apoptotic peptide is coupled to one of the homing peptides inhibited tumor growth [14]. Similar treatment also suppressed inflammation in arthritic synovium, which exhibits strong angiogenesis [19]. Moreover, combining the same pro-apoptotic peptide with a homing peptide that binds to the blood vessels of the normal prostate yielded a compound that caused partial destruction of the prostate tissue and delayed the development of prostate cancer in mice transgenic for a prostate-expressed oncogene [3]. These results, better than anything, demonstrate the power of the homing peptide targeting; the same essentially non-specific toxic peptide affects different tissues, depending on its homing peptide partner.

7.6. HOMING PEPTIDES FOR SUBCELLULAR TARGETING

Some of our homing peptides are capable of being internalized by the target cells and delivering a drug-like payload (fluorescein, rhodamine, or biotin), into the cell nucleus (Fig. 7.2; [22, 26, 28, 35]). These internalizing peptides contain numerous basic amino acid residues, which apparently are important for the internalization and may form a nuclear localization signal. Alternatively, the peptide may be binding to a transport molecule that shuttles into the nucleus. This appears to be the case with a peptide that binds to cell surface nucleolin and is internalized and transported into the nucleus [8]. These internalizing peptides may prove to be particularly useful for delivering therapeutic agents that act in the nucleus. Bhatia and coworkers have done work on quantum dots targeted to various subcellular organs [12]. Subcellular targeting, along with tissue targeting, is likely to become an important aspect of cellular level nanomedicine.

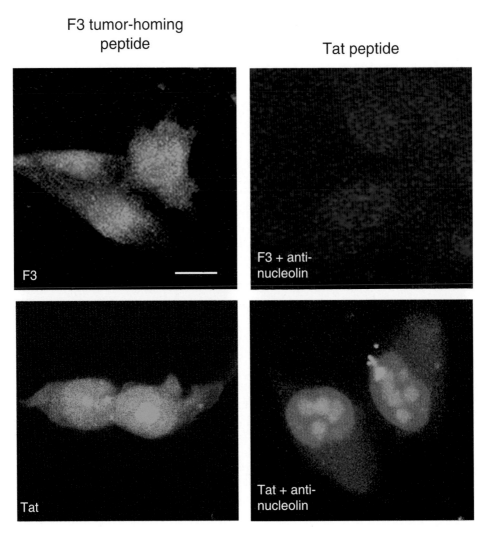

FIGURE 7.2. **Cell penetrating homing peptides**. F3 is an example of recently isolated homing peptides that are internalized by their target cells. F3 binds to angiogenic endothelial cells and certain tumor cells [8, 35]. Shown is the internalization of fluorescein-labeled F3 by cultured tumor cells (upper left panel). The F3 peptide resembles cell penetrating peptides such as the Tat peptide [30] in its ability to carry a payload to a cell, with the important difference that F3 is cell type-specifc and has a distinct receptor, cell surafe-expressed nucleolin. Anti-nucleolin inhibits the uptake of F3 (upper right panel), but has no effect on the intenalization of fluorescein-labeled Tat peptide (lower panels). Modified from [8], with permission.

7.7. NANOPARTICLE TARGETING

Vascular homing peptides can deliver particles to a specific tissue site. The phage we use to probe the vascular heterogeneity is a nanoparticle. The phage particle actually is a sophisticated nanodevice, as it functions like a nanoscale syringe that injects the phage's genome into host bacteria. Animal viruses used as gene therapy vectors have also been targeted with homing peptides. The RGD integrin recognition sequence can direct adenovirus vectors to cells expressing the $\alpha v \beta 3$ integrin, such as angiogenic endothelial cells [21, 40]. Chemical mimetics of the RGD sequence have been used to target tumor vasculature with lipid-based nanoparticles designed for imaging purposes or as gene therapy vectors [23, 31, 41].

The targeted nanoparticles used so far can seek out and concentrate at the intended target tissue. However, having reached their target, they are only capable of providing a signal or passively releasing a drug or gene payload. For the full potential of nanotechnology to be realized in medicine, it will be necessary to build more functions into such particles. One potentially important advance would be a nanoparticle in which the binding to the target acts as a switch that triggers a response such as a signal or release of a drug. Better still, the nanoparticle could incorporate a switch that can be controlled from outside, so that the functions of the particle can be activated at will and at an optimal time. Finally, the device should be capable of recognizing changes caused by the treatment and report back on the results.

The sensor aspect has essentially been solved. The homing peptides described here, as well as proteins, antibodies, and nucleic acid molecules that recognize suitable targets, can serve as sensors. Using sensors that detect vascular markers offers the considerable advantage that the endothelial marker molecule recognized by the sensor is readily available to a circulating particle. Moreover, endothelial binding is frequently followed by transendothelial transport into the tissue. In addition, some of our newest peptides recognize markers that are shared by the tumor endothelium and tumor cells [28, 35]. The dual recognition allows these peptides to deliver a payload both to tumor vessels and the tumor parenchyma. These peptides and their payload are taken up and internalized by the target cells; in some cases, the result is a tremendous concentration of the payload in tumors [29]. These peptides can also direct exquisitely specific homing of intravenously injected quantum dots into target tissues. For example, we have been able to label tumor blood vessels with quantum dots in one color and the lymphatics in another in the same tumor [2]. With the sensors in hand, the next challenge in the field is to couple the sensors to molecular switches that convert the sensor signal to a response.

7.8. FUTURE DIRECTIONS

Our homing peptide studies indicate that the endothelial cells (and possibly mural cells as well) in each vascular bed express a unique complement of cell surface molecules. Only a few of the molecules that act as receptors for homing peptides are known, but their diversity (proteases, integrins, growth factor receptors, and proteoglycans; Ruoslahti, 2002) suggests that they represent a heterogeneous group of proteins with a variety of functions. The function of any given marker molecule is likely to be important for the vascular bed that expresses it. Such molecules can also function as tissue-specific receptors for metastatic tumor cells [7]. It will be important to accelerate the discovery of homing peptide receptors; we are working on a method that allows simultaneous identification of a homing peptide and its receptor.

The homing peptides that recognize tumor vasculature fall into two categories: (1) those that bind to angiogenesis-associated molecules and recognize the vasculature of all tumors (and other angiogenic lesions) and (2) those that recognize the vasculature of some, but not all tumors. The latter kind would be potentially advantageous in therapeutic applications because sites of beneficial angiogenesis would not be affected. Our current screening efforts are directed at discovering peptides that would broadly recognize the vasculature of a given tumor type.

Homing peptide work shows that tumor lymphatics, like tumor blood vessels, differ from their normal counterpart vessels. This offers a possibility of attacking tumors from two directions, through the blood vessels and through the lymphatics, a strategy that remains to be explored.

We have described homing peptides that are internalized by the target cells and can take a payload into the cell nucleus. We are currently testing drug conjugates to utilize these properties of the peptides. We are also taking advantage of the sensor capabilities of homing peptides in designing nanodevices that can become internalized in tumor endothelial calls and tumor cells.

Being able to sense disease-associated changes in the vasculature may make it possible to develop nanodevices that serve a sentinels reporting back on early signs of a disease, such as pre-malignant lesions. We have already shown that it is possible to identify peptides that specifically recognize pre-malignant lesions by the vascular signature [22, 26]. The sentinel nanodevices would be altered upon an encounter with a specific target and would return to the general circulation to report back on their survey of the vasculature. Ingestible nanoparticles could similarly patrol the gut, and other designs could be developed for pre-malignant lesions of the urinary bladder, and so on. We are in the process of isolating peptides that recognize premalignant epithelial cells in these tissues to provide the sensors for such diagnostic nanodevices. An encounter with the target by a nanosensor for pre-malignant lesions could also induce release of a drug from the device, making it into both a diagnostic and preventative tool.

Adding responsiveness to an external signal, such a magnetic field, would allow external control of nanodevices that have identified their target in the body to effect a therapy in a highly focused manner. Finally, another set of sensors operating the same way as the homing signals, but designed to detect apoptosis, cellular distress, cell proliferation, or inflammation, could monitor the state of the tissue of interest for the physician. The design of nanodevices capable of performing these tasks will require collaboration among a diverse group of scientists with specialties ranging from tumor biology to materials science. Success in endeavors such as these is likely to bring nanomedicine to the forefront of medical practice.

ACKNOWLEDGEMENT

The author's work was supported by the NIH grants CA 82713 and Cancer Center Support Grant CA 30199, and the DAMD 17-02-1-0315 Innovator Award from the Department of Defense.

REFERENCES

[1] K. Alitalo and P. Carmeliet. Molecular mechanisms of lymphangiogenesis in health and disease. *Cancer Cell*, 1:219–227, 2002.

[2] M.E. Akerman, W.C. Chan, P. Laakkonen, S.N. Bhatia, and E. Ruoslahti. Nanocrystal targeting in vivo. *Proc. Natl. Acad. Sci. U.S.A.*, 99:12617–12621, 2002.

[3] W. Arap, W. Haedicke, M. Bernasconi,, R. Kain, D. Rajotte, S. Krajewski, H.M. Ellerby, D.E. Bredesen, R. Pasqualini, and E. Ruoslahti. Targeting the prostate for destruction through a vascular address. *Proc. Natl. Acad. Sci. U.S.A.*, 99:1527–1531, 2002.

[4] W. Arap, R. Pasqualini, and E. Ruoslahti. Cancer treatment by targeted drug delivery to tumor vasculature in a mouse model. *Science*, 279:377–380, 1998.

[5] P.J. Bates, J.B. Kahlon, S.D. Thomas, Trent, and D.M. Millet. Antiproliferative activity of G-rich Oligonucleotides correlates with protein binding. *J.Biol.Chem.*, 274:26369–26377, 1999.

[6] D.M. Brown, M. Pellecchia, and E. Ruoslahti. Drug identification through *in vivo* screening of chemical libraries. *ChemBioChemistry*, 5:871–875.

[7] D. Brown, and E. Ruoslahti. Metadherin, a novel cell-surface protein in breast tumors that mediates lung metastasis. *Cancer Cell*, 5:365–374, 2004.

[8] S. Christian, J. Pilch, K. Porkka, P. Laakkonen, and E. Ruoslahti. Nucleolin expressed at the cell surface is a marker of endothelial cells in tumor blood vessels. *J. Cell. Biol.*, 163:871–878, 2003.

[9] F. Curnis, A. Sacchi, L. Borgna, F. Magni, A. Gasparri, and A. Corti. Enhancement of tumor necrosis factor antitumor immunotherapeutic properties by targeted delivery to aminopeptidase N (CD13). *Nature Biotech.*, 18:1185–1190, 2000.

[10] F. Curnis, G. Arrigoni, A. Sacchi, L. Fischetti, W. Arap, R. Pasqualini, and A. Corti. Differential binding of drugs containing the NGR motif to CD13 isoforms in tumor vessels, epithelia, and myeloid cells. *Cancer Res.*, 62:867–874, 2002.

[11] A.M. Derfus, W.C. Chan, and S.N. Bhatia. Probing the cytotoxicity of semiconductor quantum dots. *Nano Letters*, 4:11–18, 2004.

[12] A.M. Derfus, W.C. Chan, and S.N. Bhatia, Intracellular delivery of quantum dots for live cell labeling and organelle tracking. *Adv. Mat.* published online 5/19/04, 2004.

[13] B.S. Ding, Y.J. Zhou, X. Y. Chen, J. Zhang, P.X. Zhang, Z. Y. Sun, X. Y. Tan, and J. N. Liu. Lung endothelium targeting for pulmonary embolism thrombolysis. *Circulation*, 108:2892–2898, 2003.

[14] Ellerby, W. Arap, L.M. Ellerby, R. Kain, R. Andrusiak, G. D. Rio, S. Krajewski, C. R. Lombardo, R., Rao, E. Ruoslahti. *et al.* Anti-cancer activity of targeted pro-apoptotic peptides. *Nat. Med.*, 5:1032–1038, 1999.

[15] B.P. Eliceiri and D.A. Cheresh. The role of alphav integrins during angiogenesis: insights into potential mechanisms of action and clinical development. *J. Clin. Invest.*, 103:1227–1230, 1999.

[16] B.P. Eliceiri and D.A. Cheresh. Adhesion events in angiogenesis. *Curr. Opini. Cell Biol.*, 13:563–568, 2001.

[17] M. Essler and E. Ruoslahti. Molecular specialization of breast vasculature: a breast-homing phagedisplayed peptide binds to aminopeptidase P in breast vasculature. *Proc. Natl. Acad. Sci. U.S.A.*, 99:2252–2257, 2002.

[18] M.N. Fukuda, C. Ohyama, K. Lowitz, O. Matsuo, R. Pasqualini, E., Ruoslahti, and M. Fukuda. Apeptide mimic of E-selectin ligand inhibits sialyl Lewis X-dependent lung colonization of tumor cells. *Cancer Res.*, 60:450–456, 2000.

[19] D.M. Gerlag, E. Borges, P.P. Tak, H.M. Ellerby, D.E., Bredesen, R. Pasqualini, E. Ruoslahti, and G.S. Firestein. Suppression of murine collagen-induced arthritis by targeted apoptosis of synovial neovasculature. *Arth. Res.*, 3:357–361, 2001.

[20] C. Halin, S. Rondini, F. Nilsson, A. Berndt, H. Kosmehl, L. Zardi, and D. Neri. Enhancement of the antitumor activity of interleukin-12 by targeted delivery to neovasculature. *Nat. Biotechnol.*, 20:264–269, 2002.

[21] Y.S. Haviv, J.L. Blackwell, A. Kanerva, P. Nagi,V. Krasnykh, I. Dmitriev, M. Wang, S. Naito, X. Lei, A. Hemminki, *et al.* Adenoviral gene therapy for renal cancer requires retargeting to alternative cellular receptors. *Cancer Res.*, 62:4273–4281, 2002.

[22] J.A. Hoffman, E. Giraudo M. Singh, M. Inoue, K. Porkka, D. Hanahan, and E. Ruoslahti. Progressive vascular changes in a transgenic mouse model of squamous cell carcinoma. *Cancer Cell*, 4:383–391, 2003.

[23] J. D. Hood, M. Bednarski, R. Frausto, S. Guccione, R. A. Reisfeld, R. Xiang, and D. A. Cheresh. Tumor regression by targeted gene delivery to the neovasculature. *Science*, 296:2404–2407, 2002.

[24] P. Houston, J. Goodman, A. Lewis, C. J. Campbell, and M. Braddock. Homing markers for atherosclerosis: applications for drug delivery, gene delivery and vascular imaging. *FEBS Lett.*, 492:73–77, 2001.

[25] R.C. Johnson, D. Zhu, H.G. Augustin-Voss, and B.U. Pauli. Lung endothelial dipeptidyl peptidase IV is an adhesion molecule for lung metastatic rat breast and prostate carcinoma cells. *J. Cell Biol.*, 121:1423–1432, 1993.

[26] J.A. Joyce, P. Laakkonen, M. Bernasconi, G. Bergers, E. Ruoslahti, and D. Hanahan. Stage-specific vascular markers revealed by phage display in a mouse model of pancreatic islet tumorigenesis. *Cancer Cell,* 4:393–403, 2003.

[27] S. Kim, K. Bell, S.A. Mousa, and J.A. Varner. A regulation of angiogenesis in vivo by ligation of integrin $\alpha5\beta1$ with the central cell-binding domain of fibronectin. *Am. J. Pathol.*, 156:1345–1362, 2000.

[28] P. Laakkonen, K. Porkka, J.A. Hoffman, and E. Ruoslahti. A tumor-homing peptide with a lymphatic vessel-related targeting specificity. *Nat. Med.*, 8:743–751, 2002.

[29] P. Laakkonen, M.E. Akerman, H. Biliran, M. Yang, F. Ferrer, T. Karpanen, R.M. Hoffman, and E. Ruoslahti. Antitumor activity of a homing peptide that targets tumor lymphatics and tumor cells. *Proc. Natl. Acad. Sci. U.S.A.*, 2004.

[30] Ü. Langel. *Cell-Penetrating Peptides: Processes and Applications.* Boca Raton, Florida, CRC Press, 2003.

[31] G.M. Lanza and S.A. Wickline. Targeted ultrasonic contrast agents for molecular imaging and therapy. *Curr. Prob. Cardiol.*, 28:625–653, 2003.

[32] D.P. McIntosh, X.Y. Tan, P. Oh, and J. E. Schnitzer. Targeting endothelium and its dynamic caveolae for tissue-specific transcytosis in vivo: a pathway to overcome cell barriers to drug and gene delivery. *Proc. Natl. Acad. Sci. U.S.A.*, 99:1996–2001, 2002.

[33] R. Pasqualini and E. Ruoslahti. Organ targeting *in vivo* using phage display peptide libraries. *Nature*, 380:364–366, 1996.

[34] R. Pasqualini, E. Koivunen, R. Kain, J. Lahdenranta, M. Sakamoto, A. Stryhn, R. A. Ashmun, L. H. Shapiro, W. Arap, and E. Ruoslahti. Aminopeptidase N is a receptor for tumor-homing peptides and a target for inhibiting angiogenesis. *Cancer Res.*, 60:722–727, 2000.

[35] K. Porkka, P. Laakkonen, J.A. Hoffman, M. Bernasconi, and E. Ruoslahti. Targeting of peptides to the nuclei of tumor cells and tumor endothelial cells in vivo. *Proc. Natl. Acad. Sci. U.S.A.*, 99:7444–7449, 2002.

[36] D. Rajotte and E. Ruoslahti. Membrane dipeptidase is the receptor for a lung-targeting peptide identified by in vivo phage display. *J. Biol. Chem.*, 274:11593–11598, 1999.

[37] E. Ruoslahti. Specialization of tumour vasculature. *Nat. Rev. Cancer*, 2:83–90, 2002.

[38] E. Ruoslahti, RGD story: A personal account. A landmark essay. *Matrix Biol.*, 22:459–465, 2003.

[39] E. Ruoslahti and D. Rajotte. An address system in the vasculature of normal tissues and tumors. *Annu. Rev. Immunol.*, 18:813–827, 2000.

[40] T.J. Wickham. Targeting adenovirus. *Gene Ther.*, 7:110–114, 2000.

[41] P.M. Winter, S.D. Caruthers, A. Kassner, T.D. Harris, L.K. Chinen, J.S. Allen, E.K. Lacy, H. Zhang, J.D. Robertson, S.A. Wickline, and G.M. Lanza. Molecular imaging of angiogenesis in nascent Vx-2 rabbit tumors using a novel alpha(nu)beta3-targeted nanoparticle and 1.5 tesla magnetic resonance imaging. *Cancer Res.*, 63:5838–5843, 2003.

[42] L. Zhang, J.A. Hoffman, and E. Ruoslahti. Molecular profiling of heart endothelial cell. *Circulation*, (In press) 2005.

8

Engineering Biocompatible Quantum Dots for Ultrasensitive, Real-Time Biological Imaging and Detection

Wen Jiang*, Anupam Singhal*, Hans Fischer*†, Sawitri Mardyani*, and Warren C. W. Chan*†

*Institute of Biomaterials and Biomedical Engineering, 4 Taddle Creek Rd., 408, Toronto, ON, M5S 3G9, Canada
†Department of Materials Science & Engineering, 4 Taddle Creek Rd., 408, Toronto, ON, M5S 3G9, Canada

8.1. INTRODUCTION

Advances in the design of optical probes have played a central role in the emergence of photon-based microscopy techniques for biological imaging and detection [1–6]. These advances have led to the elucidation of the biological function and activity of many proteins, nucleic acids, and other molecules in living cells, tissues, and animals. Currently, the molecular architecture of greater than 70% of all optical probes consists of an "optical emitter" attached to a "targeting molecule" [4]. The targeting molecule directs the optical emitter to specific biological sites where the optical emitter can then be used to detect the activities of biomolecules. The most popular optical probes have been traditionally designed from organic-based molecules; for instance, probes for the imaging of cellular cytoskeleton are based on the conjugation of red-fluorescent molecule Texas Red to the small targeting organic molecule phalloidin (for labeling actin fibers) and green-fluorescent Alexa Fluor 488 to a recognition antibody (for labeling microtubules) [4]. Hundreds of different types of organic-based fluorescent probes are commercially available. These probes can be used in numerous applications, including the staining of DNA and proteins, detection of subtle differences in the ionic content in living cells, or detection of protein structures [4, 7–10]. Due to their complex molecular structures, however, organic fluorophores often exhibit unfavorable absorption and emission properties, such as photobleaching, environmental quenching, broad and asymmetric emission spectra, and the inability to excite multiple fluorophores of more than 2–3 colors at a single wavelength [10, 11].

Recent developments in the field of nanotechnology have led to the fabrication of a new generation of inorganic probes that overcomes many of the problems associated with organic-based probes [6, 12–14]. Furthermore, these nanometer-sized probes provide novel and unique properties for biological applications that are not available with organic-based fluorophores. At the forefront of this new class of inorganic probes are semiconductor nanocrystals, also known as quantum dots ("QDs"). The physical properties and applications of quantum dots have been heavily investigated in many physics and engineering labs since the early 1980s [15–17]. QDs are defined as particles with physical dimensions smaller than the exciton Bohr radius; this gives rise to a unique phenomenon known as quantum-confinement. Quantum confinement, which refers to the confinement of charge carriers (i.e. electrons and holes) within a material, embarks QDs with unique optical and electronic properties that are unavailable to semiconductors in bulk solids. Although initial interest in QDs focused in physical applicatons (e.g,. making computer chips and light-emitting diodes), recent work by Nie, Alivisatos, and their coworkers have highlighted the great promise of QDs in biological applications [18, 19]. As biological probes, QDs are extremely bright (1 QD ≈ 10 to 20 organic fluorophores), have high resistance to photobleaching, have narrow spectral linewidths, and have size- and materials-tunable emission that can be excited using a single wavelength [18, 19]. Some of these properties are illustrated in figure 8.1.

When used in biological research and applications, these optical properties will lead to improved detection sensitivity for analysis, and to simplification in experimental and instrumental design. In this book chapter, we describe the development of inorganic semi-conductor QD probes for biological and medical applications.

8.2. SYNTHESIS AND SURFACE CHEMISTRY

8.2.1. Synthesis of QDs that are Soluble in Organic Solvents

Breakthroughs in the synthesis of quantum dots by Bawendi [20, 21], Guyot-Sionnest [22], Alivisatos [23] and their coworkers have paved the way for using QDs in biological applications. Prior to their work, a broad range of strategies were attempted to synthesize QDs with high optical quality (i.e., non-aggregated QDs with bright emission and stable against photodegradation). These methods include synthesis of QDs in aqueous buffers with thioglycerol as a stabilizing agent, synthesis of QDs in yeast cells with expressed proteins containing glutathione as stabilizing agents, or synthesis of QDs inside micelle systems [15, 24–27]. However, all of these approaches produced QDs with broad size distributions (relative standard deviation > 15%) and low quantum yields (< 20%). These synthetic systems produced QDs with poor optical and size properties because the reaction conditions did not provide a proper environment for rapid and even nucleation (this leads

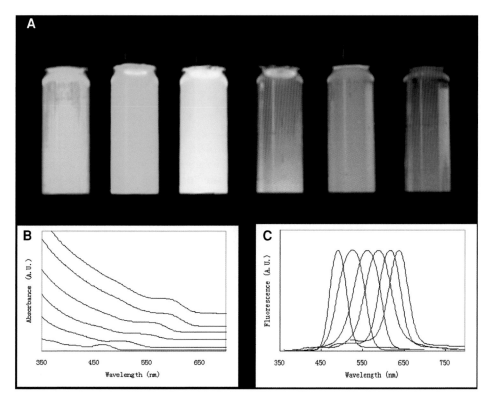

FIGURE 8.1. Size-tunable optical properties of QDs (left, 2.0 nm to right, 6.0 nm—diameter of QDs). (A) The unique emission colours of different sized QDs are observed under UV-excitation by a handheld lamp. (B) Absorbance and (C) fluorescence spectra of different-sized QDs. For the fluorescence measurement, an excitation wavelength of 400 nm is used. QDs = quantum dots.

to QDs with broad size distribution) and for careful growth of QDs (this leads to QDs with large amounts of defect sites).

In the 1990s, research by Bawendi, Guyot-Sionnest, Alivisatos and their coworkers demonstrated the key to prepare high-quality QDs is to use an organometallic reaction scheme in non-aqueous solvents. Furthermore, the reaction must take place at high temperatures to achieve rapid and even nucleation [21–23]. In one common approach high-quality QDs are synthesized via the high-temperature (e.g., 350°C) pyrolysis of organometallic and chalcogenide reagents in the solvent tri-n-octylphosphine oxide (TOPO). After rapid injection, the TOPO solution appears light yellow. As QDs grow bigger, the solution color evolves from a light yellow to a dark ruby red. At the desired size, the QDs can be isolated using solvent precipitation and centrifugation. Currently, the most established protocols for the synthesis of QDs are those from groups II-VI (e.g., CdSe, CdTe, CdS, and ZnSe) and group III-V (e.g., InP and InAs) of the periodic table [2, 19, 28–30].

La Mer and Dinegar developed theoretical model systems to explain the nucleation and growth of colloidal particles in the 1950s. Their theory has recently been adapted toward explaining QD nucleation and growth [31]. Based on their theory, the nucleation of QDs occurs when the concentration of precursor molecules are at a high enough level that they exceed the so-called "nucleation threshold". As QDs grow over time, the concentrations of precursor molecules fall below the threshold level and, ultimately, the Ostwald Ripening process replaces the nucleation process [32]. Ostwald Ripening refers to the breakdown of smaller QDs to provide the needed precursor atoms to build larger QDs. The Ostwald Ripening process is a thermal dependent process, where the growth of larger QDs requires high temperature—the breakdown of QDs in the Ostwald Ripening process is related to the temperature (i.e., larger QDs require higher temperatures). As a result, careful control of growth temperatures can be used to accurately control the average size and size distribution of QDs. During the synthesis, the size and size distributions are monitored by absorbance and fluorescence spectrophotometry. When the desired optical properties are attained, the temperature is reduced to prevent further QD growth or dissolution.

Numerous groups have developed variations to the organometallic approach for the mass-scale synthesis of QDs [30, 33]. In particular, Peng and coworkers have demonstrated that organometallic precursors (e.g. dimethyl cadmium) can be replaced with non-pyrophoric and less costly "greener" reagents (e.g. cadmium oxide, CdO, or cadmium acetate, $Cd(Ac)_2$) [34, 35]. These "alternative routes" to the synthesis of QDs in organic media can be used to reproducibly prepare high-quality CdS, CdSe, and CdTe QDs. Since QDs formed with using greener reagents exhibit slower reaction kinetics (e.g. slower nucleation), extended nucleation periods allow increased quantities of "greener" precursors to be injected at the start of the reaction; with more nucleation sites, more QDs per reaction can be synthesized. Using this synthesis technique, QDs have been successfully synthesized in quantities greater than 1 gram.

In addition to development techniques for the synthesis of high quality QDs, Bawendi, Guyot-Sionnest, Alivisatos and their coworkers demonstrated that the fluorescence efficiency of QDs could be enhanced by the formation of a secondary surface inorganic shell on top of the QD core structure [21–23]. Improvements in the fluorescence efficiency could be attributed to the removal of surface defect sites that traps excited or mobile electrons (see optical properties section below for details). The semiconductor capping material must have larger bandgap energy than the core QD and must have a bond length that is similar to that of the core. For CdSe QDs, CdS and ZnS are excellent shells. For example, the growth of a ZnS layer on CdSe at a temperature below the CdSe nucleating temperature ($\sim$100 to 310°C) has produced CdSe/ZnS core/shell QDs that exhibit luminescence yields up to 85% [21–23].

There have been great efforts recently to develop better approaches to systematically control the nucleation and growth of semiconductor QDs. A major goal in QD synthesis is to prepare QDs with a discrete size ($< 0.1\%$ in size distribution) or shape (e.g., spherical vs. rod-shaped) since the size, shape, and morphology of QDs are closely related to their optical and electronic properties [36, 37]. Although the commonly used flask synthesis has led to the synthesis of high quality QDs, there is great difficulty in controlling the kinetics and thermodynamic factors in the reaction vessel. To manipulate these factors, microfluidic technology has been used as a platform for improving the synthesis of QDs [38]. In a microfuidic chip, micrometer-sized reservoirs, channels, and flow control systems are used to regulate the amount and concentration of precursor solvents. Furthermore, temperature systems can be incorporated onto the chip; this can provide greater control and manipulation of reaction temperature. In effect, the synthesis of QDs may be better controlled and their dimensional properties dramatically improved. Chan et al. demonstrated the controlled synthesis of CdSe QDs using a microfluidic chip [38]. The synthesis of a series of CdSe QDs with average diameters of 2.44, 2.54, 2.64, and 2.69 $+/-0.06$ nm in the microfluidics chip was achieved. Although early in development, research by Chan et al. show feasibility in using microfluidic technology for manipulating the synthetic conditions of QDs; this technology holds promise for improving QD synthesis.

Other emerging areas of research in QD synthesis are the design of QDs with various shapes (e.g., rod-shaped), structures (e.g., tetrapod-branched QDs), and composition (e.g., doping CdSe QDs with Te or Mn). All of these changes can influence the optical and electronic properties of the QDs or embark new properties (e.g, magnetism) into the QDs. To date, these QDs have not found applications in biology; however, we expect this to change in the near future. We refer the interested reader to some research developments in this area of QD research [36, 37, 39].

8.2.2. Modification of Surface Chemistry of QDs for Biological Applications

Since the surfaces chemistry of both the core and core/shell QDs are typically coated with hydrophobic ligands, great efforts have been made to modify the surface chemistry of QDs in order to render them biocompatible [13, 18, 19, 40–43]. One approach to achieving QD biocompatibility is the use of displacement chemistry techniques, where the TOPO molecules on the QD surface are replaced with bifunctional molecules (e.g. mercaptoacetic acid and phospho-alcohols). One end of the bifunctional molecule contains a functional group (e.g., –SH or –P) that can interact with metal atoms on the surface of the QDs and out-compete the phosphine oxide from the TOPO molecules for binding to the metal atoms on the QD surface. The other end of the bifunctional molecules must contain a hydrophilic functional group (e.g., alcohol or carboxylic acid functional groups). Since the hydrophilic functional groups protrude from the QD surface, the QDs become charged and extremely polar. The resulting QDs are thus water-soluble and biocompatible. Furthermore, these QDs can be conjugated to biomolecules by reacting the surface alcohol and carboxylic groups with proteins, peptides, and oligonucleotides through various reaction routes.

A second approach to interface QDs with biological systems is to design amphiphilic molecules that interact with the TOPO-molecules on the QD surface. These molecules (e.g., phospholipids) typically contain both hydrophobic and hydrophilic regions. The hydrophobic end interacts with the TOPO molecule through hydrophobic-hydrophobic interactions, while the hydrophilic end, containing carboxylic acid or alcohol functional groups, protrudes from the QD surface. In this scheme, the hydrophobic stabilizing ligand TOPO do not interact with the aqueous solvent and the QDs become biocompatible because of the QD surface hydrophilic functional groups.

In either the displacement or amphiphilic techniques, the QDs may aggregate out of solution. With the displacement technique, the bifunctional stabilizing ligand on the QD surface desorbs off—this leads to the loss of surface charges that cause particle-to-particle repulsion; while in the second technique, the desorption of the amphilphile molecule exposes the hydrophobic TOPO to aqueous solvents- this leads to aggregation via hydrophobic-to-hydrophobic interactions. To alleviate this problem, cross-linking schemes to lock the organic shell from desorbing the surface of the QDs has been developed. One typical approach is the use of the amino acid lysine or short-chain of poly-lysine to cross-link the carboxylic acids on the surface of QDs via carbodiimide catalyst [41, 44, 45]. This will

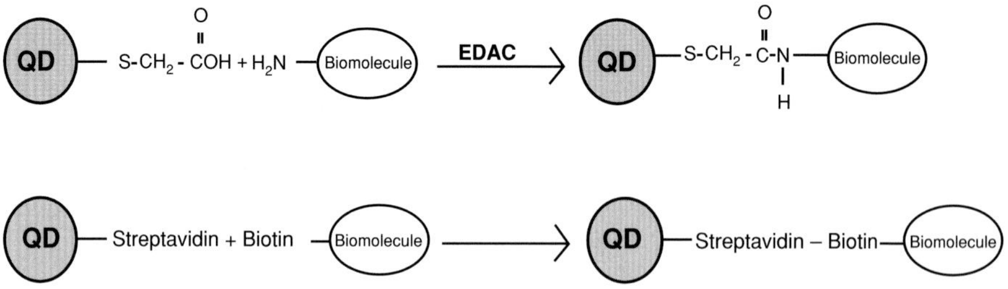

FIGURE 8.2. Methods to coat QD surface with biorecognition molecules. (A) QDs containing the carboxylic acid surface functional groups can be reacted to biomolecules containing primary amino groups through carbodiimide catalysis. EDAC (which is 1-ethyl-3-[3-(dimethylamino)propyl]carbodiimide) is a commonly used water-soluble carbodiimide. The end result of this reaction is an amide bond that links the QDs to the biomolecule. (B) QDs coated with streptavidin have also been a used for linking biotinylated biomolecules. The protein streptavidin has an extremely high binding affinity to the small organic molecule biotin.

lock the stabilizing organic ligand in place and prevent the organic ligands from desorbing the QD surface. With the organic coating in place and the appropriate surface charges, QDs will have long-term stability against aggregation.

Functional groups on QD surface provide sites to conjugate QDs to biomolecules (e.g., proteins, peptides, oligonucleotides) in order to form optical probes (figure 8.2).

One approach to bioconjugating QDs employs conventional carbodiimide chemistry [18, 46]. The addition of carbodiimide molecules to carboxylic acid-coated QDs yields an acylisourea intermediate group, which can be readily attacked by primary amines on biomolecules to form an amide linkage. Electrostatic interactions can also be used to link biomolecules onto the surface of QDs. Mattoussi and coworkers engineered a protein with a positive-charged leucine zipper and demonstrated the adsorption of such a protein onto the surface of a negatively-charged QD [13, 47, 48]. Another popular approach for linking QDs to biomolecules is through a streptavidin-biotin interaction [49, 50]. QDs are initially coated with the protein streptavidin and incubated with biotinylated biomolecules (such as biotin-conjugated antibodies). Streptavidin and biotin has a high binding affinity to each other ($K_d \sim 10^{-14}$ M, figure 8.2) [51]. A complex of QD-streptavidin-biotin-antibody is formed; when purified from uncomplexed biotin-antibody, they are ready for use as an optical probe.

8.3. OPTICAL PROPERTIES

QDs have garnered broad interests from biological and medical research communities because of their unique optical and electronic properties. Unlike organic-based fluorophores, the properties of QDs can be manipulated by simply changing their size, shape, or composition; basic semiconductor quantum chemistry and physics have been utilized to explain these properties (figure 8.3).

Bulk semiconductor materials contain two energy bands—conduction and valence band. The energy difference between the two bands is called the bandgap energy (E_v). In bulk semiconductor, the E_v is generally a fixed value that is dependent upon the composition of the material. When the semiconductor is in its normal "unexcited" state, the valence

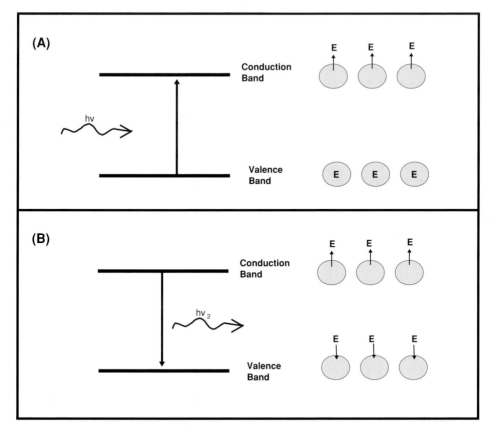

FIGURE 8.3. Schematic diagram describing the physical process of QD optical emission. (A) QDs are excited with an energy that exceeds bandgap (energy difference between a conduction and valence band, E_v). Upon optical excitation (hv), the electrons (which are in the valence band) become mobile (which enters the conduction band) and form a hole. (B) After a certain lifetime ($\sim$15 ns), the electrons recombine with the holes, producing an emission or fluorescence (hv_2). An electron-hole pair is called an exciton and the associated fluorescence is generally called an excitonic fluorescence.

band is completely filled with electrons while the conduction band is empty. When the semiconductor is optically or electrically stimulated by an energy exceeding the bandgap energy, electrons become "mobile" and are promoted from the valence to the conduction band; holes are formed in this process. The excited electrons eventually return to the valence band and recombine with the hole, forming an "exciton". The recombination of electrons with holes can be either radiative or non-radiative. Radiative recombination refers to the emission of photons while non-radiative recombination refers to the absence of photon emission.

When semiconductors are in the nanometer-scale, quantum confinement effects occur [28, 52]. The E_v's are dependent upon the size of the nanoparticles. The optical emission of QDs is a perfect example of how the optical and electronic properties of nanoscale materials are dependent upon the size. Figure 8.1a shows the emission of the ZnS-capped CdSe of a series of QDs with different sizes and their corresponding bandgap energy (which is shown as a discrete fluorescence emission). In figure 8.1a, 2.0 and 6.0 nm CdSe QDs emit blue light and red light, respectively, upon optical excitation. The classical quantum mechanical "Particle-in-a-box" model is used to explain and describe the QD's size-dependent fluorescence emission. In this model, the potential energy is considered to be infinitely high outside of the box, essentially trapping the particle inside the box. The model predicts that the wavefunction of the confined particle is dependent upon the size of the box. In QDs, the mobile carriers can be viewed as the trapped particle that is confined within the dimensions of the QD (i.e. the box).

Therefore, the wavefunction (i.e. bandgap energy and, hence, peak fluorescence emission wavelength) of the carriers is dependent upon the QD size. The bandgap energies of QDs are roughly proportional to $1/R^2$, where R is the QD radius. Due to the tunability of nanometer-sized QDs, their optical properties can be "custom-designed" to the needs of the researcher during the synthesis. In contrast, to make a series of uniquely fluorescent organic dye molecules, new synthetic procedures may be needed for each unique fluorescence emission. The simplicity of modifying the size and shape of QDs to produce a unique fluorescence emitter renders QDs advantageous over organic fluorophores for multiplexed detection and imaging.

Another advantage of QDs over organic-based fluorophores is their broad continuous absorption spectra [18, 53]; figure 8.1b shows measured absorbance spectra of a series of ZnS-capped CdSe QDs. For multiplexed analysis, a continuous absorption spectrum provides enhanced spectral separation between the fluorescence emission and excitation energy (minimizes leakage of excitation energy into detector or camera); additionally, QDs of different optical emission can all be excited using a single wavelength (figure 8.1c). In contrast, the use of organic fluorophores is limited to three-to-four colors because maximal excitation is confined to the absorbance peak (which has a specific wavelength window for each organic fluorophore). To permit the use of > four organic fluorophores, multiple optical excitation sources are generally needed; in effect, this leads to complication in instrumentation design and increase in cost of instrument.

QDs also have narrow and symmetric fluorescence spectra (i.e. narrow spectral linewidths). In bulk measurements, the excitonic fluorescence spectrum of a typical solution of ZnS-capped CdSe QDs in phosphate buffer saline (pH = 7.4) has a full-width at half maximum (FWHM) of 30 to 35 nm and is symmetric. However, the FWHM of single QDs has been shown to be approximately 13 nm, which is ~2.5 times narrower than the typical bulk measurement [18]. The broadness in the fluorescence peak can be attributed to the size distribution within the measured bulk sample. As synthetic approaches improve, the theoretical limit of 13 nm may be reachable. In contrast to QDs, the fluorescence spectra of organic dye molecules typically have FWHMs of greater than 45 nm. In addition, the fluorescence peaks of organic dyes typically have broad "tails" [4, 11]. In applications where multiple fluorophores are simultaneously employed and detected, complicated filter systems are often needed to discriminate between the unique spectra of each organic fluorophore. This overlap leads to spectral cross-talking, which can affect the overall detection efficiency. Of note, in some QD samples a fluorescence tailing or fronting is observed in the spectra; this can be attributed to non-uniform growth of QDs during the synthesis [32]. This problem can be alleviated during the synthesis where growth kinetics and thermodynamics can be altered by injection of excess precursor solvents in the reaction flask; the adjustment of reaction conditions during synthesis can lead to QDs with symmetric fluorescence

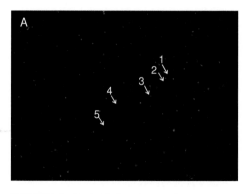
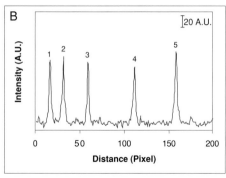

FIGURE 8.4. Single QD imaging. (A) Single QDs are imaged using a conventional epifluorescence microscope ($\square_{ex} = 430 + / - 25$, $\square_{em} = 535 + / - 25$) with a 100 W Hg lamp excitation and a 100x objective (N.A. = 1.3). These QDs are blinking on/off, a property that is common with single fluorescent molecule. (B) Linewidth measurements of five single QDs. This shows the relative brightness of each QD to background.

spectra. The narrower spectral linewidths and symmetric emission peaks of QDs bestow QDs ideal probes for multiplexed detection and analysis.

In addition to the size-tuneable and broad absorption profile, QDs exhibit large molar extinction coefficients and stability against photobleaching. Most small organic fluorophores have molar extinction coefficients (ε) of 10^3 to 10^4 M^{-1} cm^{-1}, while the ε of QDs are 10 to 100 times larger [54–56]. These relatively large molar extinction coefficients are one of the main reasons that single QDs can be easily imaged using a conventional epifluorescence microscope under ambient temperatures (see figure 8.4). A single QD is estimated to be 10 to 20 times brighter than that of a single organic fluorescent molecule [18, 41].

The fluorescence emission of QDs is extremely stable under constant photoexcitation. Figure 8.5 compares the photobleaching rate of inorganic QDs to the organic dye molecule alexa-488 in cells. Wu et al. stained a cell's microtubule with QDs and the nucleus with alexa-488. The cells were fixed with either the antifade Vectashield or glycerol [41]. Then the cells were excited using a 100W Hg lamp.

Fluorescence images were obtained at 10s intervals for 3 minutes and the fluorescence intensity measured at each of the intervals (see figure 8.5 for comparative photobleaching curve of qdots to the organic dye molecules). They showed that QD emission signal remained constant while the alexa-488's fluorescence signal decreased slowly in antifade and rapidly in glycerol. QDs are estimated to be $\sim$ 100 to 200 times more stable against photobleaching than organic dye molecules [18, 41, 57]. QDs have also been shown to be more photostable than fluorescent proteins (e.g., phycoerytherin) [58]. QDs do photobleach under extremely high power (beyond the typical power used for most biological imaging experiments). The photobleaching of QDs is due to photo-oxidation and desorption of atoms from the QD surface. Van Sark et al. showed that a 5 nm ZnS-capped CdSe QD shrank 1 nm when the QDs were excited with 468 nm line of Ar-Kr CW laser at a power density of 20 kW/ cm^2 [59]. The capping of QDs with a thick semiconductor layer (e.g., CdS or ZnS onto CdSe) to form a core/shell QD structure can slow the photodegradation process and enhance the photostability of the QDs [60].

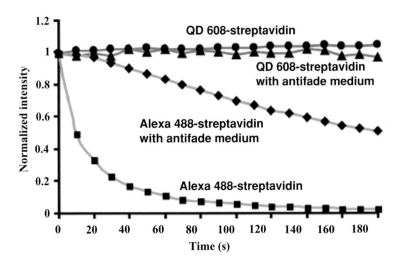

FIGURE 8.5. Photostability comparison of QDs with organic dye molecules. Fluorescence intensity measurements of cells stained with either QD-streptavidin or alexa 488-streptavidin. Within the same mammalian cell, microtubules were stained with QDs while the nucleus was stained with alexa 488-dye and the cells was mounted either with glycerol or antifade mounting medium Vectashield. The cells were excited using Hg lamp and the fluorescence intensity was measured every 10 s for 3 min. This graph shows the preservation of fluorescence intensity of QD-stained microtubules while the fluorescence intensity of alexa 488-stained nucleus slowly diminished. This experiment showed the superior photostability of QDs to organic dye molecules. Reprinted with permission from corresponding author (Dr. X. Wu). Wu et al., Nature Biotechnology, 21: 41-46. Copyright 2004 NPG.

In this section, we have briefly discussed some of the unique properties of QDs that makes them appealing in a number of biological applications. For a more in-depth review paper on the photophysics and other properties of QDs, we refer the reader to a number of excellent reviews by Alivisatos [28, 52].

8.4. APPLICATIONS

The development of optical probes for biological detection and imaging has impacted many areas of biomedical research, such as high-throughput drug screening, gene profiling, and medical diagnosis. Quantum dots (QDs) comprise a new generation of optical probes that have stirred tremendous excitement in the biological and medical community since their introduction in 1998. QDs offer novel properties that are superior to most organic dye molecules for biological research and applications, and will provide researchers with greater flexibility in experimental design. In this section, we discuss some of the successfully demonstrated biological uses of QDs.

8.4.1. In Vitro *Immunoassays & Nanosensors*

QDs have been incorporated into numerous *in vitro* assays for detecting proteins, DNA, and small organic molecules. One of the simplest immunoassay designs for biomolecule detection is the latex agglutination test, in which detection is achieved by QD-aggregation. Chan and Nie successfully demonstrated how this assay is used to detect the presence of Immunoglobulin G (IgG) [12]. By coating the surface of ZnS-capped CdSe QDs with the antibody to Immunoglobulin G (IgG), they were able to detect the presence of IgG. The IgG acted as a bridge between anti-IgG coated QDs, effectively causing aggregation of anti-IgG coated QDs. Mitchell and coworkers used a similar aggregation-based scheme for the detection of oligonucleotides [61]. While Zhu and coworkers used QD-streptavidin conjugates for the immunofluorescence labelling of *Cryptosporidium parvum* and *Giardia lamblia* cells [62].

QDs have also been employed in surface-based assays. Thompson and coworkers and Xiao and Barker extended the use of QD-oligonucleotide probes for fluorescence *in situ* hybridization, a popular approach for detecting genetic mutations in human metaphase chromosomes [63, 64]. Patolsky et al., labelled and studied the dynamics of telomerization using QDs [65]. Apart from DNA analysis, extensive work by Mattoussi and coworkers has advanced QDs as detection probes for ELISA (Enzyme Linked ImmunoAssorbant Assays)-type assays [66, 67]. Traditional ELISA detection schemes are based on an enzyme-linked color change for identifying biomolecules. Although medical and biological researchers commonly use this technique for protein detection, ELISA typically suffers from low sensitivity and cannot be used for high-throughput analyte detection. Mattoussi and et al. demonstrated the detection of proteins and toxins with a detection sensitivity of ng/mL using QD-antibody conjugates as the detection probe. Recently, they demonstrated the successful detection of four toxins (Ricin, Cholera Toxin, Shiga-like toxin 1, and staphylococcal enterotoxin B) in a single well simultaneously [43]. This was accomplished by designing four uniquely emitting QD-probes, with each probe specifically targeting toxins captured on a surface.

Nie and coworkers engineered a more sophisticated multianalyte detection system than the Mattoussi and coworkers system for the high-throughput analysis of biomolecules. They designed QD-based molecular barcodes [68–70]. Barcoding is a concept used in grocery stores to rapidly identify products during checkout. Using a similar concept, microbead barcodes are created by the infusion of QDs (at different concentrations and mixtures) into polystyrene beads; barcodes are developed based on the emission of the QDs and intensity of QDs inside the beads. For example, QDs of 6 different colours and 10 different intensity levels can be used to theoretically produce microbeads with one million unique optical signatures. Each of these optical signatures or "barcodes" can then be uniquely assigned to one million different biomolecules by tagging each uniquely-emitting microbead with a distinct targeting biomolecule (e.g., gene sequences). These QD microbeads can then be added to a solution (e.g., blood) containing both the molecule of interest and QDs conjugated to a secondary targeting molecule. The molecule of interest (or to be detected) acts as a bridge to join the QD microbeads with the QD-conjugates. Single-bead spectroscopy can then be used to identify the binding via the fluorescence spectra (i.e. if peaks for both QDs in microbeads and QD conjugates are observed, the molecule of interest has been successfully detected).

QDs can also be used to develop robust nanosensors (figure 8.6). In one example, QDs have been incorporated into fluorescence resonance energy transfer (FRET)-based nanosensors. FRET is an optical process that involves a pair of fluorescent agents—a donor and an acceptor. In a FRET assay, the energy emission from the donor excites the acceptor

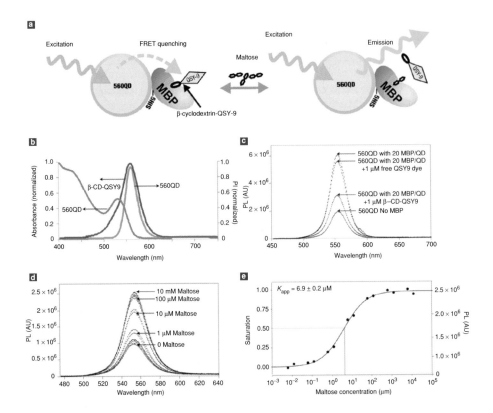

FIGURE 8.6. Function and properties of the 5600 QD-MBP nanosensor. (A) 560QD-MBP nanosensor function schematic. Each 560-nm-emitting QD is surrounded by an average of ~10 MBP moieties; a single MBP is shown for simplicity. Formation of QD-MBP-β-CD-QSY9 (maximum absorption ~565nm) results in quenching of QD emissioin. Added maltose displaces β-CD-QSY9 from the sensor assembly, resulting in an increase in direct QD emission. (B) Spectral properties of 560QD-MBP sensor. Absorption (pink) and emission spectra (green) of MBP-conjugated 560 QD are displayed along with the absorption spectra (blue) of β-CD-QSY9. Samples were excited at 400 nm and emission spectra recorded from 420 nm to 750 nm. Arrows indicate the reference axis. (C) Demonstration of 5600 QD-MBP FRET quenching. PL spectra (AU = arbitrary units) were collected from 560QDs and 560QDs mixed with an average of 20 MBP/QD (saturated for effect). Note the resulting increase in PL (~300%). These same QD-MBP conjugates were then mixed either 1 μM free QSY9 dye or 1 μM β-CD-QSY9. (D) 560 QD-10MBP maltose sensing. Titration of a 5600QD-10MBP/QD conjugate (quantum yield ~39%) preassembled with 1 μM β-CD-QSY9 with increasing concentrations of maltose. (E) Transformation of titration data. The right axis shows PL at 560 nm and fractional saturation is shown on the left axis. The point corresponding to 50% saturation was used to derive the maltose apparent dissociation constant (Kapp) value. Assuming a range of useful measurement to be determined between 10 and 90% saturation translates into a sensing range of ~500 nm to 10 μM maltose. Reprinted with permission from corresponding author (Dr. Hedi Mattoussi)—Medintz, I. et al. Nature Materials, 2003, 2: p.630–638. Both figure and caption are not modified from original publication. Copyright, 2004 NPG. CD = cyclodextrin, MBP = maltose binding protein.

when two fluorescent agents interact at a close distance. Recently, Medintz et al. designed QD-FRET nanosensors for the detection of sugar molecules. In their design, the surfaces of QDs were coated with maltose binding protein (MBP) that was initially bound to maltose molecules-conjugated to the quencher molecule QSY-7 [48]. As a result, the fluorescence emission of QDs was quenched by about 50%. Subsequently, the addition of maltose to a solution containing these QD-FRET nanosensors dislodged the maltose-quencher conjugate from the surface of the QDs, which restored the fluorescence intensities of the QDs to their prequenched state. The successful energy transfer can be used to identify a successful binding event and, hence, detection of the molecule of interest. QDs are ideal FRET donors since their emission energy can be tuned (i.e. by altering the size or composition) to match the optimal excitation energy of the acceptor.

8.4.2. Cell Labeling and Tracking Experiments

Cells are complicated systems in which biological molecules play a key role in maintaining cell viability and function. Currently, the interactions of biomolecules involved in many cellular processes (e.g., differentiation, proliferation) are not fully understood. Powerful optical probes, such as QDs, provide new tools to image and study biomolecules in action both in fixed and live cells. Early work on QD-bioconjugates demonstrated the feasibility of simultaneously identifying and imaging multiple cellular components (e.g., nuclei and cytoskeleton) within fixed mouse fibroblast cells [19]. Theoretically, by engineering a set of QDs with varying emission wavelengths, researchers could simultaneously stain and track greater than 10 cellular components within a single cell. Using a conventional epifluorescence microscope, these studies may help elucidate many of the unknown activities of biomolecules.

Various groups have also taken advantage of the long-term photostability and bright luminescence of QDs for live cell imaging [40, 57, 60]. Simon and coworkers tracked and imaged the behavior of starving D. *Dictyostelium* cells using QDs over 12 days [26]. Derfus and coworkers also used QDs to detect cellular interactions and reorganizations of co-cultures of hepatocytes (liver cells that are highly involved in the detoxification process) and fibroblast cells for 7 days [35]. Finally, Dubertret et al. injected phospholipid-coated QDs into embryos and monitored the embryogenesis process, see figure 8.7 [37].

All three of these published reports suggested that QDs do not interfere with normal cell function, and that QDs maintained their photostability and luminescence properties inside living cells. Despite these successful investigations, a great deal of work remains to be conducted to investigate the potential cytotoxicity of QDs. Derfus and coworkers demonstrated that high levels of CdSe core QDs may kill hepatocytes, but that ZnS-capped CdSe QDs cause minimal cellular toxicity [35]. Further studies on the fundamentals (including toxicity) of QD activities in cells should help optimize the parameters for using QDs for live cell imaging.

Research conducted by Dahan and coworkers opened the possibility of using QD-bioconjugates to probe the interactions of individual biomolecules in/on living cells in real-time [71]. They described the use of QD-bioconjugates to track the motion of a single glycine receptor on the surface of live neurons. This study provided kinetic information of glycine receptors that could not be obtained using 500 nm microbeads or organic dye molecules. The traditional labeling approaches are hampered by large sizes and an inability to perform long-term analysis, respectively. Similarly, Mansson et al. demonstrated the photostability and brightness of single-QDs to track the sliding of actin filaments [72]. By conducting studies similar to that of Dahan and Mansson and their coworkers, biologists may gain more in-depth information about the roles and activities of biomolecules in and on living cells and tissues.

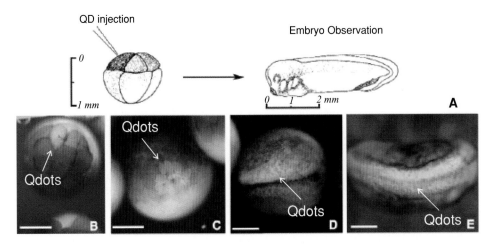

FIGURE 8.7. Long-term tracking of embryogenesis using quantum dots. (A) Schematic of quantum dot injection into embryo. Quantum dots were injected into the blastomere during early cleavage stages at a concentration of 2.1×10^9 to 4.2×10^9 particles per cell. The embryos were then monitored at each stage of development. (B) Initial injection of QDs into embryo. (C) One hour after the injection. (D) Formation of daughter cells was labeled. (E) Two neural embryos were labeled. This experiment clearly showed the ability of QDs for long-term biological lineage tracking. Reprinted with permission from AAAS (Dubertret et al., SCIENCE 298:1759-62). Copyright 2002 AAAS.

Larabel and coworkers demonstrated an interesting application of QDs for cell biology research [73, 74]. Instead of using QDs to tag biological molecules, Larabel and coworkers developed QD surfaces to track cell migration—a process that is important in cancer and tissue engineering. They coated the surface of a glass slide with a thin layer of silica-coated ZnS-capped CdSe QDs. Cells (e.g., MCF 10A, breast tumor MDA-MB-231 cells) are cultured on these QD layers. These cells can non-specifically uptake these silica-coated QDs. The optical excitation of the surface shows a non-fluorescent or dark track in the direction and path of the cell migration. Although the uptake of QDs is dependent upon their surface coating, this application provides an interesting use of QDs as a platform for cell research.

8.4.3. In Vivo *Live Animal Imaging*

Beyond simple assays and cell analysis, QDs may find use in live animal imaging. Such a technology will be useful for analyzing and detecting disease onset and progression, optically guiding surgery during tumor resection, and for elucidating biological process and kinetics in live animal models. For optimum in vivo optical imaging, traditional organic probes suffer from low tissue penetration. Optical probes emitting in the near-IR are better for in vivo imaging than visible emitting probes due to low tissue and water absorption in the near-IR emitting region [75-77]. Near-IR emitting organic fluorophores suffer from low quantum yield and weak emission; QDs can potentially overcome these problems.

As a start toward using QDs as contrast agents for in vivo animal imaging, Ruoslahti and coworkers first demonstrated the successful use of QDs for *in vivo* imaging by targeting peptide-coated QDs to the vasculatures of tumours and organs in living mice [78]. Two of these peptides were conjugated onto the surface of two QDs with distinct emission wavelengths. These peptide-conjugated QDs were then introduced into mice bearing xenograft tumours through injection into the tail vein. After 20 minutes, optical excitation revealed two distinct fluorescence signals from tumour tissues corresponding to two distinct emission wavelengths. Co-staining with markers showed red-emitting QDs localized in the blood vessels, while green-emitting QDs localized in the lymphatic vessels of the tumor. No significant QD emission was apparent in other nearby tissues and organs. Targeting molecules (e.g. peptides, antibodies) have been identified for tumour vessels and normal vessels in the brain, kidney, lungs, skin, pancreas, and other tissues [79]. In cell cultures, several groups have now shown the intracellular targeting of QDs to cellular organelles [80, 81]. This general strategy of integrating targeting molecules with nanostructures should provide a means of delivering contrast agents to specific organelles, cells, and tissues *in vivo* for ultrasensitive and multi-color optical imaging.

Recently, Gao and coworkers demonstrated the simultaneous targeting and imaging of ZnS-capped CdSe QDs [82, 83]. They conjugated prostate-specific membrane antigen (PSMA)-monoclonal antibody to ZnS-capped CdSe QDs and injected them into mouse bearing tumors in the prostate gland. Then, they placed the animal under a macroscopic imaging system and showed the accumulation of the PSMA-conjugated QDs in the prostate gland. The signal appeared as a bright orange stain that was clearly distinct from the animal's autofluorescence.

Frangioni and coworkers showed the practicality of using QDs *in vivo* to guide tumour resection, by demonstrating the use of QDs to guide the removal of the sentinel lymph node [42]. The sentinel lymph node is the first drainage site for the lymph ducts of the breast; the removal of sentinel lymph node is a popular clinical strategy for preventing the metastasis of cancer cells. As shown in figure 8.8, near-IR emitting QDs were injected into the left paw

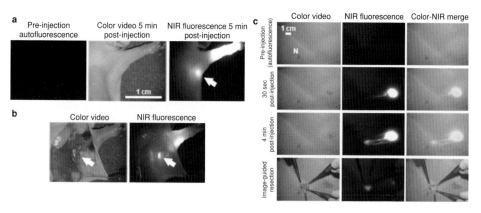

FIGURE 8.8. Sentinel lymph node mapping using near-IR QDs. (A) 10 pmol of QDs were injected into the left paw of the mouse and the flow of QDs to the lymph nodes was observed (arrow shows the axillary sentinel lymph node). (B) Isosulfan blue, a colorimetric stain, for sentinel lymph nodes was also injected into the left paw. The co-localization of the blue-stain with the QDs verifies that the QDs were located in the lymph node region. (C) QDs were then injected into 35 kg pigs and this image shows the flow of QDs in the lymphatic system. Row 4 shows the use of fluorescence to guide resection in pigs. Reprinted with permission from corresponding author (Dr. J. Frangioni). Kim et al., Nature Biotechnology, 22: 93–98. Copyright 2004 NPG.

of a mouse or the thigh of a 35 kg pig at a concentration of 400 pmol and guided the removal of the lymph node. The surgeons were able to monitor the flow of QDs to the sentinel lymph nodes by the fluorescence emission of the QDs. *In vivo* imaging depth of 1 cm using an excitation power density of 5mW/cm^2 was demonstrated using QD-contrast agents. This experiment demonstrated the feasibility of using QDs for assistance and optical guidance in surgical procedures.

8.5. FUTURE WORK

We predict that, over the next decade, a great deal of research will be conducted in the biological applications of semiconductor QDs. We have already seen advances in the synthesis and biocompatibility of QDs that have led to the commercialization of these highly luminescent probes. Proof-of-concept studies have also demonstrated the great promise of QDs for biomedical imaging and detection applications. However, before QDs reach large-scale clinical application, research will be required to elucidate the effects of surface conjugated molecules on QDs, cellular responses to the uptake of QDs, and the *in vivo* pharmacokinetics and toxicity of QDs. In the meantime, QDs will continue to be incorporated into sensors for biomolecule detection and high-throughput analytic devices. QDs will also be used to probe the activities of biomolecules in live cells and tissues, and used as ultrasensitive contrast agents for the early detection of diseases. Of note, there have been several recent publications demonstrating the use of QDs for therapeutic applications [84, 85]. Lastly, we envision that the next decade will reveal numerous unique applications and experiments using QDs that cannot be done with traditional probes and results obtained from these experiments will reveal a deeper insight into fundamental biology that can guide the design of new therapeutics and diagnostics.

ACKNOWLEDGEMENTS

W. C. W. C. would like to acknowledge the following funding sources for support of research: Canadian Foundation for Innovation (CFI) and Ontario Innovation Fund (OIT)—New Opportunity and Infrastructure Operating Fund (CFI), National Science and Engineering Council of Canada (NSERC)—Discovery Grant, Canadian Institute of Health Research (CIHR)—Regenerative Medicine and Nanomedicine Grant and Novel Technologies in Health Research Grant, and Connaught Foundation. W. J. and H. F. would like to acknowledge the Ontario Government for the Ontario Graduate Scholarship, A. S. would like to acknowledge NSERC for summer undergraduate fellowship, and S. M. would like to acknowledge NSERC for graduate fellowship.

REFERENCES

[1] J.Y. Adams et al. Visualization of advanced human prostate cancer lesions in living mice by a targeted gene transfer vector and optical imaging. *Nat. Med.*, 8(8):891–897, 2002.
[2] W.C.W. Chan et al. Luminescent quantum dots for multiplexed biological detection and imaging. *Curr. Opini. Biotechnol.*, 13:40–46, 2002.

[3] K.R. Gee et al. Detection and imaging of zinc secretion from pancreatic beta-cells using a new fluorescent zinc indicator. *J. Am. Chem. Soc.*, 124(5):776–778, 2002.

[4] R.P. Haugland. *Handbook of Fluorescent Probes and Research Products,* 9th Ed., Molecular Probes, Eugene, OR, pp. 966, 2002.

[5] M. Roederer et al. 8 color, 10-parameter flow cytometry to elucidate complex leukocyte heterogeneity. *Cytometry*, 29:328–339, 1997.

[6] S. Schultz, D.R. Smith, J.J. Mock, and D.A. Schultz. Single-Target Molecule Detection with Nonbleaching Multicolor Optical Immunolabels. *Proceedings of the National Academy of Science*, Vol. 97, no. 3, pp. 996–1001, 2000.

[7] M. Brasuel, R. Kopelman, T. Miller, R. Tjalkens, and M. A. Philbert. Fluorescent nanosensors for intracellular chemical analysis: Decyl methyacrylate liquid polymer matrix and ion-exchange-based potassium PEBBLE sensors with real-time application to viable rat C6 glioma cells. *Anal. Chem.*, 73:2221–2228, 2001.

[8] G. Gaietta, T.J. Deerinck, S.R. Adams, J. Bouwer, O. Tour, D.W. Laird, G.E. Sosinsky, R.Y. Tsien, and M.H. Ellisman. Multicolor and electron microscopic imaging of connexin trafficking. *Science*, 296:503–507, 2002.

[9] E.J. Park, M., Brasuel, C. Behrend, M.A. Philbert, and R. Kopelman. Ratiometric optical PEBBLE nanosensors for real-time magnesium ion concentrations inside viable cells. *Anal. Chem.*, 75:3784–3791, 2003.

[10] S. Weiss. Fluorescence spectroscopy of single biomolecules. *Science*, 283(5408):1676–1683, 1999.

[11] A. Waggoner. Covalent labeling of proteins and nucleic acids with fluorophores. *Meth. Enzymol.*, 246:362–373, 1995.

[12] W.C. Chan et al. Luminescent quantum dots for multiplexed biological detection and imaging. *Curr. Opin. Biotechnol.*, 13(1):40–46, 2002.

[13] H. Mattoussi et al. Self-assembly of CdSe-ZnS quantum dot bioconjugates using an engineered recombinant protein. *J. Am. Chem. Soc.*, 122:12142–12150, 2000.

[14] R. Elghanian et al. Selective colorimetric detection of polynucleotides based on the distance-dependent optical properties of gold nanoparticles. *Science*, 277(5329):1078–1081, 1997.

[15] A. Henglein. Small-particle research: physiochemical properties of extremely small colloidal metal and semiconductor particles. *Chem. Rev.*, 89:1861–1873, 1989.

[16] R. Rossetti, S. Nakahara, and L.E. Brus. Quantum size effects in the redox potentials, resonance raman spectra, and electronic spectra of CdS crystallites in aqueous solutions. *J. Chem. Phys.*, 79:1086–1087, 1983.

[17] L.E. Brus. Electronic wave functions in semiconductor clusters: experiment and theory. *J. Phys. Chem.*, 90:2555–2560, 1986.

[18] W.C. Chan and S. Nie. Quantum dot bioconjugates for ultrasensitive nonisotopic detection. *Science*, 281(5385):2016–2018, 1998.

[19] M. Bruchez, Jr. et al. Semiconductor nanocrystals as fluorescent biological labels. *Science*, 281(5385):2013–2016, 1998.

[20] C.B. Murray, D.J. Norris, and M.G. Bawendi. Synthesis and characterization of nearly monodisperse CdE (E = S, Se, Te) semiconductor nanocrystallites. *J. Am. Chem. Soc.*, 115:8706–8715, 1993.

[21] B.O. Dabbousi et al. (CdSe)ZnS Core-shell quantum dots: synthesis and characterization of a size series of highly luminescent nanocrystallites. *J. Phys. Chem. B*, 101:9463–9475, 1997.

[22] M.A. Hines and P. Guyot-Sionnest. Synthesis and characterization of strongly luminescing ZnS-capped CdSe nanocrystals. *J. Phys. Chem. B*, 100:468–471, 1996.

[23] X. Peng et al. Epitaxial growth of highly luminescent CdSe/CdS core/shell nanoocrystals with photostability and electroni accessibility. *J. Am. Chem. Soc.*, 119:7019–7029, 1997.

[24] M.L. Steigerwald, A.P. Alivisatos, J.M. Gibson, T.D. Harris, R. Kortan, A.J. Muller, A.M. Thayer, T.M. Duncan, D.C. Douglass, and L.E. Brus. Surface derivitization and isolation of semiconductor cluster molecules. *J. Am. Chem. Soc.*, 110:3046–3050, 1988.

[25] A.R. Kortan and growth of CdSe on ZnS quantum crystallite seeds, and vice versa, in inverse micelle media. *J. Am. Chem. Soc.*, 112:1327–1332, 1990.

[26] C. Dameron et al. Biosynthesis of cadmium sulfide quantum semiconductor crystallites. *Nature*, 338:596–597, 1989.

[27] A.R. Kortan, R. Hull, R.L. Opila, M.G. Bawendi, M.L. Steigerwald, P.J. Carroll, and L.E. Brus. Nucleation and growth of CdSe on ZnS quantum crystallite seeds, and vice versa, in inverse micelle media. *J. Am. Chem. Soc.*, 112:1327–1332, 1990.

[28] A.P. Alivisatos. Semiconductor clusters, nanocrystals, and quantum dots. *Science*, 271:933–937, 1996.

[29] M.A. Hines and P. Guyot-Sionnest. Bright UV-blue luminescent colloidal ZnSe nanocrystals. *J. Phys. Chem. B*, 102(19), 1998.

[30] I. Mekis et al. One-Pot synthesis of highly luminescent CdSe/CdS core-shell nanocrystals via organometallic and "greener" chemical approaches. *J. Phys. Chem. B*, 107:7454–7462, 2003.

[31] V.K. LaMer and R.H. Dinegar. Theory, production and mechanism of formation of monodispersed hydrosols. *J. Am. Chem. Soc.*, 72(11):4847–4854, 1950.

[32] X. Peng, J. Wickham, and A.P. Alivisatos. Kinetics of II-VI and III-V colloidal semiconductor nanocrystal growth: "focusing" of size distributions. *J. Am. Chem. Soc.*, 120:5343–5344, 1998.

[33] Z.A. Peng and X. Peng. Formation of high-quality CdTe, CdSe, and CdS nanocrystals using CdO as a precursor. *J. Am. Chem. Soc.*, 123:183–184, 2001.

[34] L. Qu, Z.A. Peng, and X. Peng. Alternative routes toward high quality CdSe nanocrystals. *Nanoletters*, 1(6):333–337, 2001.

[35] L. Qu and X. Peng. Control of photoluminescence properties of CdSe nanocrystals in growth. *J. Am. Chem. Soc.*, 124(9):2049–2055, 2002.

[36] L. Manna, D. Milliron, J., A. Meisel, E.C. Scher, and A.P. Alivisatos. Controlled growth of tetrapod-branched inorganic nanocrystals. *Nat. Mat.*, 2:382–385, 2003.

[37] S. Khan, T. Mokari, E. Rothenberg, and U. Banin. Synthesis and size-dependent properties of zinc-blende semiconductor quantum rods. *Nat. Mat.*, 2:155–159, 2003.

[38] E.M. Chan, R.A. Mathies, and A.P. Alivisatos. Size-controlled growth of CdSe nanocrystals in microfluidic reactors.*Nano Lett.*, 3:199–201, 2003.

[39] R.E. Bailey and S. Nie. Alloyed semiconductor quantum dots: tuning the optical properties without changing the particle size. *J. Am. Chem. Soc.*,

[40] B. Dubertret et al. In Vivo imaging of quantum dots encapsulated in phospholipid micelles. *Science*, 298:1759–1762, 2002.

[41] X. Wu, et al. Immunofluorescent labeling of cancer marker Her2 and other cellular targets with semiconductor quantum dots. *Nat. Biotechnol.*, 21:41–46, 2003.

[42] S. Kim, Y.T. Lim, E.G. Soltesz, A.M. De Grand, J. Lee, A. Nakayama, J.A. Parker, T. Mihaljevic, R.G. Laurence, D.M. Dor, L.H. Cohn, M.G. Bawendi, and J.V. Frangioni. Near-infrared fluorescent type II quantum dots for sentinel lymph node mapping. *Nat. Biotechnol.*, 22(1):93–98, 2004.

[43] S. Kim and M.G. Bawendi. Oligomeric ligands for luminescent and stable nanocrystal quantum dots. *J. Am. Chem. Soc.*, 125:14652–14653, 2003.

[44] B. Dubertret, et al. In vivo imaging of quantum dots encapsulated in phospholipid micelles. *Science*, 298:1759–1762, 2002.

[45] W. Jiang, S. Mardyani, H. Fischer, and W.C.W. Chan. Large scale surface modification of TOPO-coated quantum dots for biological applications. Submitted.

[46] G.T. Hermanson. *Bioconjugate Techniques*. Academic Press, Toronto, ON.

[47] H. Mattoussi et al. Bioconjugation of highly luminescent colloidal CdSe-ZnS quantum dots with an engineered two-domain recombinant protein. *Physica Stat. Solidi (b)*, 244(1):277–283, 2001.

[48] I.L. Medintz et al. Self-assembled nanoscale biosensors based on quantum dot FRET donors. *Nat. Mat.*, 2:630–638, 2003.

[49] J.M. Ness, R.S. Akhtar, C.B. Latham, K.A. Roth. Combined tyramide signal amplification and quantum dots for sensitive and photostable immunofluorescence detection. *J. Histochem. Cytochem.*, 51:981–987, 2003.

[50] R. Nisman, G. Dellaire, Y. Ren, R. Li, and D.P. Bazett-Jones. Applications of quantum dots as probes for correlative fluorescence, conventional, and energy-filtered transmission electron microscopy. *J. Histochem. Cytochem.*, 52:13–18, 2004.

[51] M.H. Qureshi, J.C. Yeung, S. Wu, and S. Wong. Development and characterization of a series of soluble tetrameric and monomeric streptavidin muteins with differential biotin binding affinities. *J. Biolog. Chem.*, 276:46422–46428, 2001.

[52] A.P. Alivisatos. Perspectives on the physical chemistry of semiconductor nanocrystals. *J. Phys. Chem.*, 100:13226–13239, 1996.

[53] D. Gerion et al. Synthesis and properties of biocompatible water-soluble silica-coated CdSe/ZnS semiconductor quantum dots. *J. Phys. Chem. B*, 105:8861–8871, 2001.

[54] A. Striolo, J. Ward, J.M. Prausnitz, W.J. Parak, D. Zanchet, D. Gerion, D. Milliron, and A.P. Alivisatos. Molecular weight, osmotic second virial coefficeint, and extinction coefficient of colloidal CdSe nanocrystals. *J. Phys. Chem. B*, 106:5500–5505, 2002.

[55] C.A. Leatherdale, W.K. Woo, F.V. Mikulec, and M.G. Bawendi. On the absorption cross section of CdSe nanocrystal quantum dots. *J. Phys. Chem. B*, 106:7619–7622, 2002.

[56] W.W. Yu, L. Qu, W. Guo, and X. Peng. Experimental determination of the extinction coefficient of CdTe, CdSe, and CdS nanocrystals. *Chem. Mat.*, 15:2854–2860, 2003.

[57] J.K. Jaiswal et al. Long-term multiple color imaging of live cells using quantum dot bioconjugates.[comment]. *Nat. Biotechnol.*, 21(1):47–51, 2003.

[58] W.C.W. Chan. *Semiconductor Quantum Dots for Ultrasensitive Biological Imaging and Detection, in Chemistry.* Indiana University, Bloomington. p. 153, 2001.

[59] W. Van Sark et.al. Photooxidation and photobleaching of single CdSe/ZnS quantum dots probed by roomtemperature time-resolved spectroscopy. *J. Phys. Chem. B*, 105:8281–8284, 2001.

[60] A. Derfus, W.C.W. Chan, and S. Bhatia. Probing the cytotoxicity of semiconductor quantum dots. *Nano Lett.*, 4(1):11–18, 2004.

[61] G.P. Mitchell, C.A. MIrkin, and R.L. Letsinger. Programmed assembly of DNA functionalized quantum dots. *J. Am. Chem. Soc.*, 121:8122–8123, 1999.

[62] L. Zhu, S. Ang, and W. Liu. Quantum dots as a novel immunofluorescent detection system for Cryptosporidium parvum and Giardia lamblia. *Appl. Environment. Microbiol.*, 70:597–598, 2004.

[63] S. Pathak, S.K. Choi, N. Arnheim, and M.E. Thompson. Hydroxylated quantum dots as luminescent probes for in situ hybridization. *J. Am. Chem. Soc.*, 123:4103–4104, 2001.

[64] Y. Xiao and P.E. Barker. Semiconductor nanocrystal probes for human metaphase chromosomes. *Nucleic Acids Res.*, 32:1–5, 2004.

[65] F. Patolsky, R. Gill, Y. Weizmann, T. Mokari, U. Banin, and I. WIllner. Lighting-up the dynamics of telomerization and DNA replication by CdSe-ZnS quantum dots. *J. Am. Chem. Soc.*, 125:13918–13919, 2003.

[66] E.R. Goldman, G.P. Anderson, P. T. Tran, H. Mattoussi, P.T. Charles, and M. Mauro. Conjugation of luminescent quantum dots with anitbodies using an engineered adaptor protein to provide new reagents for fluoroimmunoassays. *Anal. Chem.*, 74:841–847, 2002.

[67] E.R. Goldman, A.R. Clapp, G.P. Anderson, H.T. Uyeda, J.M. Mauro, I.L.Medintz, and H. Mattoussi. Multiplexed toxin analysis using four colors of quantum dot fluororeagents. *Anal. Chem.*, 76:684–688, 2004.

[68] X. Gao and S. Nie. Quantum dot-encoded mesoporous beads with high brightness and uniformity: rapid readout using flow cytometry. *Anal. Chem.*, 76:2406–2410, 2004.

[69] X. Gao, W.C. Chan, and S. Nie. Quantum-dot nanocrystals for ultrasensitive biological labeling and multicolor optical encoding. *J. Biomed. Opt.*, 7(4):532–537, 2002.

[70] M. Han, X. Gao, J.Z. Su, and S. Nie. Quantum-dot-tagged microbeads for multiplexed optical coding of biomolecules. *Nat. Biotechnol.*, 19:631–635, 2001.

[71] M. Dahan et al. Diffusion dynamics of glycine receptors revealed by single-quantum dot tracking. *Science*, 302:442–445, 2003.

[72] A. Mansson et al. In vitro sliding of actin filaments labelled with single quantum dots. *Biochem. Biophy. Res. Commun.*, 314:529–534, 2004.

[73] W. Parak, R. Boudreau, M.L. Gros, D. Gerion, D. Zanchet, C.M. MIcheel, S. Willimas, A.P. Alivastos, and C.Larabell. Cell motility and metastatic potential studies based on quantum dot imaging and phagokinetic tracts. *Adv. Mat.*, 14:882–885, 2002.

[74] T. Pellegrino, W. Parak, J., R. Boudreau, M. Le Gros, D. Gerion, A. P. Alivisatos, and C. Larabel. Quantum dot-based cell motility assay. *Differentiation*, 71:542–548, 2003.

[75] R. Weissleder. A clearer vision for in vivo imaging. *Nat. Biotechnol.*, 19:316–317, 2001.

[76] Y.T. Lim et al. Selection of quantum dot wavelengths for biomedical assays and imaging. *Mol. Imag.*, 2(1):50–64, 2003.

[77] W. Jiang, E. Papa, H. Fischer, S. Mardyani, and W.C.W. Chan. Semiconductor quantum dots as contrast agents for whole animal imaging. *Trends Biotechnol.*, 22:607–609, 2004.

[78] M.E. Akerman et al. Nanocrystal Targeting In Vivo. *Proceedings of the National Academy of Sciences of the United States of America*, Vol. 99, no. 20, pp. 12617–12621, 2002.

[79] E. Ruoslahti. Specialization of tumor vasculature. *Nat. Rev. Cancer*, 21:84–90, 2002.

[80] A. Derfus, W.C.W. Chan, and S. Bhatia. Intracellular delivery of semiconductor quantum dots. *Adv. Mat.*, 16:961–966, 2004.

[81] F. Chen and D. Gerion. Fluorescent CdSe/ZnS nanocrystal-peptide conjugates for long-term nontoxic imaging and nuclear targeting in living cells. *Nano Letters*, 4:1827–1832, 2004.

[82] X. Gao, Y. Cui, R.M. Levenson, L.W.K. Chung, and S. Nie. In vivo cancer targeting and imaging with semiconductor quantum dots. *Nat. Biotechnol.*, 22:969–976, 2004.

[83] R.K. Jain and M. Stroh. Zooming in and out with quantum dots. *Nat. Biotechnol.*, 22:959–960, 2004.

[84] R. Bakalova, H. Ohba, Z. Zhelev, T. Nagase, R. Jose, M. Ishikawa, and Y. Baba, Quantum dot anti-CD conjugates: are they potential photosensitizers or potentiators of classical photosensitizing agents in photodynamic therapy of cancer? *Nanoletters*, ASAP, 2004.

[85] A.C. Sarmia, X. Chen, and C. Burda. Semiconductor quantum dots for photodynamic therapy. *J. Am. Chem. Soc.*, 125:15736–15737, 2003.

9

Diagnostic and Therapeutic Applications of Metal Nanoshells

Leon R. Hirsch, Rebekah A. Drezek, Naomi J. Halas, and
Jennifer L. West

*Rice University, Department of Bioengineering, 6100 Main St., MS 142,
Houston, TX, 77005*

9.1. METAL NANOSHELLS

Nanoshells are a new class of nanoparticles with tunable optical properties that have shown promise in a number of diagnostic and therapeutic applications. Conceived of over 50 years ago [1] but not realized until the 1990's [3, 25, 48], metal nanoshells consist of spherical dielectric nanoparticles surrounded by a thin, conductive, metallic layer. By varying the composition and dimensions of the nanoparticles, nanoshells can be designed and fabricated with plasmon resonances from the visible to the mid-infrared regions of the spectrum [26]. For a given composition of core and metal shell, the plasmon resonances of the nanoparticle, which determines the particle's optical absorption and scattering, may be tuned by changing the ratio of the nanoparticle's core size to its shell thickness. Essentially, for a core of given diameter, metal shells of diminishing thickness (or higher core:shell ratios) will result in red shifting of the peak plasmon frequency (Figure 9.1).

$$\alpha = 4\pi\varepsilon_0 r_2{}^3 \left[\frac{\varepsilon_2\varepsilon_a - \varepsilon_3\varepsilon_b}{\varepsilon_2\varepsilon_a + 2\varepsilon_3\varepsilon_b} \right] \tag{9.1}$$

$$\varepsilon_a = \varepsilon_1(3 - 2P) + 2\varepsilon_2 P \tag{9.2}$$

$$\varepsilon_b = \varepsilon_1 P + \varepsilon_2(3 - P) \tag{9.3}$$

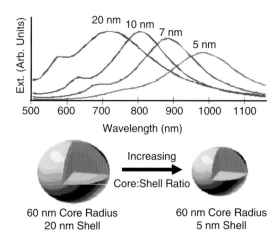

FIGURE 9.1. Optical tenability is demonstrated for nanoshells with a 60 nm silica core redius and gold shells 5, 7, 10, and 20 nm thick. Observe that the plasmon resonance (extinction) of the particles red shifts with decreasing thickness of the gold shell (or an increasing core:shell ration).

$$P = 1 - (r_1/r_2)^3 \tag{9.4}$$

$$\frac{r_1}{r_2} = \left[1 + \frac{3}{2} \frac{\varepsilon'_2(\lambda)(\varepsilon_1 + 2\varepsilon_3)}{[\varepsilon'_2(\lambda)]^2 - \varepsilon'_2(\lambda)(\varepsilon_1 + \varepsilon_3) + \{\varepsilon_1\varepsilon_3 - [\varepsilon''_2(\lambda)]^2\}} \right]^{1/3} \tag{9.5}$$

Medieval alchemists were some of the first to discover the plasmon resonance phenomenon when they successfully reduced gold from a salt solution into its reddish colloidal form. The reddish color arises from the metal colloid's extinction at approximately 520 nm, which results from optical resonances of surface plasmons (or oscillating conducting electrons) in the metal induced by the incident light. Many bulk metals (i.e. Au, Ag, Ni, Pt) demonstrate a plasmon resonance with each metal having a characteristic peak within a defined region of the visible spectrum [5]. However, unlike solid metal colloid, core-shell metal nanoparticles possess a plasmon resonance which can be tuned in resonant frequency by varying the magnitude of the core and shell layers, a unique property of this topology [30].

Experimentally, the first metal nanoshell, developed by Zhou, et al. [48], consisted of an Au_2S dielectric core surrounded by a gold shell. Depending upon the size of the nanoparticles, it was possible to shift the plasmon resonance to longer wavelengths of light (red shift), ranging from the standard gold colloid peak of $\sim$520 nm out to $\sim$900 nm. Originally, it was proposed that this red shift in the extinction peak was due to quantum confinement within the Au shell. However, Averitt, et al. later contested this theory and concluded that neither quanum confinement nor electron scattering was the cause of this red shift in the particle's extinction [3], rather this tunability was related to the size and shape dependent plasmon response of the metallic shell layer. By applying the boundary conditions to the geometry depicted in Figure 9.2, Averitt, and colleagues were able to solve for the electromagnetic fields within each layer of the nanoparticle for an incident plane wave and determined the polarizability (α) of the nanoshell (Equation 9.1), where ε_0 is the permittivity of free space, ε_1, ε_2, ε_3, are the dielectric constants of the core, shell, and embedding medium

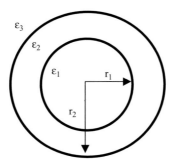

FIGURE 9.2. Geometry for metal nonoshell consisting of dielectric functions for embedding medium (ε_3), shell (ε_2), and core (ε_1), as well as core and total radii (r_1) and (r_2) respectively. These dimensions and constants are used to define the resonance conditions outlined in Equations 9.1–9.5.

respectively; ε_a and ε_b are defined by Equations 9.2 and 9.3. P (Equation 9.4) is simply the ratio of the shell volume to the core volume [4]. For mathematical simplicity, these models use quasi-static calculations (also known as the dipole limit), which assume that the incident light wave is uniform over the spatial extent of the nanoparticle. As long as the particle radius is much smaller than the incident wavelength, the quasi-static assumption is appropriate.

Plasmon resonance occurs at the wavelength of maximum polarizability, which is when the real component of the denominator in Equation 9.1 approaches zero. Given this condition, and the fact that the imaginary components of the dielectric function for the core and embedding medium are zero, one can solve for the resonance condition in Equation 9.5. Here, we see the relationship between the core/shell ratio (r_1/r_2) and the plasmon resonance at λ that was discussed earlier in this section.

Gold-gold sulfide (Au-Au$_2$S) nanoshells have limits to their size (≤ 40 nm) and plasmon tunability due their synthesis chemistry. These nanoparticles are grown in a one step process where chloroauric acid (HAuCl$_4$) and sodium sulfide (Na$_2$S) are mixed. Depending upon the ratios of HAuCl$_4$ and Na$_2$S added (with an excess of Au), Au-Au$_2$S nanoshells are grown with different core and shell thicknesses. Due to the kinetics of the core and shell growth (Averitt, 1997) this synthesis lacks the facility for independent control over nanoparticle core and shell dimensions.

Oldenberg, et al. developed a new nanoparticle, a silica-gold core-shell nanoshell, which overcame many of the limitations of the Au-Au$_2$S particles [25]. To synthesize these particles, a dielectric silica core is grown by the Stöber method [41], where tetraethylorthosilicate is reduced in ethanol under basic conditions resulting in the nucleation and growth of highly monodisperse and spherical silica colloid. Particle diameters ranging from 50 to 500 nm can be synthesized by this method. The surfaces of the silica core particles are then functionalized with amine groups by reaction with aminopropyltriethoxysilane (APTES). Small gold colloid (1–2 nm) is then adsorbed onto their aminated surfaces. This disperse surface Au colloid layer serves as nucleation sites for further reduction of gold onto the silica nanoparticle core by reduction of Au in a HAuCl$_4$ solution. As more gold is reduced, the surface coating grows and coalesces into a complete gold shell (Figure 9.3). The amount of HAuCl$_4$ added determines the final thickness of the gold shell, which can typically range between

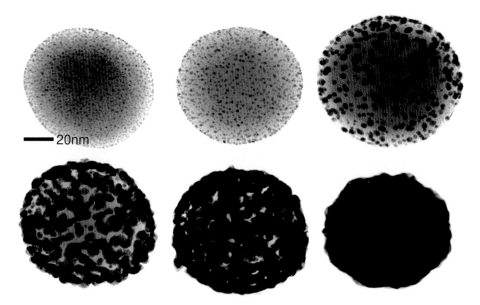

FIGURE 9.3. Series of TEM images showing gold colloid growth into a complete shell on silica core particle surface. Beginning form the upper left, the gold colloid (black dots) serve as nucleation sites for additional electroless plating of gold. As addition gold is plated onto the gold islands, the gold grows until coalescing with neighboring colloid, finally forming a complete metal shell (bottom right).

5 and 30 nm. By changing the ratio of the core diameter to shell thickness, the plasmon resonance peak may be placed anywhere within the visible to mid-infrared region of light. More recently, this geometry has been extended into multilayered, concentric nanoshells, producing more complex hybridized resonances whose spectral profiles span the infrared spectrum to thermal wavelengths and further illustrate the properties of the tunable plasmon in the context of a Plasmon Hybridization model. This theoretical picture exploits the similarity between plasmon resonances and molecular orbital theory, and explains why the core-shell geometry of the nanoshell uniquely supports such a structurally sensitive tunable plasmon [30].

Gold nanoshells are attractive for use in biological applications due in part to the fact that their outer shell is composed of reduced gold. This noble metal is resistant to corrosion, demonstrates low toxicity, and is a popular material for medical applications such as dental prosthetics, which utilize its inert chemical properties and conformational flexibility [6, 9, 28]. Biomedical applications also use gold in electrodes for amperometric detection of analytes (i.e. O_2, H_2O_2, urea, glucose) in long term sensing applications [15, 22, 33].

Gold also facilitates easy conjugation of various proteins onto its surface. Proteins spontaneously chemisorb to gold surfaces when incubated with the metal at or slightly basic to their isoelectric point. This technique is commonly used for immunogold labeling techniques [16]. However, most current methods prefer a more irreversible, if not covalent, method of molecule immobilization onto gold surfaces. Proteins and other thiol/disulfide-possessing molecules are capable of spontaneously self-assembling into dense networks on gold surfaces. This strong dative interaction between the sulfur and gold is reversible, but strongly favors binding in the forward direction with a bond enthalpy of 126–146 kJ/mol—a strength approaching that of other covalent bonds [24]. Proteins and other synthetic molecules are easily functionalized with these sulfahydryl moities to create a robust, simple method of tailoring the gold nanoshell surface chemistry, and concomitantly, its bioreactivty/inertness for biosensing or therapeutic applications.

Using this type of chemistry, metal nanoshells can be chemically modified with either hydrophobic or hydrophilic species. This provides the advantage of using nanoshells in solution-based systems without compromising solubility or colloidal stability in either organic or aqueous solution. For *in vivo* applications, nanoshells can be immoblized within "stealthing" polymers or microecapsulated to provide steric stabilization and reduce the risk of opsonization [13].

9.2. BIOMEDICAL APPLICATIONS OF GOLD NANOSHELLS

Optical methods for diagnosis and treatment in medicine and biology are attractive due to their potential for non-invasive and minimally-invasive applications. Near infrared light between 700–1100 nm (the so-called "water window") is particularly interesting due to the absence of any significant absorption from either biological chromophores or water within this region, permitting deep optical penetration into biological samples such as tissue or whole blood [46]. Numerous diagnostic systems employing NIR optical probes are under investigation. Tomographic techniques, such as optical coherence tomography (OCT), use backscattered light to reconstruct high-resolution images (≤ 10 μm) of tissue morphology [18]. While OCT has been used most extensively in ophthalmology, it has also examined features such as intimal thickening in human aortas [7]. On a larger scale, tomographic reconstructions of whole brain specimens have been made with NIR light. Although this technique does not approach the spatial resolution of conventional methods (i.e., MRI, CT), it does contain spectral information as well; NIR spectroscopy in conjunction with tomography may provide both space- and time-resolved information about cerebral oxygenation, which is of particular interest in neonates [17]. Some other NIR applications worth mentioning are confocal imaging, iridotomy, and photothermal coagulation—all of which take advantage of tissue's increased transparency within this region [32, 43]. As discussed above, nanoshells can be easily tuned to have strong scattering or absorption properties in the NIR, enabling many new biomedical applications in this interesting spectral region.

9.2.1. Nanoshells for Immunoassays

Immunoassays utilize antibody-antigen interactions to detect a specific antigen within a complex mixture. These systems are used in both clinical and research settings to detect a variety of components, including: toxins (i.e. snake venom), drugs, viruses (i.e. HIV, hepatitis), hormones, DNA, and many other molecules [14, 20, 21]. The Sandwich-Type Enzyme Linked Immunosorbent Assay (ELISA) is the most widely used immunoassay. These assays take place in a microtiter plate where a primary capture antibody is first adsorbed onto the surfaces of the well. The immobilized capture antibody will then bind with its complimentary antigenic analyte when exposed to the test solution of interest. After rinsing, a secondary enzyme-labeled antibody is added which has an affinity for the analyte as well, forming the sandwich complexes. The enzymatic activity remaining on the surface of the well after rinsing is indicative of the amount of antigen present in the analyte solution. Although these methods are very effective at detecting small amounts of analyte, they suffer a few limitations. These systems rely upon a colorimetric change in solution to determine antigen concentrations, making it necessary to use purified, cell-free specimens and perform multiple rinsing steps in order to minimize optical interference and obtain a pure signal. For the analysis of a blood specimen, these additional purification steps may lengthen the time necessary to complete an assay (4–24 hrs)—time which the patient may not be able to afford. In addition, these assays are performed on a solid, macroscopic substrate, limiting the application to *in vitro* use only. The ELISA's dependence upon enzymatic activity for detection is another problem, due to its dependence upon temperature, denaturation, pH, and other factors. To attempt to overcome these issues, a new immunoassay technique has been developed utilizing antibody-conjugated, near-infrared resonant nanoshells. This assay can be performed in whole blood and provides results within several minutes.

The underlying mechanism behind the nanoshell-based immunoassay is similar to that of conventional latex agglutination (LA) and sol particle immunoassays (Singer and Plotz, 1956, Leuvering et al., 1980). When antibody conjugated particles are exposed to a multivalent analyte, multiple particles will bind to the analyte, forming particle dimers and higher order assemblies of particles. Unfortunately, LA and sol particle assays are difficult to perform in whole blood due to blood's high turbidity and strong visible extinction. The aggregation of nanoshells gives rise to additional optical resonances at longer wavelengths for the aggregate structure. This appears in the optical signal as a net shift in the nanoshell resonance to longer wavelengths in the near infrared extinction spectrum, allowing detection of analyte in blood samples [12]. The plasmon resonant spectral response of nanoparticle masses differs significantly from that of isolated, dispersed plasmon resonant nanoparticles [31]. Nanoshell dimers and larger aggregates produce a significantly red-shifted plasmon resonance. The red-shifted plasmon from aggregation is simultaneously accompanied by a decrease in the amplitude of the single nanoshell plasmon resonance in the overall spectral response [31]. This provides a straightforward method for detecting nanoshell-bioconjugate aggregation via monitoring of the decrease in extinction at the original nanoshell plasmon resonance peak, as shown in Figure 9.4 [12].

The performance of the nanoshell-based immunoassay has been evaluated in saline, serum, and blood samples for a variety of analytes [12]. For these assays, antibodies are conjugated to nanoshells via a polyethylene glycol linker that is derivatized with thiol groups. Quantitative analyte detection could be achieved over the concentration range of 0.4–400 ng/ml, with completion of the assay in 10–30 minutes. Figure 9.5 shows the detection of varying concentrations of rabbit IgG in saline, bovine serum, and human blood. The availability of this type of rapid, *in situ*, whole blood assay with the capacity to detect a variety of analytes would greatly benefit point-of-care or public health applications where there is a strong demand for rapid, high throughput screening of blood-borne species such as bacteria, viruses, or proteins.

9.2.2. *Photothermally-modulated Drug Delivery Using Nanoshell-Hydrogel Composites*

Over the past several decades, the field of controlled drug delivery has faced two major challenges; sustained zero-order release of a therapeutic agent and deliver of a therapeutic

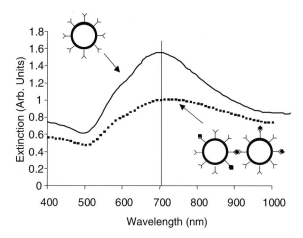

FIGURE 9.4. The principle of the nonoshell immunoasay. Nanoshells conjugated with antibody specific for a particular analyte are well dispersed in an analyte free environment (—), and possess a spectrum with a resonance in the near infrared. In the presence of analyte, however, antibody-antigen interactions induce nanoshell agglutination, a pheonomenon that is easily detected by a reduction in the nanoshell solutions' original peak resonance (- -).

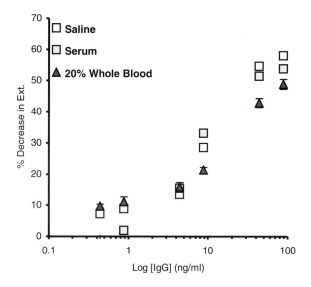

FIGURE 9.5. Rabbit IgG (analyte) induced aggregation of nanoshells in saline, serum, and blood at 30 min. Successful detection of analyte in serum reveals good sensitivity of the assay, as serum contains a multitude of proteins that could potentially promote nonspecific antibody binding. Likewise, nanoshell performance in whole blood proceeded with only a slight reduction in sensitivity. Whole blood performance demonstrates that nanoshells possess the optical contrast required for detection.

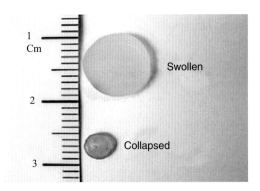

FIGURE 9.6. Nanoshells entrapped within the NIPAAm-co-acrylamide hydrogels absorb near infrared light to drive the phase change of the thermally responsive polymer. Exceeding the LCST caused collapse of the hydrogel material, due to expulsion of water and collapse of the polymer chains.

agent in a pulsatile or staggered fashion. The first goal has been largely addressed by a variety of delivery systems, including osmotically driven pumps and biodegradable matrices (reviewed in [39]). The second goal, controlled modulation of drug delivery, has proved more difficult. Thermally responsive hydrogels and membranes have been extensively evaluated as platforms for the pulsatile delivery of drugs. One of the characteristics of temperature-responsive hydrogels is the presence of a lower critical solution temperature (LCST), a temperature at which the hydrogel material will undergo a dramatic phase change. The driving force for this phase change is based on the interactions between the polymer and the surrounding water [34, 42]. Below the LCST, the most thermodynamically stable configuration is for the water molecules to remain clustered around the polymer chains. Above the LCST, the polymer chains collapse upon each other and minimize interaction with water. Due to this phase change, a macroscopic hydrogel will undergo a drastic change in dimensions, collapsing as the temperature exceeds the LCST, with expulsion of water (and dissolved drug) from the matrix.

N-isopropylacrylamide (NIPAAm) is a commonly used thermoresponsive polymer, and copolymers of NIPAAm and acrylamide form hydrogel materials with LCSTs ranging from 32–60°C, depending on the composition of the copolymer [29]. Pulsatile delivery of a variety of drugs has been demonstrated from NIPAAm-co-acrylamide hydrogels by subjecting the materials to changes in temperature [29, 47]. While very nice results have been achieved with this system, the practical implementation of the system has been difficult due to issues with inducing temperature changes in implanted materials. In order to develop more easily manipulated implants for modulated drug delivery, composites hydrogels formed from NIPAAm-co-acrylamide and near-infrared absorbing nanoshells have been developed [38]. The composites were fabricated by mixing nanoshells into the monomer mixture. After polymerization, the nanoshells were physically entrapped in the hydrogel matrix. As shown in Figure 9.6, the composite hydrogels undergo a pronounced collapse in response to near infrared light; the nanoshells absorb the light, generating heat within the composite to exceed the LCST of the copolymer.

The collapse of the hydrogel material is completely reversible. When the temperature falls below the LCST, the polymer chains extend and interact with water, causing the

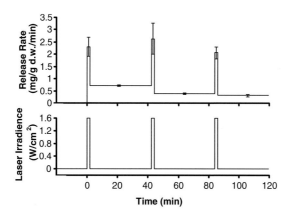

FIGURE 9.7. Pulsatile release of protein (top panel) from composite hydrogels was achieved through pulsatile near infrared irradiation (bottom panel).

material to swell. The reversible nature of this phenomenon allows one to partially collapse the material, re-swell it, cycling above and below the LCST repeatedly. If a drug has been incorporated into the hydrogel matrix, each time the hydrogel collapses, a burst of drug will be expelled from the material, as demonstrated in Figure 9.7 [38]. Because of the deep penetration of near infrared light through tissue, the composite hydrogels may be implanted subcutaneously, with the phase change behavior easily manipulated by externally applied light.

9.2.3. Photothermal Ablation

As discussed above for the drug delivery application, nanoshells can be designed to strongly absorb near infrared light and thus generate localized heating, potentially enabling nanoshell-mediated thermal ablation therapies for applications such as cancer treatment. Thermal ablation therapies can provide a minimally invasive alternative to surgical excision of tumors and are particularly attractive for situations where surgery is not possible. Thermal delivery methods under investigation for local tissue ablation include lasers [2, 44], microwave and radio frequency energy [8, 35], magnetic thermal ablation [11], and focused ultrasound [19]. The goal of thermal ablation is to conform a lethal dose of heat to a prescribed tissue volume with as little damage to intervening and surrounding normal tissue as possible, which has been difficult with the majority of techniques currently under investigation. Due to the lack of absorption of near infrared light by tissue components, use of this type of light source, with nanoshells at the desired tissue locations, should minimize collateral tissue damage. In vitro studies with nanoshells bound to breast carcinoma cells have demonstrated effective destruction of the cancerous cells upon exposure to near infrared light [12], with cell damage limited to the laser treatment spot (Figure 9.8).

The efficacy of nanoshell-mediated photothermal ablation has also been assessed in several in vivo studies. Initial studies involved directly injecting nanoshell suspensions into tumor sites and utilizing MRI thermal imaging to monitor temperature profiles during heating [12]. These studies demonstrated rapid heating of nanoshell-laden tissues upon exposure to the near infrared light. Evaluation of the gross pathology and histology demonstrated

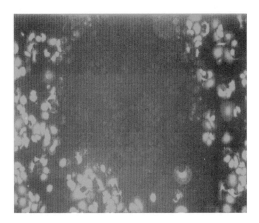

FIGURE 9.8. Breast carcinoma cells treated with nanoshells and near infrared light (821 nm). Cells were stained with the fluorescent viability stain, calcein AM. All cells within the circular laser spot were destroyed. The region of cell death corresponds with the diameter of the laser.

marked tissue damage at the treatment sites, with little or no damage to surrounding tissue. This initial work also provided information about the relationships between nanoshell dosages, light intensity, and duration of illumination with the ultimate thermal profile and resultant tissue damage. However, for the majority of applications, direct injection into the tumor site may not be feasible.

An alternative approach is to inject nanoshells intravenously, allowing them to circulate and accumulate at the tumor site before near infrared treatment. The size of nanoshells is critical to the success of this type of approach. Substantial prior research has investigated the delivery of macromolecules and small particles through the tumor vasculature. These efforts have demonstrated that particles in the 60–400 nm size range will extravasate and accumulate in many tumor types via a passive mechanism referred to as the "enhanced permeability and retention" (EPR) effect [23]. This effect has been attributed to the highly proliferative vasculature within neoplastic tumors. During rapid angiogenesis, defects in the vascular architecture are often present, resulting in leaky vessels. Nanoshells fall within the range applicable for the EPR effect, and thus should accumulate in most tumor types following intravenous injection. The efficacy of photothermal ablation following systemic delivery of nanoshells has been evaluated in a mouse model [27]. Complete regression of tumors was observed following treatment with nanoshells and near infrared light, with no tumor re-growth over at least 60 days. Survival times for mice with the nanoshell treatment in this study were significantly improved compared to untreated mice or those receiving laser treatment alone. Additionally, it is possible to conjugate nanoshells to antibodies to oncoproteins or endothelial markers, which may improve the accumulation of nanoshells in the tumor tissue and further localize nanoshells to targeted cells at the treatment site.

9.2.4. Nanoshells for Molecular Imaging

The drug delivery and thermal ablation applications described in the preceding sections used nanoshells designed to strongly absorb light in the NIR spectral region. By fabricating nanoshells designed to preferentially scatter rather than absorb light, nanoshells can serve as strong optical contrast agents for a variety of biomedical optical imaging applications. Photonics-based imaging technologies offer the potential for non-invasive, high-resolution in vivo imaging at competitive costs. However, the clinical utility of optical imaging strategies has been significantly constrained both by the limited variety of endogenous chromophores present in tissue and by relatively low levels of optical contrast between normal and diseased tissue. Furthermore, in the case of cancer, where early detection is critical to reducing morbidity and mortality, it is often desirable to image specific molecular biomarkers which are present long before pathologic changes occur at the anatomic level. Imaging biomarkers of interest requires development of targetable optical contrast agents. A recent demonstration of scattering-based molecular imaging used gold colloid conjugated to antibodies to the epidermal growth factor receptor (EGFR) as an optical contrast agent for imaging early cervical precancers [40]. While gold colloid bioconjugates are valuable as contrast agents for detecting superficial epithelial cancers with visible light, a primary challenge in optical contrast agent development has been the need for optical contrast agents at the multiple laser wavelengths within in the NIR spectral region commonly used in optical imaging applications. The facile tunability of nanoshells facilitates their use as NIR contrast agents. In addition, nanoshells offer other advantages relative to conventional imaging agents including more favorable optical scattering properties, enhanced biocompatibility, and reduced susceptibility to chemical/thermal denaturation. Furthermore, as described earlier in this chapter, nanoshells are readily conjugated to antibodies or other targeting moieties of interest enabling molecular specific imaging.

Initial in vitro studies were conducted to demonstrate the potential of nanoshell bioconjugates for molecular imaging applications. These experiments used nanoshell designed to strongly scatter light throughout the NIR "optical window" of 700–1200 nm. To enable molecular targeting, antibodies were conjugated onto nanoshell surfaces using ortho-pyridyl-disulfide-polyethylene-glycol-n-hydroxy-succinimide (OPSS-PEG-NHS) as a linker. Following antibody conjugation, nanoshell surfaces were further modified with PEG-thiol in order to block non-specific adsorption sites. Cells incubated with scattering nanoshell bioconjugates were viewed under darkfield microscopy, a form of microscopy sensitive only to scattered light. Significantly increased optical contrast due to expression of HER2, a clinically relevant cancer biomarker, was observed in HER2-positive breast carnicoma cells targeted with HER2-labeled nanoshells compared to the contrast observed in cells targeted by either nanoshells non-specifically labeled with IgG or control cells which were not exposed to nanoshell conjugates (Figure 9.9). Using a qualitative silver stain capable of detecting the presence of gold on cell surfaces, greater staining intensity was seen in HER2-targeted cells providing additional evidence that the increased contrast seen under darkfield was specifically attributable to nanoshell targeting of the HER2 receptor.

Nanoshell-based molecular contrast agents offer advantages including tunability, size flexibility, and systematic control of optical scattering and absorption properties. While darkfield microscopy is appropriate for in vitro imaging applications, use of nanoshell conjugates in vivo will require more sophisticated imaging techniques. Work currently underway is assessing nanoshell contrast agents in vivo in animal models using scattering-based optical systems including reflectance confocal microscopy and optical coherence microscopy. Furthermore, the high level of control over nanoshell properties achievable through systematic manipulation of design parameters suggests the potential for biomedical

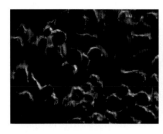  

FIGURE 9.9. Molecular imaging in living cells using nanoshell bioconjugates. Scattering contrast is evident in HER2+ breast carcinoma cells incubated with HER2 nanoshell bioconjugates (left). Scattering signal is not present when the cells are exposed to nanoshells conjugated to a non-specific Ab (middle) or in control cells not exposed to nanoshells (right).

applications requiring more complex functionalities including integrated imaging and therapy of cancer.

REFERENCES

[1] A.L. Aden and M. Kerker. *J. Appl. Phys.*, 22:1242–1246, 1951.

[2] Z. Amin, W. Thurrell, G.M. Spencer, S.A. Harries,W.E. Grant, S.G. Bown, and W.R. Lees. *Invest. Radiol.*, 28:1148–1154, 1993.

[3] R.D. Averitt, D. Sarkar, and N.J. Halas. *Phys. Rev. Lett.*, 78:4217–4220, 1997.

[4] R.D. Averitt, S.L. Westcott, and N.J. Halas. *J. Opt. Soc. Am. B*, 16:1824–1832, 1999.

[5] C.F. Bohren and D.R. Huffman. *A bsorption and Scattering of Light by Small Particles.* Wiley-Interscience, New York, pp. 379, 1983.

[6] K. Eichner. *Int. Dent. J.*, 33:1–10, 1983.

[7] J.G. Fujimoto, M.E. Brezinski, B.J. Tearney, S.A. Boppart, B. Bouma, M.R. Hee, J.F. Southern, and E.A. Swanson. *Nat. Med.*, 1:970–972, 1995.

[8] G.S.Gazelle, S.N. Goldberg, L. Solbiati, and T. Livraghi. *Radiology*, 217:633–646, 2000.

[9] H. Hagman. *Dent. Lab. Rev.*, 54:28–30, 1979.

[10] G.D. Hale, J.B. Jackson, T.R. Lee, and N.J. Halas. *Appl. Phys. Lett.*, 78:1502–1504, 2000.

[11] I. Hilger, R. Hiergeist, R. Hergt, K. Winnefeld, H. Schubert,and W.A. Kaiser. *Invest. Radiol.*, 37:580–586, 2002.

[12] L.R. Hirsch, J.B. Jackson, A. Lee, N.J. Halas, and J.L. West. *Analyt. Chem.*, 75:2377–2381, 2003.

[13] L.R. Hirsch, R.J. Stafford, J.A. Bankson, S.R. Sershen, B. Rivera, R.E. Price, J.D. Hazle, N.J. Halas,, and J.L. West. *Proc. Natl. Acad. Sci. U.S.A.*, 100:13549–13554, 2003.

[14] M. Ho, M.J. Warrell, D.A. Warrel, D. Bidwell, and A. Voller. *Toxicon*, 24:211–221, 1986.

[15] N. Holstrum, P. Nilsson, J. Carlsten, and S. Bowland. *Biosens. Bioelectron.*, 13:1287–1295, 1998.

[16] M. Horisberger and J. Rosset. *J. Histochem. Cytochem.*, 25:295–305, 1977.

[17] J.V. Houten, D. Benaron, S. Spilman, and D.K. Stevenson. *Ped. Res.*, 39:470–476, 1996.

[18] D. Huang, E.A. Swanson, C.P. Lin, J.S. Schuman,W.G. Stinson,W. Chang, M.R. Hee, T. Flotte, K. Gregory, C.A. Puliafito, and J.G. Fujimoto. *Science*, 254:1178–1181, 1991.

[19] F.A. Jolesz and K. Hynynen. *Cancer J.*, S1:100–112, 2002.

[20] B.M. Kapur. *Bull. Narcot.*, 45:116–154, 1993.

[21] E. Kuun, M. Brashaw, and A.d.P. Heyns. *Vox Sanguinis*, 72:11–15, 1997.

[22] B. Lindholm-Sethson, J.C. Gonzalez, and G. Puu. *Langmuir*, 14:6705–6708, 1998.

[23] H. Maeda, T. Sawa, and T. Konno, *J. Control. Rel.*, 74:46–61, 2001.

[24] R.G. Nuzzo, F.A. Fusco, and D.L. Allara. *J. Am. Chem. Soc.*, 109:2358–2368, 1987.

[25] S.J. Oldenberg, R.D. Averitt, S.L. Westcott, and N.J. Halas. *Chem. Phys. Lett.*, 28:243–247, 1998.

[26] S.J. Oldenberg, S.L. Westcott, R.D. Averitt, and N.J. Halas. *J. Chem. Phys.*, 111:4729–4735, 1999.

[27] D.P. O'Neal, L.R. Hirsch, N.J. Halas, J.D. Payne, J.L. West. *Cancer Lett.*, (In Press), 2004.
[28] M. Pourbaix. *Biomaterials*, 5:122–134, 1984.
[29] J.H. Priest, S.L. Murray, R.J. Nelson, and A.S. Hoffman. *Revers. Polym. Gels Related Syst.*, 350:255–264, 1987.
[30] E. Prodan, C. Radloff, N. Halas, and P. Nordlander. *Science*, 302:419–422, 2003.
[31] M. Quinten. *Appl. Phys. B*, 73:317–326, 2001.
[32] M. Rajadhyaksha, S. González, J.M. Zavislan, R.R. Anderson, and R.H. Webb. *J. Investig. Dermatol.*, 113:203–303, 1999.
[33] C. Ruan, R. Yang, X. Chen, and J. Deng. *J. Electroanalyt. Chem.*, 455:121–125, 1998.
[34] S. Sasaki, H. Kawasaki, and H. Maeda. *Macromolecules*, 30:1847–1848.
[35] T. Seki, M. Wakabayashi, T. Nakagawa, M. Imamura, T. Tamai, A. Nishimura, N. Yamashiki, A. Okamura, and K. Inoue. *Cancer*, 85:1694–1702, 1999.
[36] S. Sershen and J. West. *Adv. Drug Del. Rev.*, 54:1225–1235, 2002.
[37] S.R. Sershen, J.L. Westcott, J.L. West, and N.J. Halas. *Appl. Phys. B*, 73:379–381, 2001.
[38] S.R. Sershen, S.L. Westcott, N.J. Halas, and J.L. West, *J. Biomed. Mater. Res.*, 51:293–298, 2000.
[39] S.R.Sershen, S.L. Westcott, N.J. Halas, and J.L. West, *Appl. Phys. Lett.*, 80:4609–4611, 2002.
[40] K. Sokolov, M. Follen, I. Pavolva, A. Malpica, R. Lotan, and R. Richards-Kortum. *Cancer Res.*, 63:1999–2004, 2003.
[41] W. Stober and A. Fink. *J. Colloid. Interf. Sci.*, 26:62–69, 1968.
[42] N. Tanaka, S. Matsukawa, H. Kuroso, and I. Ando. *Polymer*, 39:4703–4706, 1998.
[43] A. Vogel. *Phys. Med. Biol.*, 42:895–912, 1997.
[44] T.J. Vogl, M.G. Mack, R. Straub, K. Engelmann, S. Zangos, and K. Eichler. *Radiologe*, 39:764–771, 1999.
[45] R. Weissleder. *Nat. Biotech.*, 19:316–317, 2001.
[46] A. Welch and M. van Gemert (eds.). *Optical-Thermal Response of Laser-Irradiated Tissue*, Plenum Press, New York, 1995.
[47] R. Yoshida, K. Sakai, T. Okano, and Y. Sakurai. *J. Biomat. Sci. Polym. Ed.*, 6:585–598, 1994.
[48] H.S. Zhou, I. Honma, and H. Komiyama. *Phys. Rev. B*, 50:12052–12056, 1994

10

Nanoporous Microsystems for Islet Cell Replacement

Tejal A. Desai[1], Teri West[2], Michael Cohen[2], Tony Boiarski[2], and Arfaan Rampersaud[2]

[1]*Department of Bioengineering and Physiology, University of California, San Francisco, CA*
[2]*IMEDD Inc., Columbus, OH 43212*

10.1. INTRODUCTION

10.1.1. The Science of Miniaturization (MEMS and BioMEMS)

Micro-Electro-Mechanical Systems technology, commonly known with the acronym MEMS, refers to the fabrication of devices with dimensions on the micrometer scale. For comparison, a human hair is about 80 μm in diameter. The most essential elements of MEMS consist of miniaturized, highly precise, and repeatable structures that can be stationary or moving. These structures are created via fabrication processes and equipment developed for the integrated circuit (IC) industry. The fabrication of MEMS commonly involves bulk or surface machining. Bulk machining defines microstructures by etching directly into the bulk material such as single crystal silicon. The advantage of bulk machining is that it allows the integration of active devices and the use of integrated circuit technology. Surface machining defines the release and movable structure in a polysilicon film or sacrificial layer of silicon dioxide, both deposited on bulk silicon. More complex microchips including multilayer interconnections can be obtained by bonding together and laser drilling of several layers of the components. Typically, microfabrication has a limit of resolution on the order of microns. However, specialized techniques can be used to create features in the nanoscale, as it will be shown later in the fabrication process of the nanoporous membrane. More importantly, the incorporation of new materials and the range of processes now extend far beyond just those found in the IC industry.

Medicine and biology are among the most promising, although most challenging, fields of application for MEMS. This does not come as a surprise considering that the technology has the capability to fabricate minimally invasive yet highly functional microdevices that match the size range of many structures found in the human body. Examples include pressure sensors that are small enough to fit through 1 mm catheters, but are priced to be disposable, and pacemakers that have incorporated microscale accelerometers to pace the heart in proportion to the patient activity. These tiny devices, also referred to as biomedical microsystems, hold promise for precision surgery with micrometer control, rapid screening of common diseases and genetic predispositions, and autonomous therapeutic management of allergies, pain, and neurodegenerative diseases. The health care implications predicted by successful development of this technology are enormous, including early identification of disease and risk conditions, less trauma and shorter recovery times, and more accessible health care delivery at a lower total cost.

The development of new, affordable, disposable analytic microchips are changing diagnostics. Examples of analytical functions that are benefiting from such developments include blood supply screening, analysis of biopsy samples and body fluids, minimally invasive and noninvasive diagnostic procedures, rapid identification of disease, and early screening. These systems will eventually perform diagnostic procedures in a multiplexed format that incorporates multiple complementary methods. Ultimately these systems will be combined with other devices to create completely integrated analysis and treatment systems.

New drug delivery methods seek to develop tool capable of delivering precise quantities of a drug at the right time and as close as possible to the treatment site. Sustained drug delivery provides the same medicinal effect with higher efficiency, longer duration and less side effects than traditional delivery methods. There are a number of mechanisms to provide timed release of drugs, such as microencapsulation, transdermal patches, and implants. Transdermal release is an attractive alternative for formulations which cannot be effectively delivered using pills and injection because of limitations related to gastrointestinal drug degradation and the inconvenience and pain related to intramuscular and intravenous injections. Implantable devices are preferred for therapies that require several daily injections, such as for diabetes treatment. If the drug level is monitored in real-time, it could also be adapted to metabolic variations. In treatments like chemotherapy, the device can be implanted where the drug is most needed. The current state-of-the art includes systems that are approximately the size of a hockey puck, have a limited battery lifetime of 3–7 years, and rely on the use of power-consumptive electromagnetic dispensing of fixed amounts of medication at programmed intervals, regardless of body need. Among these techniques, implantable pumps have the advantage that the drug therapy can be delivered at the optimal time and concentration to a specific site. MEMS systems combine miniature size, which is amenable to implantability, low power requirements, and the potential to precisely meter fluid samples.

10.1.2. Cellular Delivery and Encapsulation

The immunoisolation of transplanted cells and tissue by size-based semipermeable membranes has emerged as an extremely promising method of treating hormone deficiencies arising from such diseases as Type I diabetes, Alzheimer's, and hemophilia [1–5]. It has been demonstrated that cellular transplants, such as isolated pancreatic islets of Langerhans or hepatocytes, respond physiologically both in vitro and in vivo by secreting bioactive substances in response to appropriate stimuli, as long as they are immunoprotected. However, with the exception of autologous cells and tissue, overcoming immunologic rejection of the transplanted cells is still the greatest obstacle. Although approaches involving polymeric microcapsules have yielded promising results for autologous and allogeneic cell transplantation without immunosuppression [15–18], few approaches have been effective for non-immunosuppressed xenogeneic cell encapsulation [19–23], due to mechanical rupture of the membrane, biochemical instability, compatibility with islet cell heterogeneity, and broad pore size distributions [1, 5, 24–26]. While Duvivier et al. showed that size based exclusion is not critical for auto or allografts [64], several groups have found that the protection of porcine islets and xenografts, in general, will be more difficult to protect than allografts, as suggested by studies performed with permeable polymer membranes [30,69]. Their studies also confirmed that cytokines can cross polymeric alginate membranes and damage the islet cells contained inside the capsules. This may not be an issue for allografts but will certainly play a role in xenotransplantation as discussed below.

Immunoisolation of xenogeneic cells requires stringent biological and physical criteria to be met. The immunological response to a xenograft has various features that distinguish it from the alloresponse. The first difference depends on the species crossed, in that the recipient may have natural cytotoxic antibodies against the xenoantigens in the donor tissue. In addition, the major method of recognition of xenoantigens is via the indirect pathway, in which xenoantigen is processed and presented by the host antigen-presenting cells to T_h cells. The reaction to xenogeneic cells is very complex and involves not only cells and antibodies, but also complement and a host of cytokines such as tumor necrosis factor, which can inflict cell damage. Thus, the clinical success of encapsulated islet transplantation is still minimal, with less than 30 documented cases of insulin independence occurring from over 250 attempts at clinical islet allo-transplantation since 1983 and no *clinically* successful cases of xenogeneic islet encapsulation without immunosuppression [27–29]. In light of this, it is critical to turn to new capsule materials, designs, and fabrication approaches, focusing on such fundamental issues as the immunoisolation membrane material, membrane pore parameters, and cellular arrangement within the device. Only then can we begin to develop cell encapsulation devices with long-term therapeutic efficacy and safety.

The immunoisolation membrane should allow permeability of glucose, insulin, oxygen and other metabolically active products, to insure islet functionality and therapeutic effectiveness. It must also prevent the passage of cytotoxic cells, macrophages, antibodies and complement to remain effective. Previous studies indicated that immunoisolation could be attained if C1q and IgG were completely retained, such as the case with membranes having pore diameters between 30 to 50 nm [1]. In fact, Hirotani and Ohgawara measured complement permeability for a membrane with pore size 0.1 to 0.2 microns and reported that complement C3 became inactivated upon the passage through the membrane. Thus, membrane thickness may also be a critical parameter. More recently, it has been elucidated that true immunoisolation also requires the blockage of cytokines such as TNF and cell-secreted antigens. For example, Brauker et al. [33] found that xenografts (CF1 mouse embryonic lung implanted into Lewis rats) were destroyed within 3 weeks even when implanted in devices with intact membranes. The death of the xenogeneic tissues was accompanied by a severe local accumulation of inflammatory cells and a decrease in local vascularization. Studies by Kulseng et al. [7] have also suggested the possibility that xenografts are killed by local accumulation of inflammatory cells, mediated by the release of antigens from encapsulated cells which bind nonspecifically to the outer capsule surface and trigger complement activation and a subsequent local inflammatory response. Weber and colleagues [31–35] also presented evidence that leakage of antigens out of the capsule incited macrophage activation and cytokine release resulting in pericapsular inflammation and fibrosis and cytokine-mediated islet destruction.

Based on these studies, it seems that successful immunoisolation will require membranes that not only provide protection of the encapsulated tissues from the host immune system but also have properties that diminish the release of xenogeneic antigens through reduced pore size and surface properties that prevent nonspecific adsorption. It is thought that current membrane technology has reached a plateau and molecular weight cut-offs cannot be made tighter without compromising the health of the implants [66]. Although it may be technically possible to eliminate the escape of shed antigens, such a diminished pore size would most probably limit the diffusion of essential nutrients to the encapsulated cells. Some have suggested that we must combine new membrane immunoisolation technologies (having tighter cut-offs) with the immunomodulation of cell sources themselves [67]. In addition, we believe that the combination of new membrane architectures with surface modification strategies that resist non-specific adsorption will also be effective.

Current polymeric biocapsules have not been able to achieve uniform pore size membranes in the tens of nanometer range. These membranes, due to their polymeric nature, have found that meeting cut-off requirements is quite difficult, due to the broad pore size distribution of real membranes. Even if only 1% of pores were larger than the cut-off goal, the pores would allow the passage of antibodies, complement, and cytokines in sufficient amounts to initiate immunorejection pathways [1]. The immune system is quite specific and as such, a membrane that attempts to circumvent the immune system should be specific and precise as well. Moreover, effective methods for encapsulation are needed which do not damage cells during the process. For example, processes involving vacuum, high shear, organic solvent, or free radicals are not compatible with cell viability. Processes that produce less than complete coverage of the cells will result in host immune sensitization by cellular means followed by humoral destruction.

10.1.3. Microfabricated Nanoporous Biocapsule

The silicon-based nanoporous biocapsule is achieved by applying fabrication techniques originally developed for Micro Electro Mechanical Systems (MEMS), and represents one of first therapeutic applications of micro and nanotechnology in biomedicine (bioMEMS) [36–41]. Utilizing bulk and surface micromachining and microfabrication, membrane-based biocapsules can be engineered to have uniform and well-controlled pore sizes, channel lengths, and surface properties [42]. We have developed several variants of microfabricated diffusion barriers, containing pores with uniform dimensions as small as 7 nanometers [43].

The control of pore size down to tens of nanometers, coupled with modification of the outer biocapsule wall to prevent protein binding may be able to overcome some of these immunological challenges and at least hinder the passage of small cytokines and cell-secreted antigens. This control over membrane parameters has been suggested by other groups as being the only way to achieve immunoisolation [44–45]. Due to the fact that we can control the geometry and length of the diffusion path (straight vs. L-shaped), it is thought that we can have greater control over the diffusion rate of small molecules. Furthermore, improved dynamic response of islets tissue can be obtained due to the reduced membrane thickness (6–9 μm) of microfabricated membranes compared to polymeric membranes (100–200 μm) [1]. It is important to retain rapid intrinsic secretion kinetics, in particular first phase insulin release, so as to provide physiological feedback control of blood glucose concentrations. Moreover, we have shown that our nanoporous membranes display little non-specific protein/peptide adsorption or fibrosis, which may be critical for limiting the local inflammation due to a cell-secreted antigen response [54].

Another important consideration is the ability to insert islets into the device in situ. By incorporating a refilling/recharging port into the design, we have the unique ability to introduce islet after the devices have been implanted and after neovascularization has taken place. Moreover, the ability to remove cells after implantation allows for recharging of the device if new cells are needed long term. This is also a critical aspect in the safety of our device, in that cells can be added/removed in the case of too much or too little insulin output.

In addition, the design of the microfabricated membrane may be advantageous due to the ability to create a bi-level pore structure. The outer openings can be 2 by 2 microns while the inner diffusion channels can be designed to have a minimum dimension of tens of nanometers. Studies by Brauker et al. [46] revealed that neovascularization at the membrane-tissue interface occurred in membranes that had pore sizes large enough to allow complete penetration by host cells (0.8–8 microns pore size). When the vascularization of the membrane-tissue interface of 5-microns-pore-size polytetrafluoroethylene (PTFE) membranes was compared to 0.02-microns- pore-size PTFE membranes, it was found that the larger pore membranes had 80–100-fold more vascular structures. The increased vascularization was observed even though the larger pore membrane was laminated to a smaller pore inner membrane to prevent cell entry into the prototype immunoisolation device. We have observed similar phenomena with our microfabricated membranes.

Compared to alginate systems, micro and nanofabrication technology offers control of pore dimensions, short distance through pores, durable materials, and is refillable and removable. Compared to existing membrane implants, microfabricated nanoporous membranes can allow for precise control of pore dimensions (small and uniform) and the potential for active diffusion. Moreover, the ability to integrate other "smart" capabilities such as multicompartmental structure; local release of immunosuppressive drugs; biosensor incorporation, self-cleaning capabilities; and modulation of angiogenesis via surface architecture or immobilized growth factors, is extremely attractive.

10.2. FABRICATION OF NANOPOROUS MEMBRANES

The micromachined immunoisolation biocapsule concept started with the development of a membrane with highly defined pores into a structure that would allow the microencapsulation of cells for immunoisolation. The basic technology that was developed for the nanopores themselves was the use of a sacrificial oxide sandwiched between silicon layers, thus defining a space that could be opened by a subsequent etching of the oxide in

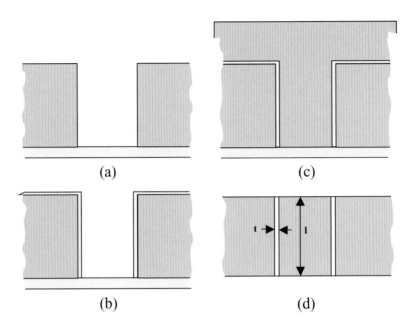

FIGURE 10.1. Process flow diagram for creating nanochannels using microfabrication technology.

HF. To make the complete immunoisolation capsule, the silicon substrate is etched up to the membrane, leaving a cavity in the wafer with an encapsulating immunoisolation membrane [11]. The design has been refined with the help of iMEDD, Inc. for the past several years to optimize and scale up the production of the nanoporous biocapsules. The biocapsule, in its current state, consists of a refillable housing with the desired cells contained within the housing. In general, photolithography is not amenable to the fabrication of pores with dimensions smaller than 0.35–1 micrometer. To reach a desired pore size in the tens of nanometers range, we have developed strategies based on the use of a sacrificial oxide layer, sandwiched between two structural layers, for the definition of the pore pathways [43]. This strategy encompasses a multitude of viable embodiments for the biocapsules [36, 43].

The overall process is shown schematically in Figure 10.1 (Desai et al., 2000a). The first step is the etching of the support ridge structure into the substrate (not shown in the figure). These ridges provide mechanical rigidity to the final membrane structure. A low stress silicon nitride layer (nitride), which functions as an etch-stop, is then deposited. A polysilicon film, that acts as the base structural layer (base layer), was deposited on top of the etch-stop layer. Because the etch-stop layer is very thin, the structural layer gets deposited down into the support ridge, which will remain after the membrane is released and the etch stop layer is removed.

The next etch step is the etching of holes in the base layer (10.1a), which defines the overall shape of the pores. The holes are etched through the polysilicon by chlorine plasma, with a thermally grown oxide layer used as a mask. After the pore holes are defined and etched through the base layer, the pore sacrificial oxide is grown on the base layer (10.1b). The sacrificial oxide thickness determines the pore size in the final membrane, so control

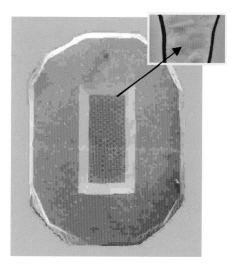

FIGURE 10.2. Top view of nanoporous membrane with close-up of pores.

of this step is critical to reproducible pores in the membranes. The necessary control is accomplished by the thermal oxidation of the silicon in dry oxygen. Thermal oxidation of polysilicon allows the control of the sacrificial layer thickness of less than 0.5 nm across the entire wafer. Limitations on this control can come from local inhomogeneities in the polysilicon, such as the initial thickness of the native oxide (especially for polysilicon), the grain size or density, and the impurity concentrations. To mechanically connect the base polysilicon with the plug polysilicon, which is necessary to maintain the pore spacing between layers, anchor points are defined in the sacrificial oxide layer. After the anchor points are etched through the sacrificial oxide, the plug polysilicon is deposited to fill in the holes (10.1c). The plug layer is then planarized down to the base layer, leaving the final structure with the plug layer only in the base layer openings. Chemical mechanical polishing is used to planarize the polysilicon, leaving a very smooth surface with the pores exposed.

A protective nitride layer is then deposited on the wafer (completely covering both sides of the wafer). This layer is completely impervious to the KOH chemical etch used to release the membranes from the bulk silicon. The backside etch windows are etched in the protective nitride, exposing the silicon wafer in the desired areas, and the wafer is placed in an 80°C KOH bath to etch. After the silicon is completely removed up to the membrane (as evidenced by the smooth buried etch stop layer), the protective, sacrificial, and etch stop layers are removed by etching in HF (10.1d).

The structure and pore size of these membranes were verified via scanning electron microscopy (figure 10.2). The dimension of the membrane, including the support ridge, are 6×8 mm. The active area measures 3.5×2 mm and has a thickness, or channel length, equal to 5 μm. The overall porous surface is equal to 7 mm^2. Pores can be annular (C shaped) or linear, organized in parallel arrays along the membrane major dimension, separated by 5 μm long anchor points. Regardless of the pore width, the length is fixed at 45 μm and there are 10,000 pores per mm^2. Hence, the total pore area increases linearly with the pore size.

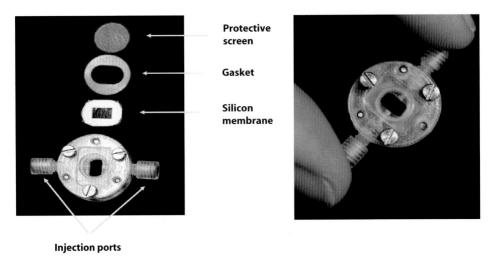

FIGURE 10.3. Schematic of Biocapsule Assembly.

10.3. BIOCAPSULE ASSEMBLY AND LOADING

A schematic of the biocapsule assembly is shown in figure 10.3. The biocapsule is assembled with latex gaskets between the membrane and plastic body. To eliminate the introduction of air bubbles, assembly is done underwater in a glass 125 mm × 75 mm dish. It is important to loosely assemble the biocapsule to allow for expansion of plastic parts during sterilization. The submerged biocapsule is sterilized in the dish by autoclaving at 121°C for 15 minutes. The biocapsule is allowed to cool in a tissue culture hood, and a sterile filling tube will be attached to one port. Approximately 40 ul of islet/alginate/complete

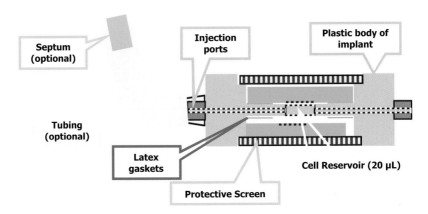

FIGURE 10.4. Components of biocapsule including refilling ports used to load cells and recharge system.

medium media is carefully delivered into the capsule through a tuberculin syringe attached to the filling tube. Then the tube and syringe will be removed and the ports are covered with caps (figure 10.4). The alginate is allowed to polymerize overnight in complete media containing $CaCl_2$ (37°C in humidified 5% CO_2). The following day the filled capsule is evaluated in vitro or implanted.

One concern that is often brought up in related to encapsulation technologies has to do with the escape of very small antigenic peptides or the entrance of small inflammatory cytokines from the device, thus leading to an inflammatory response despite protection of the cells from larger antibodies and complement molecules. This issue has been of great interest in the encapsulation community since it brings up serious concerns about the potential value of any immunobarrier regardless of its molecular weight cut-of. The requirements of a membrane for immune protection are related to its diffusive permeability and will be primarily dependent on the donor-host mismatch: allogeneic vs. xenogeneic [61,30]. The indirect pathway, where shed antigens from the graft stimulate system CD4 T-cells, is secondary and less significant for allografts [33]. In either case, however, it is the goal to minimize the contact between the host's immune cells and any component of the encapsulated cells, because any activation of the local immune cell population may result in the local production of inflammatory cytotoxic agents such as IL-1β, IFNγ, and TNFα. Therefore, we believe that the issue of cell-secreted antigens and cytokines needs to be further investigated in the context of our immunoprotective membranes. We have already used surface modification strategies for our biocapsule which involve chemical vapor deposition of polyethylene glycol to diminish non-specific binding [65]. These membranes have shown to have significantly decreased peptide adsorption and fibroblast adhesion, and may be advantageous in terms of limiting the inflammatory response due to non-specific adhesion of cell-secreted antigens.

10.4. BIOCOMPATIBILITY OF NANOPOROUS MEMBRANES AND BIOCAPSULAR ENVIRONMENT

Early studies showed that silicon microdevices were biocompatible in vitro and in vivo [55]. For these microimplants, there appeared to be no changes in the mechanical properties of the implants and no corrosion was observed. The filtration channels appeared clear and free from any obstructions. No gross abnormalities of color or consistency were observed in the tissue surrounding the implant. No necrosis, calcification, tumorgenesis, or infection was observed at any of the implant sites, suggesting that silicon substrates were well-tolerated and non-toxic both in vitro and in vivo, leading to our further studies on cell encapsulation within biocapsules.

The behavior of different cell types in three-dimensional silicon microstructures was studied using microfabricated half-capsules [10]. All cells had normal growth characteristics, morphology, and greater than 90% viability. Overall, islets in microfabricated silicon pockets and the control dishes appeared to have similar morphology and viability. Glucose-supplemented medium was allowed to diffuse to the islets, from underneath the membrane, to stimulate insulin production and monitor cell functionality. The concentration of insulin, secreted by the islets through the membrane, into the surrounding medium was compared

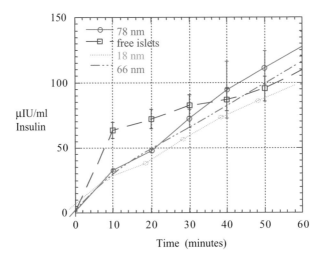

FIGURE 10.5. Insulin secretory profile through differing pore sizes.

between the unencapsulated islets and the islets on micromachined membranes [11, 14]. The amounts were similar in concentration and time release suggesting that glucose was able to sufficiently pass through the pores of the wafer pockets to stimulate islets for insulin production. Figures 10.5 and 10.6 show the typical insulin release profile in response to stimulatory (16.7 mM) glucose medium over 1 hour under static incubation for 78 nm, 66 nm, and 18 nm pore-sized membranes and in the presence of an alginate matrix within the biocapsule. This profile indicated that insulin and glucose diffusion occurred at sufficiently high rates through the microfabricated membrane to ensure nutrient exchange for encapsulated islet cells. These experiments show that no diffusion barrier is formed by the membrane for glucose and insulin, while taking into account the effect of rotation on mass transfer.

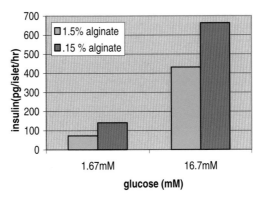

FIGURE 10.6. Glucose stimulation of rat islets in alginate filled biocapsules.

10.5. MICROFABRICATED BIOCAPSULE MEMBRANE DIFFUSION STUDIES

Since biocapsules are diffusion based devices, understanding the diffusion properties of the semipermeable membrane employed is critical to optimizing the passage of relevant biomolecules, and to tailoring immunoisolation capabilities. Commercial polymeric membranes utilized for cell encapsulation purposes have been poorly characterized in terms of large and small molecule diffusion properties. Indeed, it is often complicated to determine the diffusive transport properties of these membranes due to their variability and asymmetry. Typically, they are based on the nominal molecular weight cut off of the membrane, however, the non-uniformity in membrane architectures makes it difficult to define the absolute cut-off dimensions as well as the channel length, pore sizes and distribution. Conversely, microfabrication technology allows to precisely control pore size, pore distribution and diffusion path length. Therefore, due to the uniformity and known geometry of the micromachined membrane, it is much more straightforward to determine important diffusion parameters. Diffusion studies described here focused on the parallel pore design and different pore sizes were taken into consideration. In this case, immunoisolative properties were characterized in addition to glucose and albumin diffusion.

The mass transport properties of an encapsulation membrane are critical since the influx and outflux of relevant molecules will determine the extent of encapsulated cell viability. At the same time, the membrane must be able to provide considerably greater impedance to the diffusive transport of large molecular weight immunomolecules. In addition, when the device functions as an implantable homeostatic sensor-release system, it is fundamental for the entrapped cells to be able to respond promptly to fluctuations in solute concentrations of the interstitial fluid in order to retain a physiologic dynamic response.

Glucose, albumin, and IgG diffusion through parallel pore membranes with 7, 13, 20, and 49 nm was characterized. Because the initial concentrations in compartment A is significantly high, it is reasonable to assume that the donor cell acts as an infinite source and the concentration gradient between the two cells does not change significantly over the course of the experiment. Therefore, the flux across the membrane can be considered in steady state, and zero-order kinetics is expected from such system. Deff was calculated from concentrations experimentally measured. The diffusion coefficients were then normalized by the diffusion coefficient in water, calculated according to the Stokes-Einstein equation.

In the current fabrication protocol, the pore cross-sectional area is a linear function of the pore size, thus Aeff increases linearly with the pore size (Table 10.1). The measured effective diffusion coefficients were all lower than that for free diffusion in water (Table 10.2). For glucose and albumin, the relative diffusivity (Deff/DH2O) values increased with pore size,

TABLE 10.1. Pore size versus pore area.

Pore Size nm	Porosity %	Pore Area mm^2
7	0.30	0.0208
13	0.55	0.0387
20	0.85	0.0595
49	2.08	0.1458

TABLE 10.2. Measured effective diffusion coefficients.

Molecule	MW Da	Stokes Radius Nm	D_{H2O} cm²/s	D_{eff} 7 nm pore size	D_{eff} 13 nm pore size	D_{eff} 20 nm pore size	D_{eff} 49 nm pore size
Glucose Short-term	180	0.37	6.14E-06	6.34E-07	2.06E-06	3.20E-06	4.24E-06
Glucose Long-term	180	0.37	6.14E-06	N/A	1.98E-06	N/A	3.18E-06
Albumin	67,000	3.55	6.40E-07	2.10E-08	1.90E-07	3.09E-07	4.3E-07
IgG	150,000	5.90	3.85E-07	1.33E-10	2.19E-10	N/A	2.71E-09

nearing 1.0 for 49 nm pore sizes (Figure 10.7). As the nanopore channel width decreased, approaching several times the molecular dimensions of the solute molecules themselves, the rates of diffusion deviated appreciably from those predicted by Fick's first law. For example, the diffusion coefficient of glucose measured using a 7 nm nanopore membrane is substantially less than that seen with a 13 nm or 20 nm nanopore membrane (figure 10.8).

These data suggest that at channel widths below about 13 nm, the diffusion of glucose is constrained by the geometry of the nanopore channels, resulting in a slower than expected efflux through the nanopore membrane. A similar phenomenon was seen for a higher molecular weight solute, albumin (MW 67,000 daltons), except the channel widths at which non-Fickian diffusion is seen is in the range below 20 nm. Note that at channel widths below about 7 nm, virtually no efflux of albumin is seen. At a channel width above 20 nm, the measured diffusion constant is about equal to that predicted by Fick's law. The Stokes radius of albumin determined by gel exclusion chromatography is about 4 nm, suggesting that its molecular diameter is about 8 nm. The lack of albumin diffusion through the nanopore with the 7 nm membrane is understandable, as the channel thickness would physically exclude passage of the molecule.

These data suggest a relationship between the rate of diffusion of molecules through nanopore membranes and the size of the nanopore channels. On one extreme, as expected, if the channel width is below the molecular dimensions of the solute, virtually no diffusion is observed. On the other extreme, for channel widths that are approximately 2–5 fold greater

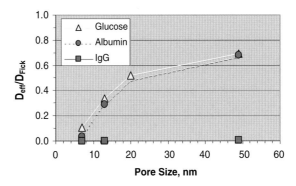

FIGURE 10.7. Effective diffusion versus molecular size and pore size of the membrane.

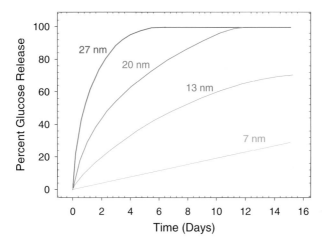

FIGURE 10.8. Diffusion profiles of glucose through Microfabricated membranes. At small pore sizes, molecules exhibit non-Fickian behavior.

in size than molecular dimensions of the solute, Fickian diffusion behavior is observed, that is, the geometry of the nanopore membranes does not constrain diffusion. For channel widths ranging from approximately the molecular dimensions of the solute to several- fold larger than this value, hindered diffusion is observed.

The origin of the constrained diffusion seen for solutes passing through nanopore membrane channels may be somewhat related to the phenomenon of single- file. Single-file diffusion is a one-dimensional transport process involving particles that cannot pass each other due to constrained pathways [56–59]. However, in our case, we actually have slit-like diffusion, since our channels are rectangular. Similar effects occur when a colloidal particle is confined in a small gap between two flat walls such as is the case for solute molecules diffusing through nanopore channels [60].

Another possible explanation for the constrained diffusion seen here are drag effects. When a colloidal particle suspended in a quiescent fluid is close to a flat wall, the Stokes drag force acting on it is increased relative to that when far from the wall and, therefore, its diffusion coefficient is smaller than that when far from the wall. The increase of the drag force is attributed to the alteration of the hydrodynamic interaction between the particle and the fluid generated by the boundary condition imposed by the nearby wall.

10.5.1. IgG *Diffusion*

Our studies have indicated that microfabricated biocapsule membranes can be tailor-made to attain desired IgG diffusion kinetics [14, 43]. At the same time, we have shown that the deselection of IgG requires absolute pore dimensions of 13 nm or less (Figure 10.9). This refines the previous understanding that pores in the range 30–50 nm would suffice to provide membrane-based immunoisolation [5]. We found that the percent of IgG diffusion (concentration of IgG that passes through the membrane) was 2% after over 150 hours through the 18 nm membranes and less than 0.001% in 13 nm membranes after 4 days. Compared to commonly used polymeric membranes, this rate was several times smaller

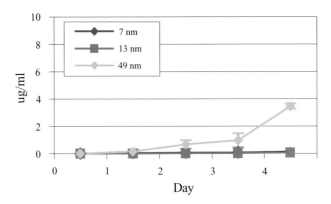

FIGURE 10.9. Diffusion of IgG through nanoporous biocapsule. Note that 7 and 13 nm membranes do not allow passage of IgG.

indicating superior immunoprotection. For example, Dionne et al. measured an IgG concentration of 1% after 24 hours through poly(acrylonitrile-co-vinyl chloride) membranes with a molecular weight cut-off of $\approx 80,000$ MW [61]. Similarly, Wang and colleagues (1997) investigated permeability of relevant immune molecules to sodium alginate/poly-L-lysine capsules and found that significant amounts of IgG (close to 40%) passed through both 230 kD and 110 kD membranes in 24 hours [4].

Immunoglobulin G molecules together with IgM are the most abundant immunomolecules involved in the humoral host response. Once they are bound to the grafted tissue, their interaction with the C1q component of the complement cascade activates a pathway which will lead to the destruction of the implanted cells. It is estimated that pore sized between 30 and 50 nm should be able to exclude IgG [1]. Although the effective Deff obtained for the microfabricated membrane are significantly lower than other membranes reported in literature total immunoisolation is difficult to achieve even at lower pore sizes. A few possibilities may explain this behavior. Proteins do not have a fixed conformation, rather they resonate among several energetically allowed ones. With this in mind, it needs to be considered that unlike albumin, immunoglobulins are not globular, but that they have a less compact, more flexible characteristic Y shape which may allow for greater conformational changes. Partial or total unfolding of the protein three-dimensional structure might also occur, in which case diffusion would be further eased. Therefore, a slow diffusion of IgG may be expected even for pore sizes below 20 nm. Small defects in the membranes cannot also be completely excluded. A change in the pores geometry, namely a decrease in their major lateral dimension, could prove useful in further reducing IgG effective diffusivity through the microfabricated membrane. However, Iwata et al. showed that complement components are rapidly inactivated and therefore it is enough to hinder IgG diffusion in the first days following implantation, rather than prevent it. In fact, they found that 7.5% agarose microbeads with IgG Deff in the order of 10–7 were effective in maintaining xenograft viability for over to one hundred days [8]. Similarly, Lanza et al. [5,6] reported that alginate microcapsules permeable to IgG with a MWCO in excess of 70 kDa can accept islet xenografts.

The microfabricated nanoporous membrane represents an exciting alternative to the traditional permselective membranes employed in cell encapsulation applications. With the advantage that microfabrication can offer in term of achieving uniform and well controlled pore width, length and configuration, the membrane can potentially be engineered and optimized to satisfy precise specifications, as required in size-based immunoisolating devices. Progress has been made in overall membrane design, and performance has been augmented with the parallel pore arrangement. Diffusivities for glucose and albumin have approached commercial polymeric membranes, while strict controlled over IgG permeation has been achieved.

10.6. MATRIX MATERIALS INSIDE THE BIOCAPSULE

Oxygen is known to be a critical factor that influences insulin secretion by islets. The diffusional limitations imposed by a biocapsule construct can potentially make it a hypoxic environment for the cells. In the attempt to partially overcome this constraint, cells were seeded in various extracellular matrix materials within the biocapsule. Their behavior was compared with free cells, which form aggregates. We showed that the addition of a matrix, particularly alginate or collagen, within the micromachined biocapsule proved to augment the functionality islets in microfabricated biocapsules [62] (Figure 10.10). Supporting our studies, Tatarkiewicz et al. also demonstrated that the use of alginate three-dimensional matrix allowed cell clusters to be cultured at least two times higher density compared with culture in suspension. The clusters immobilized in a matrix resulted with 3-fold increase in insulin content and 9-fold increase in insulin/DNA ratio [58]. This suggests the importance of optimizing cell density in within a matrix in the encapsulation device.

It is important to remember that that inclusion of matrix materials does promote a more homogeneous distribution of cells, but can also limit the density of cells that can be attained in the device. The importance of cell density in a practical encapsulation device is increasingly being recognized. It has been noted that we must better understand of the role of packing density on cell function and then incorporate those considerations into our devices. King et al. suggested that for planar devices, the goal should be a device less than 3 times the volume of the tissue inside and with the center of every islet less than 250 microns from the outer surface of the capsule. Due to the micron thickness of our membrane and an

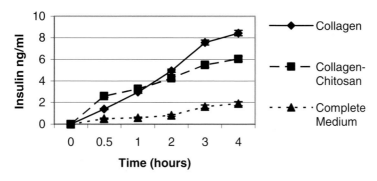

FIGURE 10.10. Insulin secretion for cells loaded in different matrix materials.

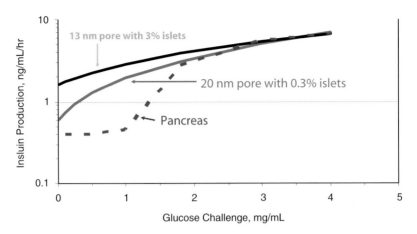

FIGURE 10.11. Model of insulin production for different loading conditions.

overall device thickness of less than 500 microns, we can achieve this. Moreover, because we use microfabrication approaches, the thicknesses of the device and length/width can be easily changed. Recently, we have conducted studies looking at viability and insulin secretion of cell populations of differing densities in our proposed device. As shown in figure 10.11, functionality of our devices can be controlled based on packing density and insulin levels are comparable to that seen from a normal pancreas. The insulin output for glucose stimulated rat islets should be somewhere around 2 ng/min/100 islets or about 120 pg/islet/hour. This is usually 10 times more than islets in low glucose. For a mouse implant, the number of islets that would fit into a 10 ul biocapsule would be 200 to 400 islets and would represent a packing density of 2–6%. This is reasonable achievable in our current designs.

In our experience, single islets in culture tend to develop a non-viable core, while islets in high density culture also show loss of viability and functionality. In fact, insulin secretory capacity may increase very little or even decrease as loading increases. As studied extensively by Colton and Colleagues, the packing density directly relates to the ability of the device to provide adequate oxygenation of the cells inside. Suzuki et al. performed several studies related to macrocapsule loading and islet density. After being mixed with a 1% alginate solution, a total of 250, 500, 750 or 1000 islets were loaded into the devices, which were implanted into the epididymal fat pad(s) of streptozocin diabetic mice [70]. The success rate for restoration of normoglycemia at week 4 was highest for the recipients receiving two devices, each with 500 islets. Loading 750 or 1000 islets provided no improvement over loading 500 islets in a single device. Devices containing 250 islets were rarely successful. In a related study, they found that islet cell volumes of .20 microliters restored normoglycemia in STZ-mice [71]. However, they also showed that islet necrosis in such devices was 10–15% within two weeks. These studies defined important limitations in the requirements for islet packing density in macroencapsulation. They also pointed to the fact that new approaches for improving islet packing density must be developed to make diffusion-dependent macroencapsulation more practical. Thus, we have focused on two aspects: choice of matrix materials and refilling capabilities after transplantation. It is

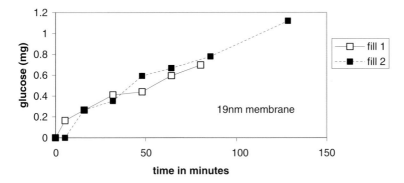

FIGURE 10.12. Glucose diffusion is similar before and after refilling the capsule in situ.

important to note that using our devices, it is possible to prevascularize prior to loading with islets. This could make a large difference in enhancing the packing density because early loss of tissue due to hypoxia might be reduced. In fact, we have shown that our biocapsules can be easily refilled with matrix containing cells in a reproducible manner as shown in figure 10.12.

10.6.1. In-Vivo Studies

We have conducted animal studies to determine the feasibility of implanting micro-fabricated biocapsules for cell delivery. These studies have served as a basis for the proposed studies described in this grant application. For these studies, 2 male Lewis rats (Harlan Sprague-Dawley Inc., Indianapolis, Indiana) weighing 200–230 g, were used in the first group, and 8 in the second. For Group 1, one animal received a total of 120 μl of 6.7 μg/ml mouse insulin solution equally aliquoted into three capsules with 49 nm pores. The second animal received insulinoma cells. The first animal was sacrificed at day 10 POD and the devices were retrieved. Capsulectomy was performed on the second animal at day 14 POD. The rat receiving pure insulin solution showed a decrease in blood glucose levels at day 1 and 2 POD. On day 3 and the following days, blood glucose levels were over 300 mg/dl. At day 10 POD the animal was sacrificed and the capsule retrieved for histology. The capsule wrapped into the omentum was intact, thus suggesting this as an optimal place to implant the device. Biocapsules released their insulin content over 24 hours, and in an amount insufficient to restore euglycemia. However, the animal responded to the treatment, even though for a short time. The second animal received encapsulated insulinoma cells. The rat successfully reverted diabetes and his blood glucose levels were in the normal range from day 1 POD (figure 10.13). On day 14, POD capsules were retrieved. Additional experiments performed have shown that microfabricated capsules can retain insulin stimulatory capacity of islet cells in vivo, given a pore size of less than 20 nm (Desai et al., 1999). Whereas 66 nm capsules led to loss of cell function in vivo, the 20 nm capsules maintained similar secretory output compared to in vitro levels after being implanted for two weeks. We obviously need to enlarge and extend these studies in vivo to validate our system.

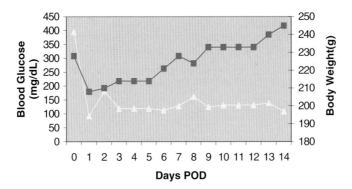

FIGURE 10.13. Non fasting blood glucose concentration and body weight in STZ-diabetic Lewis rat after bio-capsule implantation.

10.6.2. Histology

At a gross examination, the capsules seemed free of fibrotic tissue and clean. A rich network of blood vessels surrounded the microfabricated membrane in proximity of the diffusion area, minimizing possible limitations of glucose-insulin exchange due to the lack of a well developed vascular system surrounding the biocapsule (Figure 10.14). Microscopic analysis of tissue sampled from the biocapsule located in the omentum revealed no evidence of macrophages or lymphocytic infiltration. Small vessels characterized by a thin layer of elongated cells were also dispersed in the tissue, typical of the lining endothelium in capillaries. Several issues regarding biocapsule implantation and effectiveness were revealed in these experiments. Cells did maintain their viability within microfabricated biocapsules over the studied period. This demonstrated the feasibility of this approach. Just as important, the capsules were shown not to elicit a deleterious tissue response.

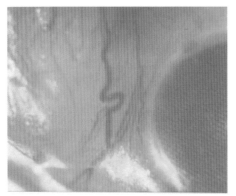 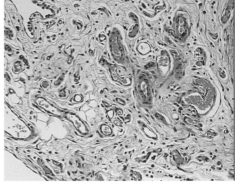

FIGURE 10.14. (left) Tissue surrounding capsule retrieved from the peritoneal cavity; (right) H–E stained tissue (×20). Several blood vessels are visible.

CONCLUSIONS

A method to create precise nanoporous biocapsules for cell encapsulation via microfabrication technology has been described. Membranes can be fabricated to present uniform and well-controlled pore sizes as small as 7 nm, tailored surface chemistries, and precise microarchitecture. These platforms can be interfaced with living cells to allow for biomolecular separation and immunoisolation. Ideally a membrane in contact with cells should be biocompatible and allow for the free exchange of nutrients, waste products, and secreted therapeutic proteins. Furthermore, where nutrients and time sensitive compounds are diffusing across a membrane it is highly desirable to be able to control the diffusion characteristic precisely in order to retain the dynamic response of seeded cells to external stimuli. Membranes were shown to be sufficiently permeable to support the viability of pancreatic islets and insulinoma. Applications of these nanoporous membranes range from cellular delivery to cell-based biosensing to in vitro cell-based assays.

ACKNOWLEDGEMENTS

Portions of this project were funded by NSF and The Whitaker Foundation. Special thanks to iMEDD, Inc. for their technical support.

REFERENCES

[1] C.K. Colton. Implantable hybrid artificial organs. *Cell Transplantat.*, 4(4):415–436, 1995.

[2] R.P. Lanza, J.L. Hayes, and W.L. Chick Encapsulated cell technology. *Nat. Biotechnol.*, 14(9):1107–1111, Sept. 1996.

[3] R.P. Lanza and W. Chick. Encapsulated cell therapy. *Sci. Am. Sci. Med.*, 2(4):16–25, 1995.

[4] T.Wang, I. Lacik, M. Brissova, A. Anilkumar, A. Prokop, D. Hundeler, R. Green, K. Shahrokhi, and A. Powers. An encapsulation system for the immunoisolation of pancreatic islets, *Nat. Biotechnol.*, 358–362, 15 April, 1997.

[5] R.P. Lanza et al. Xenotransplantation and cell therapy: progress and controversy. *Mol. Med. Today*, 5(3):105–106, Mar 1999.

[6] R.P. Lanza et al. Xenogenic humoral responses to islets transplanted in biohybrid diffusion chambers. *Transplantation*, 57(9):1371–1375, 15 May, 1994

[7] B. Kulseng, S. Gudmund, L. Ryan, A. Andersson, A. King, A. Faxvaag, and T. Espevik. Transplantation of alginate microcapsules: generation of antibodies against alginates and encapsulated porcine islet-like cell clusters. *Transplantation*, 67(7):978–984, 1999.

[8] H. Iwata, N. Morikawa, T. Fujii, T. Takagi, T. Samejima, and Y. Ikada. Does Immunoisolation Need to Prevent the Passage of Antibodies and Complement? *Transplantation Proceedings*, Vol. 27(6), pp. 3224–3226, December 1995.

[9] T.A. Desai, M. Ferrari, and G. Mazzoni. Silicon microimplants: fabrication and biocompatibility. In T. Kozik (ed.). *Materials and Design Technology* 1995. ASME, pp. 97–103, 1995.

[10] M. Ferrari, W.H. Chu, T.A. Desai, D. Hansford, T. Huen, G. Mazzoni G, and M. Zhang. Silicon nanotechnology for biofiltration and immunoisolated cell xenografts. C.M Cotell, A.E. Meyer, S.M. Gorbatkin, and G.L. Grobe. *Thin Films and Surfaces for Bioactivity and Biomedical Application* (eds.), MRS, Vol. 414, pp. 101–106, 1996.

[11] T.A. Desai, W.H. Chu, J.K. Tu, G.M. Beattie, A. Hayek, and M. Ferrari. Microfabricated immunoisolating biocapsules. *Biotechnol. Bioeng.*, 57:118–120, 1998.

[12] T.A. Desai, J. Tu, G. Rasi, P. Borboni, and M. Ferrari. Microfabricated biocapsules provide short-term immunoisolation of insulinoma xenografts. *Biomed. Microdev.*, 1(2):1999.

[13] M. Zhang, T.A. Desai, and M. Ferrari. Proteins and cells on PEG immobilized silicon surfaces. *Biomaterials*, 19:953–960, 1998.

[14] T.A. Desai, D. Hansford, and M. Ferrari. Characterization of micromachined membranes for immunoisolation and bioseparation applications. *J. Memb. Sci.*, 4132:1–11, 1999.

[15] R.P. Lanza et al. Transplantation of islets using microencapsulation: studies in diabetic rodents and dogs. *J. Mol. Med.*, 77(1):206–210, Jan 1999.

[16] U. Siebers et al. Alginate-based microcapsules for immunoprotected islet transplantation. *Ann. NY Acad. Sci.*, 831:304–312, Dec 31, 1997.

[17] R. Calafiore et al. Transplantation of allogeneic/xenogeneic pancreatic islets containing coherent microcapsules in adult pigs. *Transplant Proc.*, 30(2):482–483, Mar 1998 .

[18] C.J. Weber et al. Encapsulated islet iso-, allo-, and xenografts in diabetic NOD mice. *Transplant Proc.*, 27(6):3308–3311, Dec 1995.

[19] R.P. Lanza et al. Xenotransplantation. *Sci. Am.*, 277(1):54–59, Jul 1997.

[20] C.J. Weber et al. Xenografts of microencapsulated rat, canine, porcine, and human islets. In C. Ricordi (ed.), *Pancreatic Islet Cell Transplantation*, pp. 177–189, 1991.

[21] P. Lacy, O.D. Hegre, A. Gerasimidi-Vazeou, F.T. Gentile, and K.E. Dionne. Maintenance of normoglycemia in diabetic mice by subcutaneous xenograft of encapsulated islets. *Science*, 254(5039):1728–1784, 1991.

[22] P. Marchetti et al. Prolonged survival of discordant porcine islet xenografts. *Transplantation*, 61(7):1100–1102, Apr 15, 1996.

[23] R.P. Lanza, W. Kuhtreiber, K. Ecker, W.M. Beyer, and W.L. Chick. Xenotransplantation of porcine and bovine islets without immunosuppression using uncoated alginate microspheres. *Transplantation*, 59:1366–1385, 1995.

[24] P. Soon-Shiong et al. An immunologic basis for the fibrotic reaction to implanted microcapsules. *Transplant. Proc.*, 23(1 Pt 1):758–759, Feb 1991.

[25] C.K. Colton and E. Avgoustiniatos. Bioengineering in development of the hybrid artificial pancreas. *Trans. ASME*, 113:152–170, 1991.

[26] D. Chicheportiche et al. In vivo activation of peritoneal macrophages by the implantation of alginatepolylysine microcapsules in the BB/E rat. *Diabetologia*, 34:A170, 1991.

[27] P. Soon-Shiong, R. Feldman, R. Nelson, Q. Heintz, Z. Yao, T. Yao, N. Zheng, G. Merideth, T. Skjak-Braek, T. Espevik et al. Long-term reversal of diabetes by the injection of immunoprotected islets. *Proc. Natl. Acad. Sci. U.S.A.*, 90(12):5843–5847, Jun 15, 1993.

[28] P. Soon-Shiong, R.E. Heintz, N. Merideth, Q.X. Yao, Z. Yao, T. Zheng, M. Murphy, M.K. Moloney, M. Schmehl, M. Harris et al. Insulin independence in a type 1 diabetic patient after encapsulated islet transplantation. *Lancet*, 343(8903):950–951, Apr 16, 1994.

[29] R. Calafiore, G. Basta, G. Luca, C. Boselli, A. Bufalari, G.M. Giustozzi, L. Moggi, and P. Brunetti. Alginate/polyaminoacidic coherent microcapsules for pancreatic islet graft immunoisolation in diabetic recipients. *Ann. NY Acad. Sci.*, 831:313, Dec 31, 1997

[30] J. Brauker, L.A., Martinson S.K., Young and R.C. Johnson Local inflammatory response around diffusion chambers containing xenografts. Nonspecific destruction of tissues and decreased local vascularization. *Transplantation*, 61(12):1671–1677, Jun 27, 1996.

[31] C.J. Weber et al. The role of CD4+ helper T cells in the destruction of microencapsulated islet xenografts in nod mice. *Transplantation*, 49(2):396–404, Feb 1990.

[32] C.J. Weber et al. CTLA4-Ig prolongs survival of microencapsulated neonatal porcine islet xenografts in diabetic NOD mice. *Cell Transplant.*, 6(5):505–508, Sep-Oct, 1997.

[33] D.W. Gray. Encapsulated islet cells: the role of direct and indirect presentation and the relevance to xenotransplantation and autoimmune recurrence. *Br. Med. Bull.*, 53(4):777–788, 1997.

[34] K.E. Ellerman et al. Islet cell membrane antigens activate diabetogenic CD4+ T-cells in the BB/Wor rat. *Diabetes*, 48(5):975–982, May 1999.

[35] U. Siebers, A. Horcher, H. Brandhorst, D. Brandhorst, B. Hering, K. Federlin, R.G. Bretzel, and T. Zekorn. Analysis of the cellular reaction towards microencapsulated xenogeneic islets after intraperitoneal transplantation. *J. Mol. Med.*, 77(1):215–218, Jan 1999.

[36] D.J. Edell, V. Van Toi, V.M. McNeil, and L.D. Clark. Factors influencing the biocompatibility of insertable silicon microshafts in cerebral cortex. *IEEE Trans. Biomed. Eng.*, 39(6):635–643, 1992.

[37] K.D. Wise et al. Micromachined Silicon Microprobes for CNS Recording and Stimulation. *Annual Conference of the IEEE Engineering in Medicine and Biology Society*, Vol. 12, no. 5, pp. 2334–2335, 1990.

[38] D.J. Edell, J.N. Churchill, and I.M. Gourley. Biocompatibility of a silicon based peripheral nerve electrode. *Biomat. Med. Dev. Artif. Org.*, 10(2):103–122, 1982.

[39] T. Akin, K. Najafi et al. A micromachined silicon sieve electrode for nerve regeneration applications. *IEEE Trans. Biomed. Eng.*, 41(4):305–313, Apr 1994.

[40] G.T.A. Kovacs, C.W. Storment, M. Halks-Miller, C.R. Belczynski, C.C. Della Santina, E.R. Lewis, and N.I. Maluf. Silicon-substrate microelectrode arrays for parallel recording of neural activity in peripheral nerves. *IEEE Trans. Biomed. Eng.*, 41:567–577, 1994.

[41] P.L. Gourley. Semiconductor microlasers: a new approach to cell-structure analysis. *Nat. Med.*, 2(8):942–944, Aug 1996.

[42] W. Chu, T. Huen, J. Tu, and M. Ferrari, Silicon-micromachined, direct-pore filters for ultrafiltration. In P.L. Gourley (ed.), *SPIE Proc. of Micro- and Nanofabricated Structures and Devices for Biomedical Environmental Applications*. Vol. 2978, pp. 111–122, 1996.

[43] T.A. Desai. Microfabricated interfaces: new approaches in tissue engineering and biomolecular separation.*Biomol. Eng.*, 17(1):23–36, Oct 2000.

[44] M. Brissova et al. Control and measurement of permeability for design of microcapsule cell delivery system. *J. Biomed. Mater. Res.*, 39(1):61–70, Jan 1998.

[45] I. Lacik et al. New capsule with tailored properties for the encapsulation of living cells. *J. Biomed. Mater. Res.*, 39(1):52–60, Jan 1998.

[46] J.H. Brauker, V.E. Carr-Brendel, L.A. Martinson, J. Crudele, W.D. Johnston, and R.C. Johnson Neovascularization of synthetic membranes directed by membrane microarchitecture. *J. Biomed. Mater. Res.*, 29(12):1517–1524, Dec 1995.

[47] T.A. Desai, W.H. Chu, J. Tu, P. Shrewsbury, and M. Ferrari. Microfabricated biocapsules for cell xenografts: a review. *Proc. SPIE*, 2978:216–226, 1996.

[48] D. Hansford, T. Desai, J. Tu, and M. Ferrari. Biocompatible siliconwafer bonding for biomedical microdevices. *Micro and Nanofabricated Electro-Optical-Mechanical Systems for Biomedical and Environmental Application*. Vol. 3258, pp. 164–168, 1998.

[49] N. Hellerstrom, J. Lewis, H. Borg, R. Johnson, and N. Freunkel. Method for large scale isolation of pancreatic islets by tissue culture of fetal rat pancreas. *Diabetes*, 28:766–769, 1979.

[50] S. Efrat, S. Linde, H. Kofod, D. Spector, M. Delannoy, S. grant, D. Hanahan, and S. Baekkeskov. Beta-cell lines derived from transgenic mice expressing a hybrid insulin gene-oncogene. *Proc. Natl. Acad. Sci. U.S.A.*, 85:9037–9041, 1988.

[51] W. Tan, R. Krishnaraj, and T. Desai. Evaluation of composite collagen-chitosan matrices for tissue engineering. *Tissue Eng.*, 7(2):2001.

[52] C.J. Weber and K. Reemstma. Microencapsulation in small animals—Xenografts. In R.P. Lanza and W.L. Chick (eds.), *Pancreatic Islet transplantation: Vol. III. Immunoisolation of Pancreatic Islets*. New York, RG Landes Co., pp. 59–79, 1994.

[53] T.N. Salthouse and B.F. Matlaga. An approach to the numerical quantification of acute tissue response to biomaterials. *J. Biomatls. Med. Dev. Art. Orgs.*, 3(1):47–56, 1975.

[54] T.A. Desai, D.J. Hansford, L. Leoni, M. Essenpreis, and M. Ferrari. Nanoporous antifouling silicon membranes for implantable biosensor applications. *Biosen. Bioelect.*, 15(9–10):453–462, 2000.

[55] T.A. Desai, M. Ferrari and G. Mazzoni. Silicon microimplants: fabrication and biocompatibility. In T. Kozik (ed.), *Materials and Design Technology* 1995. ASME, pp. 97–103, 1995.

[56] S.K. Aityan and V.I. Portnov. *Gen. Physiol. Biophys.*, 5(4):351–364, 1986.

[57] S.K. Aityan and V.I. Portnov. *Gen. Physiol. Biophys.*, 7(6):591–611, 1988.

[58] K. Hahn, J. Karger, and Kukla. *Phys. Rev. Lett.*, 76(15):2762–2765, 1996.

[59] Q.C. Wei, Bechinger, and P. Leiderer. *Science*, 287(5453):625–627, 2000.

[60] J.A. Hernandez and J. Fischbarg. *Biophy. J.*, 67(3):996–1006, 1994.

[61] K. Dionne, B.M. Cain, R.H. Li, W.J. Bell, E.J. Doherty, D.H. Rein, M.J. Lysaght, and F.T. Gentile. Transport in immunoisolation membranes. *Biomaterials*, 17(3):1996.

[62] L. Leoni and T.A. Desai. Nanoporous biocapsules for the encapsulation of insulinoma cells: biotransport and biocompatibility considerations. *IEEE Trans. Biomed. Eng.*, 48(11):Nov 2001.

[63] -K. Tatarkiewicz, -M. Garcia,-M. Lopez-Avalos,-S. Bonner-Weir, and -G.-C. Weir, Porcine neonatal pancreatic cell clusters in tissue culture: benefits of serum and immobilization in alginate hydrogel. *Transplantation*, 71(11):1518–1526, Jun 15, 2001.

[64] G.M. Beattie, J.S. Rubin M.I. Mally, T. Otonkoski, and A. Hayek. Regulation of proliferation and differentiation of human fetal pancreatic islet cells by extracellular matrix, hepatocyte growth factor, and cell-cell contact. *Diabetes*, 45(9):1223–1228, 1996.

[65] S.N. Bhatia, M.L. Yarmush and M. Toner Controlling cell interactions by micropatterning in co-cultures: hepatocytes and 3T3 fibroblasts fibroblasts. *J. Biomed. Mater. Res.*, 34:189–199, 1997

[66] T. Loudovaris, B. Charlton, R.J. Hodgson, and T.E. Mandel. Destruction of xenografts but not allografts within cell impermeable membranes. *Transplant. Proc.*, 24:2291, 1992.

[67] T. Zekorn, U. Siebers, R.J. Bretzel, M. Renardy, H. Planck, P. Zshcocke, and K. Federlin. Protection of islets of langerhans from IL-1 toxicity by artificial membranes. *Transplantation*, 50:391–394, 1990.

[68] D.R. Cole, M. Waterfall, M. McIntyre, and J.D. Baird. Microencapsulated islet grafts in the BB/E rat: a possible role for cytokines in graft failure. *Diabetologia*, 35:231–237, 1992.

[69] J.A. Hunt, P.J. McLaughlin, and B.F. Flanagan. Techniques to investigate the cellular and molecular interactions in the host response to implanted biomaterials. *Biomaterials*, 18:1449–1459, 1997.

[70] K. Popat, Robert W. Johnson, and T.A. Desai. Vapor deposited thin silane films on silicon substrates for biomedical microdevices. *Surf. Coat. Technol.*, Accepted Dec 2001.

[71] D.W.R. Gray. Pancreatic islet transplantation—open issues. *NY Acad. Sci.*, 2001.

[72] A. Prokop. Bioartificial pancreas: materials, devices, function, and limitations. *Diabetes Technol. Therapeut.*, 3(3):2001.

[73] F. Valerie, -K. Duvivier, O. Abdulkadir, R.J. Parent, J.J. O'Neil, and G.C. Weir. Completeprotection of islets against allorejection and autoimmunity by a simple barium-alginate membrane *Diabetes*, 50:1698–1705, 2001.

[74] T. Loudovaris, S. Jacobs, S. Young, D. Maryanov, J. Brauker, and R.C. Johnson. Correction of diabetic NOD mice with insulinomas implanted within Baxter immunoisolation devices. *J. Mol. Med.*, 77:219–222, 1999

[75] K. Suzuki, S. Bonner-Weir, N. Trivedi, K.H. Yoon, J. Hollister-Lock, C.K. Colton, and G.C. Weir. Function and survival of macroencapsulated syngeneic islets transplanted into streptozocin-diabetic mice. *Transplantation*, 66(1):21–28, Jul 15, 1998.

[76] K. Suzuki, S. Bonner-Weir, J. Hollister-Lock, C.K. Colton, and G.C. Weir. Number and volume of islets transplanted in immunobarrier devices. *Cell Transplant.*, 7(1):47–52,Jan 2, 1998.

11

Medical Nanotechnology and Pulmonary Pathology

Amy Pope-Harman[1] and Mauro Ferrari[2]

[1]*Department of Internal Medicine, Division of Pulmonary, Critical Care, and Sleep Medicine, The Ohio State University College of Medicine and Public Health*
[2]*Professor, Brown Institute of Molecular Medicine Chairman, Department of Biomedical Engineering, University of Texas Health Science Center, Houston, TX; Professor of Experimental Therapeutics, University of Texas M.D. Anderson Cancer Center, Houston, TX; Professor of Bioengineering, Rice University, Houston, TX; Professor of Biochemistry and Molecular Biology, University of Texas Medical Branch, Galveston, TX; President, the Texas Alliance for NanoHealth, Houston, TX*

ABSTRACT

Diseases of the lungs are common and potentially devastating. Though advances in medical science have been significant, there is yet substantial need for improvement in the ability to determine precisely what is occurring in the body on a local level during disease and to intervene in a timely and targeted manner Despite the needs of medicine and of patients, there are forces that tend to slow the progress of medical innovation and the incorporation of new practices into common medical care.

The lungs as a therapeutic and diagnostic site provide particular challenges. In spite of this, physicians are already commonly using medications and therapeutics that came about through the application of molecular chemistry and evolving nanotechnology. There is need for continued development of nanotechnology toward lung applications. Several of these opportunities are discussed. We encourage the cooperation between bedside physicians and those of the technologic creativity and knowledge in the identification and delineation of unsolved clinical problems toward identification of means to overcome those clinical hurdles.

11.1. INTRODUCTION

Diseases that affect the lungs are among the most frequently-encountered and deadly we face as a population. During the period from 1982 to 1992 (the most recent decade for which data are available), the overall annual age-adjusted prevalence rate of self-reported asthma increased to 49.4 per 1000, and we have seen an age adjusted mortality due to asthma increase to nearly 5 in 1 million in the United States [90, 91]. Lung cancer is the deadliest of the solid tumors, leading to the deaths of more individuals than the combination of the three next most lethal cancers [78]. Emphysema and other chronic obstructive pulmonary diseases affect an estimated 2 million Americans [88]. Venous thrombosis becomes symptomatic in approximately 100 per 100,000 Americans per year, with thirty to forty percent of those resulting in pulmonary embolism [3, 17, 40, 59, 83]. Cystic fibrosis, a genetic disease that causes abnormally thick lung secretions with subsequent recurrent infections and extensive scarring of lung tissues, affects one in four thousand Caucasian births. The lungs are the delicate casualty of many inflammatory insults, from sepsis to environmental antigen exposure. These represent only the most common of diseases that manifest primarily in the lungs.

Clinical medicine has seen tremendous improvements that have allowed individuals to survive many acute illnesses to which they would have succumbed in the past. The body's efforts to heal, however, often lead paradoxically to further injury, abnormality, and disability. Physicians are frustrated in their inability to monitor the body's injury and repair processes as they occur on a sub-clinical basis in the lungs and elsewhere, prior to the time at which they may currently be practically detected by means of screening, onset of symptoms, or acute deteriorations. Doctors are often incapable of supervising directly or intervening specifically and locally as injuries occur and afterward—able instead only to make gross assessments of global dysfunctions and either stand by as damage accumulates or to offer therapeutics that may not address specific pathogenic events in a timely manner.

On the other hand, the worldwide scientific community has achieved great advances toward understanding many of the basic mechanisms behind lung and other diseases. It is increasingly apparent that most disease states come about not as a result of merely a single insult or abnormality, but as complex matrices of events and responses that evolve over time. As scientists gain specific knowledge of isolated relationships that define the complexities of our inner biology [47], it becomes ever less satisfying to approach problems by blanket assumptions, offering interventions that may or may not have beneficial effects for a specific individual at a particular site and point in time, and that in fact could even be detrimental in the given situation.

It is clear, then, that we must develop methods to thoughtfully unite our varied and burgeoning pieces of medical knowledge with practical means to precisely apply it clinically. We must span the gaps between our needs, knowledge, and expectations and the medical reality. Medical nanotechnology promises just such a bridge. By working on a nanometer, molecular scale, there is potential to simultaneously assess and intervene with accuracy in space, time, and activity that we have never before enjoyed.

The public imagination has been piqued by works such as Asimov's *Fantastic Voyage* and Michael Crichton's *Prey*. Despite this interest, however, the entry of nanotechnology into the medical realm has seemed to advance at an our-world pace that reflects the dimensions of its products. Well, there is no teeny Raquel Welch in a skin-tight inner-space suit, but many physicians are quietly incorporating advances made through nanotechnology into their practices even now, and even if they might not be aware of the origins or nomenclature. We will explore such current applications in this chapter, particularly in relation to diseases of the lungs, after discussing a few of the challenges faced in introducing new technology into the general medical environment, discussing particularly dangers that must be anticipated in interventions that target the lungs. The bulk of this chapter will encompass a few of the theoretical applications of nanotechnology to diseases of the lungs and conclude with a call to continued cooperation and communication between physicians who identify the unsolved clinical frustrations and failures at the bedside and the biologists, chemists, engineers, and other visionaries who may provide the tools to solve them.

11.1.1. Today's Medical Environment

While often frustrated with current limitations in diagnostic and therapeutic options, the majority of physicians do not have sufficient free time to dream of, pursue, or develop new technologies for many reasons. Medical practice generally has little room for creativity, no potential downtime to re-think and re-configure. Much of the literature on nanotechnology traditionally has not appeared in medical journals, but instead in those of chemistry, engineering, and physics and in the newer nanotechnology and microtechnology literature. Thus, information regarding new technology is not readily accessible nor comprehensible to physician clinicians, even were they capable of determining which of the developments may hold potential usefulness in their practices, for the diseases they treat. The technology then often lacks advocates for its rapid advancement and direction toward meaningful medical applications.

Also, emphasis on existing management strategies and treatments, coming from outcomes research and consensus-derived practice guidelines that go on to dictate current patient care protocols, builds the perception that there is no room for revolutionary improvements in existing practices. The legal and third party payer environments inflict similar constraints on straying from routines of diagnosis and disease management. Most changes creep slowly into medical practice, relative to the general explosion of information and innovation outside of medicine, and inch into common use based upon awareness of the medical community, availability, proven safety record, improved efficacy over existing treatments, demand of the public and clinicians, and affordability. Further, regulatory bodies have also slowed the progress of advancement and implementation of nanotechnology in medicine, in part because they have not known how to proceed. Experts are few and the potential dangers have not all been explored. Caution is clearly and understandably the prevailing attitude. Predicting future developments with some degree of realism is difficult. No one can guarantee what is possible or likely, thus complicating the planning process. Still, there is exponential growth in our knowledge and capabilities that should continue to produce exciting medical applications.

11.1.2. Challenges for Pulmonary Disease-Directed Nanotechnology Devices

The respiratory system is open to the environment, and thus is armed with means to protect against infections and other environmental threats. When considering inhaled access, the pattern of laminar flow of air within the upper airways and lungs; branching of

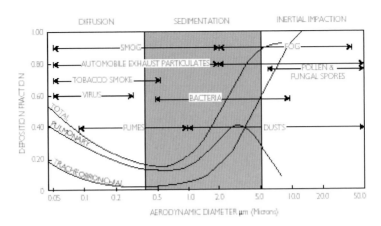

FIGURE 11.1. Relationship between particle size and location of deposition in the respiratory tract. [25]

the airways; and moisture, mucous, and ciliation of the epithelial lining make it difficult for larger particles (greater than approximately 5 microns, depending upon shape, composition, charge, etc.) (Figure 11.1) to travel as far as and remain in the lower reaches (alveolar spaces) of the lungs. The lungs are rich in macrophages that are readily capable of engulfing and neutralizing foreign material (including intentionally-introduced medical devices) and initiating an aggressive acute inflammatory response. While this capability may have allowed our forebears the opportunity to survive infectious onslaughts that might otherwise have been fatal, it may impair the access of devices that are to employ in the lungs and could also present a threat to the entire organism in the potential triggering of overwhelming systemic inflammatory responses. This is of special concern were there unintentional contamination of inhalable devices with infectious material, lipopolysaccharide, or other toxins. The intricacies of the immune system response are not fully understood, and concerns that inhalable interventions may either de-sensitize to true pathogens or "overwhelm" the first-line defense mechanisms have still not been completely allayed. The lungs are rich in lymphatic vessels and lymph tissue, assuring that subsequent exposures to materials the body perceives as "undesirable" will be met with even more immediate and vigorous humoral protection. The lung is the initiating site of many allergic responses as well—with unpleasant effects running the gamut from runny nose through rash to severe, life-threatening bronchospasm and anaphylaxis. Devices that remain in the lungs, accessing via airways, vasculature, or otherwise, are potentially subject to local healing/walling-off granulomatous or fibrosis responses that may of themselves be potential threats (as might occur with occlusion of the pulmonary vasculature leading to pulmonary hypertension and consequent right heart failure) or may impede the device function. Also, materials and devices that are sufficiently small to achieve the alveoli and traverse the extensive alveolar-capillary membrane can enter the circulation in a very rapid manner, as the surface area over which diffusion may take place is large and such diffusion is generally quite efficient, with the diffusion gradient maintained by the rapid movement of the entirety of the cardiac output through the lungs. Materials may also migrate in an unpredictable manner through the tissues of the lungs and into the entire body, where local response may cause secondary and unintended effects [9]. Implantable drug delivery devices have the potential for toxic effects, secondary to the device itself, its carrier, or the contents. Special care must be taken to assure that any toxic contents will not leak unexpectedly and that the device is impervious to potential traumas. Also, any material that is foreign to the body serves as a potential focus for infection, so care must be taken to minimize this risk.

11.2. CURRENT APPLICATIONS OF MEDICAL TECHNOLOGY IN THE LUNGS

11.2.1. Molecularly-derived Therapeutics

The lungs, directly accessible via the airways to the outside environment, were the target of one of the first molecularly-based medical therapies. Dornase alpha recombinant human deoxyribonuclease I (rhDNase), launched in 1994, is a proteinaceous enzyme that selectively cleaves DNA [43]. Cystic fibrosis patients use rhDNAase, administered by inhalation of an aerosol mist, to cleave the sticky free DNA that has been released from destroyed inflammatory cells and that comprises a significant portion of mucous. This lysis of the DNA facilitates expectoration. DNAase therapy has contributed to improvements in the multi-component management of an autosomal recessive inherited disease that had previously resulted in near uniform mortality by the early teens to result in a life expectancy for these patients that may reach now into the mid-thirties and even beyond [81].

After having been originally applied to treat acute coronary syndromes in the late 1980's, tissue plasminogen activators have come to use in the lungs as treatment for large and life-threatening pulmonary emboli. Their value in this setting, though, continues to be questioned, plagued by complications related to inability to target their site of action, resulting in the potential for bleeding complications, and the sheer volume of clot that is often seen in acute thromboembolic disease. Thus, the role of tissue plasminogen activators in this setting is still under exploration [43].

11.2.2. Liposomes

Liposomes in many sizes and compositions are in fairly common use in cosmetic and medical applications and are continuing to undergo further refinement and testing for additional medical uses [49]. The liposome refinements attempt to overcome early problems related to variability in size, amount of medication incorporation into the particles, integrity of payload encapsulation, and targeting to diseased tissue and organs. Encapsulation of drugs in liposomes has often resulted in improved therapeutic efficacy over that of the drug alone following administration by multiple routes, including topical, injection, and inhalation. Physicians are accustomed especially to intravenous use of liposomal formulations of toxic azole antifungal medications [39].

Additional non-cosmetic liposomal formulations are in various stages of testing. Lung deposition of the antibiotics gentamicin, amikacin, and tobramycin markedly improves when these antibiotics are administered in inhalable liposomal forms, compared to non-liposomal inhalation [18, 26, 35, 36, 41, 66, 71]. A liposomal form of ketotifen fumarate, an experimental asthma bronchodilator, is in human trials [55], as is liposomal cyclosporine. Cyclosporin is currently most often utilized as an anti-rejection medication for transplant patients in oral or intravenous form, where its potential for system-wide side effects is significant. When given via liposomal inhalation for lung transplants, the peak serum level of cyclosporin is on the order of 110 nanograms per milliliter, a fraction of that necessary if the drug is administered by other routes, thus reducing its toxic potential [89]. Interleukin 2 has been in use for many years to combat various cancers. Its administration, usually intravenously or subcutaneously, is often complicated by uncomfortable and occasionally life-threatening infection-like symptoms. Liposome-encapsulated Interleukin 2 via inhalation has been tested with promising results in animal models of several cancers that metastasize to the lungs [60]. Clodronate disodium encapsulated within liposomes administered via the airway in a murine model of Pseudomonas bacterial pneumonia and sepsis results in depletion of alveolar macrophages, thus reducing the initial inflammatory response and bringing about improvements in mortality [46]. Evaluation of gene therapies with the genetic material introduced via liposomal inhalation is underway. Liposomal formulations of the gene for α-tocopherol, a naturally-occurring anti-oxidant, have been administered in rats in anticipation of severe hypoxia (low oxygen levels). After hypoxia, the treated rats show downregulation of caspases, resulting in a limitation of apoptosis in the lungs. Their mortality is reduced from 60 to 30 percent [67, 86]. This approach also may be useful to protect the lungs during time-limited risks such as radiation therapy through the expression of manganese superoxide dismutase plasmids [28]. These are but a few of the many liposomal delivery systems that are under current use and study.

11.2.3. Devices with Nanometer-scale Features

Filters with pores of 15–40 nanometers have seen increased use in the manufacture of plasma-derived biopharmaceuticals as a means of size-excluding viral and other potentially infectious elements from therapeutic protein mixtures [13, 93]. Microarrays of many types and varieties are under development to aide in the identification of disease states or monitoring of disease activity. Many of the emerging systems utilize micron and nanometer scale manufactured features. Though these are still primarily a research tool, as the results become more reliable and the array results are correlated with particular disease states, we can anticipate more common use in the patient population.

11.3. POTENTIAL USES OF NANOTECHNOLOGY IN PULMONARY DISEASES

It is truly impossible for us to conceive or predict how all of the achievements in science may eventually channel into advances in medicine. It is, though, a useful pleasure to make the attempt. We can anticipate that particular tasks will continue to need to be accomplished. We will frame this discussion using those basic medical tasks. These immutable tasks are: 1) diagnosis and 2) therapeutics or management of disease. Therapeutics may be divided further into the administration of therapeutic agents versus provision of mechanical or structural improvements. We will begin with a description of the needs and potentials within each of these areas, then elaborate based upon available technology or technology in near development, as applied in specific examples to common pulmonary diseases. There are thousands of potential medical applications and nearly as many individual technology platforms. It is impossible to achieve complete comprehensiveness, but we will attempt to be as inclusive as space and reason will allow. We will not mention anything along the lines of "bioreactors," or implantation or transplantation of cell colonies or tissues designed or programmed to act in a particular biologic fashion. We will make minimal mention of nanotechnologic approaches to systemic, body-wide, delivery via the inhalational route, which could encompass a chapter in itself.

11.3.1. Diagnostics

The first broad category of need in pulmonary medicine, as in all of medicine, is in the area of diagnostics. This topic is huge, and the potential for improvement in existing practice is significant. Diagnosis encompasses identification of indicators, or markers, of disease states; various forms of imaging; localization, staging, and quantification of disease; prognostics; and monitoring of the progression of disease. It is conceivable that many techniques that may be used to localize and identify a disease may also be applied toward targeting for therapeutic intervention.

11.3.1.1. Disease Markers and Localization
A disease state may be recognized or localized by the identification of markers of disease. These may include any of several potential biomolecular clues. Proteins may manifest disease in the form of either completely abnormal proteins / peptides, chemical or electrical changes that are encountered in places and at times where they should not normally be located, or proteins that are processed abnormally. Similarly, a disease state (or predisposition to it) may be predicted by the presence of particular genes within an individual's total genome. In fact, nanotechnology provides a bridging step toward "personal medicine," where an individual's response to a planned therapeutic intervention may be predicted by the presence of particular genetic polymorphisms, as may their risks for certain disorders. Also, abnormalities in the structure of genetic material may come about as a result of local injury or other insult (structural genetic aberrations), or abnormal patterns of gene expression for the given conditions (up- or down-regulations). These changes may be associated with particular diseases or may predispose to disease states such as cancer.

Disease markers may be identifiable in exceedingly small amounts or found only in relatively inaccessible anatomic locales. Nanotechnology may permit less invasive, less uncomfortable means of making diagnoses. Additionally, nanotechnology may allow for improvement in the lower limits of detection of disease, and thus improve the chances for early diagnosis as well as providing more accurate monitoring of therapeutic response or progression of disease. This can be accomplished by several means. Local chemical, electrical or physical property-changes in cells or tissues may be monitored by means of nanometer scale tubes and wires in the form of field effect transistors [23, 56]. Concentration of markers that are present only in small amounts, either by attraction of markers of interest to a stationary binding site or by the binding of biomarkers remotely onto particulates, with subsequent harvesting of bound particulates for query may be accomplished. Alternatively, identification of biomarkers found in low concentrations may be improved through amplification in cascade reactions or by tagging known biomarkers with indicators that may be more easily detected. Imaging, or localization of abnormalities, may take advantage of the presence of markers, the function or behavior of organs or tissues, or the architectural structure of the tissue or organ.

11.3.1.2. Imaging
There are many means of imaging that may be used to detect local abnormalities. Positron emission tomography (PET) scanning demonstrates differences in metabolism between diseased tissues from that of normal tissue. Nanotechnology should

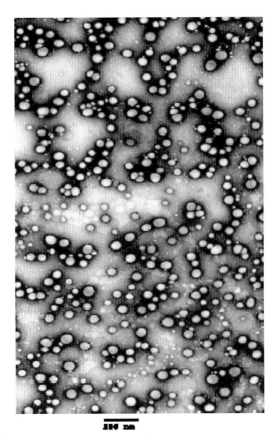

FIGURE 11.2. Transmission electron micrograph (TEM) showing the size and morphology of folate-coated gadolinium nanoparticles [75].

permit increased sensitivity through amplification of these detected chemical changes or by increasing their specificity, allowing improved differentiation between "normal" areas of increased metabolic activity and areas of true disease, or even potentially permitting discrimination among tumor types and of biologic behavior in the case of cancer. Contrast agents have been used for many years to improve the diagnostic capabilities in magnetic resonance imaging (MRI), with more nanometer-scale contrast agents in development. Electron Paramagnetic Resonance (EPR) is an evolving imaging technique that may allow an even greater understanding of the regional metabolic activity within diseased tissues, also enhanced through administration of nanometer-scale contrasts [74] (Figure 11.2). Nanotechnology may allow improvement in the sensitivity of detection or even allow differentiation between similar disease states (i.e. distinguish between different types of infectious pathogens) using traditional radiographic techniques such as computerized tomography (CT) or plain radiographs along with labeling of organisms or features of organisms. Mechanical differences between normal and diseased tissue may be identified through the employment of specialized precision ultrastructural ultrasound techniques [64], and these characteristics further enhanced by employing nanoscale contrast agents. Disease-related abnormalities that may not be otherwise directly visually identifiable can be made manifest directly through targeting of color- or fluorescent-contrast agents or even expression of indicator genes that are targeted to the diseased tissue of interest [1].

11.3.2. Therapeutics

11.3.2.1. Therapeutic Agent Delivery   Nanotechnology will offer significant and increasing advantages in the administration of therapeutic agents. Materials that now require injection could potentially be inhaled or swallowed through the use of nano-engineered delivery devices, thus improving patient comfort and compliance [31, 85]. Even through-the-skin administration of some agents such as vaccines can be made more comfortable by employing multitudes of micro- or nanometer-scale needles, to which human nerves are insensitive, rather than a single painful injection [57] (Figure 11.3). Compliance may be improved with longer dosing intervals [61], which is possible with the use of nanoparticulate delivery systems. Nanoparticulates may permit both the improvement in targeting of drugs to diseased tissue as well as an increase in the amount of drug that reaches sites of activity within abnormal cells. This site-specific (and also potentially metabolism-specific) targeting, nanoparticle-induced seclusion of potentially-toxic therapies, and possibly the requirement for less drug overall may significantly improve the toxicity profile of some drugs that are effective but have at this point an unacceptable toxicity profile. Also, activation of devices or release of drug may be made dependent upon a site- or tissue-specific second stimulus or even via mechanical means as in the localized application of ultrasound or a magnetic field.

Nanometer-scale features may allow the practical construction and use of fully implantable and controllable drug delivery devices. Implantable nanochannel drug delivery devices can provide further improvement of control and optimization of pharmacodynamics, allowing prolonged steady-state drug (or individually-tailored drug mixture) release without potentially-dangerous peaks, start / stop of drug administration, adjustment in level of dosing, or pulsatile release, depending upon the requirements for the particular therapeutic agent [31]. Mathematical modeling gives precise predictions of the rate of drug release over

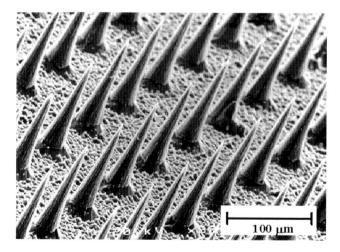

FIGURE 11.3. Microneedles that are sufficiently long to penetrate through the stratum corneum, but short enough not to stimulate nerves in deeper tissue. These arrays are created using an SF6/O2 plasma [2].

time [24]. These devices may eventually be able to be controlled from externally (telemetry) or integrated into a feedback loop system using incorporated or associated diagnostics and monitoring. Many of the initial difficulties in the use of indwelling sensors such as fouling, signal alteration, etc. are becoming better understood [16], and may allow adjustment of dosing based upon direct feedback from monitoring devices in the local bodily environment. The inner activities of as few as one cell at a time may be monitored and therapeutics adjusted accordingly [27, 44, 65, 82] (Figure 11.3).

11.3.2.2. Mechanical/Structural Interventions   There are several areas in which the application of nanotechnology may in the future provide significant help with more structural problems in the pulmonary system. Currently, we are capable of installing, via bronchoscope, tubular stents to lessen the collapse of larger airways and blood vessels secondary to loss of structural integrity—as in the case of chronic inflammation of cartilage, scarring, or ingrowth of tumor (Figure 11.3). These stents are limited in their efficacy and usefulness by the branching of the airways and vasculature (confining their practical use essentially to large and centrally-located airways and vessels), by local tissue reactions (i.e. granulation tissue formation, or keloid), by the tendency to mucous accumulation at the stent interfaces, and by progression of the original pathology. Nanotechnology may provide means of delivering structural support without the limitations of existing technology, possibly through self-assembly of inhaled components. This goal of course will require significant work and is not likely to be achieved until long in the future.

The lung, as is the case with several organs, even when badly damaged, may repair itself to a state of healthy normalcy if the underlying architecture of the tissue is intact. It is not possible at this point, aside from attempts to grossly limit damage secondary to the acute manifestations of disease, to assure that the lung tissue architecture will support normal healing after an insult. In fact, the foundation is being laid for such future work. Elastin polypeptides have been induced to self-assemble into something that resembles elastin [7, 11] (Figure 11.4). Extensive work has been accomplished in the area of tissue scaffolding and repair. [14, 87] (Figure 11.5).

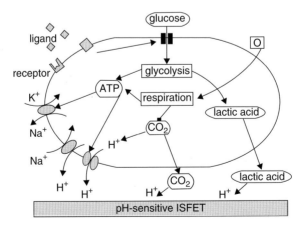

FIGURE 11.4. Extracellular acidification can be measured by field-effect sensors [44].

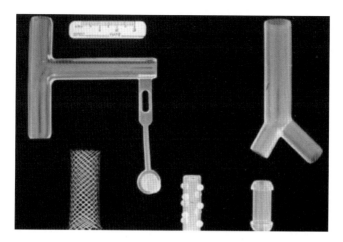

FIGURE 11.5. Commonly used silastic and expandable wire stents for airway obstruction. (Kaplan P)

Possibly one day in the future we will be able to utilize nanotechnologic self-assembling components to support airways and provide infrastructure for damaged alveolar lung tissue so that the patient may then be able to complete the healing process with her own cells and tissues. Such an intervention would extend tremendous hope for those who suffer from emphysema, whose only option at this point may be lung transplantation.

One of the major difficulties of patients who suffer from many lung diseases is that of cough. Damaged or inflamed airway lining cells and glands respond in part through increased mucous and other secretion production. Cilia, normally responsible for sweeping out the material lining the inner airways, are sometimes damaged by exposure to toxins or may be genetically pre-destined to malfunction. The nanotechnology of our children and grandchildren could provide direct help with the breakdown and clearance of secretions. One might envision in the distant future the inhalation of single-task nano-robots designated to break apart mucous and carry it out along the oxygen gradient. Damaged or disordered cilia components could assemble in-situ, within pulmonary epithelial lining cells, to assist in mucociliary clearance.

11.3.3. Evolving Nanotechnology in Pulmonary Diseases

11.3.3.1. Lung Cancer   With lung cancer as most cancers, one of the primary goals is to determine earlier, at a time when intervention may be more effective, that an individual may have cancer. We would like to identify pre-cancerous lesions to avert the risk of malignancy. We want to use more accurate means to identify metastatic spread or recurrence of previously-treated cancer so that it may be addressed. We want to provide more effective means of targeting therapeutics to diseased tissues and cells. Nanotechnology offers many potential tools to achieve these goals [32], first in the form of nanoparticulates. There are numerous targets that may be utilized to identify and localize cancerous lesions with nanoparticulates. Detection of abnormal areas of endothelial growth factors may be associated with (and provide a target against) pulmonary metastasis [54]. Methylation of $p16^{INK4a}$, DAPK, MGMT, and FHIT is seen in lung epithelial tissue at risk for the development of cancer [6, 75]. Even single nucleotide polymorphisms are recognizable using current biochip technology [12]. We know that the number of identifiable genetic abnormalities increases with progression to cancer from pre-cancerous risk and that early intervention through re-instatement of absent or malfunctioning genetic material is effective in averting the transition to cancer [52, 73].

Nanoparticulates can be paired with existing imaging modalities to aide in visualization of cancerous tissue and its characteristics. The redox state of specific biologic molecules in living systems [20, 22, 50] can be determined. Additionally, conjugation of folate and polyethylene glycol onto superparamagnetic particles results in increased uptake in cancer cells [95]. These may be imaged by MRI. Areas of tumor heterogeneity can be identified by contrast-enhanced MRI [19]. Gadolinium nanoparticles (folate and polyethylene glycol-coated) may be useful for both imaging and treatment through neutron capture therapy, and have been demonstrated to localize to tumors in vivo [74]. Iodinated nanoparticles have successfully localized to lymph nodes after bronchoscopic instillation, and may be imaged by CT [58]. Quantum dots, visible on stimulation with discrete light wavelengths, can be targeted to protein markers in the lung endothelium and to cancers [1]. Tissue proteomic techniques may be employed to identify the early protein markers of cancer [38].

Many types of biodegradable nanoparticles for the purpose of therapeutic agent delivery have been manufactured [79, 96]. Polymeric nanoparticles have been used to achieve gene transfection within tumor cells [77]. Nanoparticles may serve to localize drug delivery and to avoid first-pass metabolism [4]. In fact, certain nanoparticles may be able to partially evade the immediate recognition of the recipient's immune system, allowing the potential for a longer time of efficacy and improved arrival at target tissues [68]. Nigavekar and colleagues have taken initial steps toward targeting nanodendrimers, which may serve as drug delivery devices, to tumors [70]. Polymer nanoparticles have brought about enhanced uptake of immuno-active drugs (cyclosporin) into macrophages [89]. These are just a few examples of many promising technologies in current development in the fight against the pain and suffering due to lung and other cancers.

11.3.3.2. Pulmonary Thromboembolic Disease Pulmonary thromboembolic (PTE) disease is a potential killer that is common in both the hospitalized and non-hospitalized patient. It involves formation of clot, usually in the large slow-flowing veins of the legs and pelvis, followed by dislodgement of recently-formed clot into venous flow to travel through the right heart and into the pulmonary vasculature. PTEs kill by both occluding the pulmonary vasculature, thus impairing the ability of the normally-weak right heart to provide sufficient blood return to the systemic (left) heart and by causing profound decreases in blood oxygen levels. The diagnosis of PTE is fraught with difficulty. The "gold standard" is still considered to be pulmonary angiogram, a procedure that involves threading a catheter into the pulmonary arteries in question, which requires a skilled invasive radiologist to perform and interpret. Smaller clots are not easily detected by this or other means. With the more recently-recognized CT angiogram, accurate results depend upon correct timing and technique of intravenous dye injection and are still subject to misinterpretation by radiologists of lesser familiarity with the technique.

Nanotechnology may provide help in this realm as well. Nanoparticles conjugated to proteins that recognize mature clot (i.e. fibrin) could be constructed to be identified by common radiographic modalities, thus allowing more accurate diagnosis. Were thrombolytic

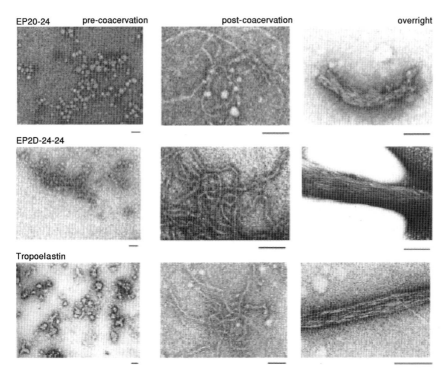

FIGURE 11.6. Comparison of structures formed from EP20-24, EP20-24-24, and tropoelastin at three stages of coacervation. Precoacervation and immediately (10 min) following coacervation ("postcoacervation")the structures formed by the polypeptides are similar to those formed by tropoelastin. However, after overnight incubation above the coacervation temperature ("overnight") the structures formed from EP20-24-24 closely resemble that of tropoelastin, whereas structures formed from EP20-24 are less compact and well-organized. All scale bars represent 100 nm [7, 11].

medications to be packaged within fibrin-targeting nanoliposomes, the clot-lysis medication could be released specifically at or near the site of the obstructing clot by a technique as simple as ultrasound directed at the area of clot. The nanoliposomes containing the potentially-dangerous thrombolytics are disrupted by the ultrasound at the site of the clot and re-form when they are not under the influence of ultrasound. Thrombolytics are released from the disrupted liposomes only into the area around the clot. Liposomes that target clots that are not involved in the embolic event do not release their payload. Such site specificity could allow repeated treatments and more concentrated thrombolysis at the site of the clot burden, which could provide greater success in resolution of particularly large and mature clots [33]. Kim Hamad-Schifferli's innovation of on-off regulation of chemical events by application of localized magnetic fields to achieve temporary denaturation of proteins after administration of protein-conjugated metallic nanoparticulates may be applicable to this and other, more precise biologic localization and control applications [45]. Certain polymers change their configuration with changes in pH. Were these polymers configured to contain thrombolytics or other medication while in a higher-pH environment and release in areas of lower pH, the medication could be released at the site of clot (with its subsequent

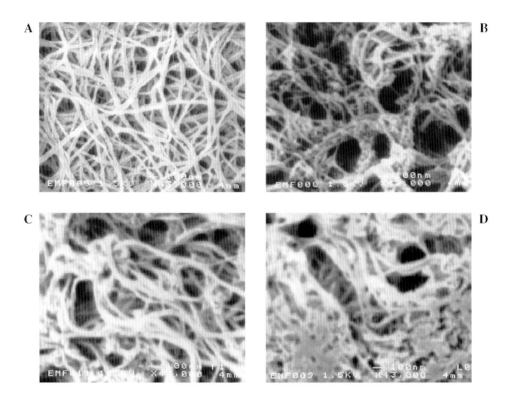

FIGURE 11.7. SEMs of collagen with chitosan of different proportions where collagen concentration is always 8.0 mg/ml: **(A)** pure collagen matrices; **(B)** collagen–chitosan composite matrices with 1:1 proportion; **(C)** collagen–chitosan matrices with 1:3 proportion; and **(D)** collagen-chitosan matrices with 1:3 proportion and with K562 cells in three-dimensional gel. (Original magnification: 343,000.) [87]

ischemia), where the pH would be expected to be relatively lower owing to the anaerobic environment. Were there difficulty in re-establishing adequate circulation or oxygenation after or during PTE, nanoparticles may be able to enhance oxygen delivery in thrombotic states [15].

11.3.3.3. Asthma Simple asthma may be diagnosed by pulmonary function testing, which demonstrates diminished exhalation ("obstruction") during exacerbations. This abnormality usually improves to normal between episodes. Some patients, though, who have severe or chronic asthma, never return to normal breathing. Although there may be many causes of this lack of normalization between asthma exacerbations, it has been recognized over the past several years that airway remodeling, or a change in the structure of the underlying tissues that form the air passages, may be responsible. Airway remodeling is characterized by thickened smooth muscle and subepithelium, increases in collagens III and V, tenascin, laminin, and fibrin in the sub-basement membrane, as well as matrix metaloproteins and secretion by the smooth muscle of immunomodulating cytokines [53, 76]. Currently, there is no way to quantify this response aside from intra-airway biopsies, which are difficult to perform and pose some risks. Ultrastructure-Accounting Characterization

Mode Ultrasound [64] via bronchoscopy may provide information regarding both the underlying tissue structure of the airways as well as the particular composition and presence of inflammatory indicators through the use of targeted contrast agents.

11.3.3.4. Pulmonary Infections   Pulmonary infections are common and account for significant morbidity and mortality [34, 72]. Even under the most ideal circumstances, we can determine culprit organism(s) less than 50% of the time [69]. This lack of information regarding responsible infectious agents not only prevents patients from receiving optimal antibiotics sooner (especially in the case of antibiotic-resistant organisms), but also contributes to the problem of antibiotic resistance by forcing empiric initiation of available antibiotics in patients who are too ill to wait for results of cultures and sensitivities. A means by which we could both quickly identify specific organisms and determine their sensitivities to antibiotics would likely result in significant improvement in pneumonia-related morbidity and mortality [51]. The specific organisms as well as the presence of any resistance-conferring genetic material could be quickly identified by use of lab-on-a-chip microsystems [90], among other means.

Prevention of pulmonary infections, making use of the body's inherent mechanisms of directed humorally-mediated protection, may be even more appealing than attempting to treat an already-established problem. Vaccines hold promise as well for the policing of abnormal cells that may develop into cancers. The immunogenicity of biologic polymers may be optimized either for use as vaccines [34] or, as in the case of conjugated nanoparticulates for drug delivery, for avoidance of immune response [62]. G Ferrari with the author and SC Lee, in unpublished calculations, have developed programming by which optimal sequences of proteins may be manufactured to allow immunologic response to multiple antigenic sites at once. Polymer nanoparticulates in the form of chitosan nanoparticles may provide a means of bringing about unique topical (epithelial) genetic immunization [21].

11.3.3.5. Pulmonary Fibrosis   Idiopathic pulmonary fibrosis is a devastating pulmonary disorder that often results in death within a few years [29]. The only recourse that some patients have is lung transplantation, which is not an ideal solution, and one that is not open to many of this generally-older patient population. Among other therapies, interferon gamma has been utilized with some success to delay the inevitable deterioration of this disease [94]. Interferon gamma results in stimulation of the immune system and is poorly tolerated by some patients, especially the elderly, causing malaise, fevers, and occasional low blood pressure. An implantable drug delivery device for the administration of interferon is currently under development for use in anti-cancer therapy [63] and may similarly be applicable to use in idiopathic pulmonary fibrosis, allowing therapy to maintain efficacy without as many of the side effects.

11.4. CONCLUSION

In conclusion, we have briefly introduced many of the common lung diseases we face as a population. Potential barriers to the development of new medical practices were discussed. Pitfalls that may be encountered when accessing and manipulating the lungs were clarified. We have discussed nanotechnology as it is already introduced into pulmonary medical practice. Lastly, we have presented a few of the many potential applications—both near at hand and as projections for the future—of nanotechnology toward the most common and devastating problems of the pulmonary system.

There are many opportunities in medicine, advances yet to be made, problems to be tackled. Nanotechnology and its offshoots may provide the means to greatly improve patient care and life in general. It will require, though, continued hard work and collaboration. People with specialized technologic knowledge are not necessarily present to see the needs of the patients. Clinicians must remain alert to the problems that they see at the bedside rather than merely accepting the status quo. The clinical problems must then be brought to the engineers and the chemists and the physicists, a dialogue established so that they may begin to apply their expertise to together find acceptable solutions. We are fully encouraging of these collaborations. This interaction is most critical in bringing about further advances, though is sometimes challenging as well, requiring that differences in language and constraints of time both be overcome. Interdisciplinary training for the future professionals in these areas will be crucial in maintaining the momentum toward further advances.

REFERENCES

[1] M.E. Akerman, W.C. Chan, P. Laakkonen, S.N. Bhatia, and E. Ruoslahti. Nanocrystal targeting in vivo. *Proc. Natl. Acad. Sci. U.S.A.*, 99(20):12617–12621, 2002.

[2] M.G. Allen. www.cmmt.gatech.edu, Center for MEMS and Microsystems Technologies, Georgia Institute of Technology, 2002.

[3] F.A. Anderson Jr., H.B. Wheeler, and R.J. Goldberg. A population-based perspective of the hospital incidence and case-fatality rates of deep vein thrombosis and pulmonary embolism: theworcester DVT study. *Arch. Intern. Med.*, 151:933–938, 1991.

[4] P. Arbos, M.A. Campanero, M.A. Arangoa, and J.M. Irache. Nanoparticles with specific bioadhesive properties to circumvent the pre-systemic degradation of fluorinated pyrimidines. *J. Control Rel.*, 96(1):55–65, 2004.

[5] C. Beaulac, S. Clement-Major, J. Hawari, and J. Lagace. Eradication of mucoid *Pseudomonas aeruginosa* with fluid liposome-encapsulated tobramycin in an animal model of chronic pulmonary infection. *Antimicrob. Agents Chemother.*, 40:665–669, 1996.

[6] S.A. Belinsky, K.J. Nikula, W.A. Palmisano, R. Michels, G. Saccomanno, E. Gabrielson, S.B. Baylin, and J.G. Herman. Aberrant methylation of p16(INK4a) is an early event in lung cancer and a potential biomarker for early diagnosis. *Proc. Natl. Acad. Sci. U.S.A.*, 95(20):11891–11896, 1998.

[7] C.M. Bellingham, M.A. Lillie, J.M. Gosline, G.M. Wright, B.C. Starcher, A.J. Bailey, K.A. Woodhouse, and F.W. Keeley. Recombinant human elastin polypeptides self-assemble into biomaterials with elastin-like properties, *Biopolymers*, 70(4):445–455, 2003.

[8] Y. Benenson, B. Gil, U. Ben-Dor, R. Adar, and E. Shapiro. An autonomous molecular computer for logical control of gene expression. *Nature*, 429(6990):423–429, 2004.

[9] P.J. Borm. Particle toxicology: from coal mining to nanotechnology. *Inhal. Toxicol.*, 14(3):311–324, 2002.

[10] H. Bounameaux, L. Hicklin, and S. Desmarais. Seasonal variation in deep vein thrombosis. *BMJ*, 312:284–285, 1996.

[11] G. Bressan, I. Pasquali-Ronchetti, C. Fornieri, F. Mattioli, I. Castellani, and D. Volpin. *J. Ultrastruct. Res.*, 94:209–216, 1986.

[12] J. Burmeister, V. Bazilyanska, K. Grothe, B. Koehler, I. Dorn, B.D. Warner, and E. Diessel. Single nucleotide-polymorphism analysis by chip-based hybridization and direct current electrical detection of gold-labeled DNA. *Anal. Bioanal. Chem.*, 379(3):391–398, 2004.

[13] T. Burnouf, Radosevich. Nanofiltration of plasma-derived biopharmaceutical products. *Haemophilia*, 9:24–37, 2003.

[14] E.L. Chaikof, H. Matthew, J. Kohn, A.G. Mikos, G.D. Prestwich, and C.M. Yip. Biomaterials and scaffolds in reparative medicine. *Ann. NY Acad. Sci.*, 961:96–105, 2002.

[15] C. Chauvierre, M.C. Marden, C. Vauthier, D. Labarre, P. Couvreur, and L. Leclerc. Heparin coated poly(alkylcyanoacrylate) nanoparticles coupled to hemoglobin: a new oxygen carrier. *Biomaterials*, 25(15):3081–3086, 2004.

[16] R.J. Chen H.C. Choi, S. Bangsaruntip, E. Yenilmez, X. Tang, Q. Wang, Y.L. Chang, and H. Dai. An investigation of the mechanisms of electronic sensing of protein adsorption on carbon nanotube devices. *J. Am. Chem. Soc.*, 126(5):1563–1568, 2004.

[17] W.W. Coon. Epidemiology of venous thromboembolism. *Ann. Surg.*, 186:149–164, 1977.

[18] C. Cordeiro, D.J. Wiseman, P. Lutwyche, M. Uh, J.C. Evans, B.B. Finlay, and M.S. Webb. Antibacterial efficacy of gentamicin encapsulated in pH-sensitive liposomes against an in vivo *Salmonella enterica* serovar Typhimurium intracellular infection model. *Antimicrob. Agents Chemother.*, 44:533–539, 2000.

[19] N.G. Costouros, D. Lorang, Y. Zhang, M.S. Miller, F.E. Diehn, S.M. Hewitt, M.V. Knopp, K.C. Li, P.L. Choyke, H.R. Alexander, and S.K. Libutti. Microarray gene expression analysis of murine tumor heterogeneity defined by dynamic contrast-enhanced MRI. *Mol. Imaging*, 1(3):301–308, 2002.

[20] J.E. Crowther, V.K. Kutala, P. Kuppusamy, J.S. Ferguson, A.A. Beharka, J.L. Zweier, F.X. McCormack, and L.S. Schlesinger. Pulmonary surfactant protein a inhibits macrophage reactive oxygen intermediate production in response to stimuli by reducing NADPH oxidase activity. *J. Immunol.*, 172(11):6866–6874, 2004.

[21] Z. Cui and R.J. Mumper. Chitosan-based nanoparticles for topical genetic immunization. *J. Control Rel.*, 75(3):409–419, 2001.

[22] M.C. Daniel, J. Ruiz, S. Nlate, J.C. Blais, and D. Astruc. Nanoscopic assemblies between supramolecular redox active metallodendrons and gold nanoparticles: synthesis, characterization, and selective recognition of H2PO4-, HSO4-, and adenosine-5′-triphosphate (ATP2-) anions. *J. Am. Chem. Soc.*, 125(9):2617–2628, 2003.

[23] J.J. Davis, K.S. Coleman, B.R. Azamian, C.B. Bagshaw, and M.L. Green. Chemical and biochemical sensing with modified single walled carbon nanotubes. *Chemistry*, 9(16):3732–3739, 2003.

[24] P. Decuzzi, S. Lee, M. Decuzzi, and M. Ferrari. Adhesion of microfabricated particles on vascular endothelium: a parametric analysis. *Ann. Biomed. Eng.*, 32(6):793–802, 2004.

[25] M.B. Dolovich and M.T. Newhouse. Aerosols. Generation, methods of administration, and therapeutic applications in asthma. In E. Middleton Jr., C.E. Reed, E.F. Ellis, N.F. Adkinson Jr, J.W. Yunginger, and W.W. Busse(eds.), *Allergy. Principles and practice*, (4th Edn.), St Louis, Mosby Year Book, Inc., pp. 712–739, 1993.

[26] P. Demaeyer, E.M. Akodad, E. Gravet, P. Schietecat, J.P. Van Vooren, A. Drowart, J.C. Yernault, and F.J. Legros. Disposition of liposomal gentamicin following intrabronchial administration in rabbits. *J. Microencapsul.*, 10:77–88, 1993.

[27] S.V. Dzyadevych, A.P. Soldatkin, Y.I. Korpan, V.N. Arkhypova, A.V. El'skaya, J.M. Chovelon, C. Martelet, and N. Jaffrezic-Renault. Biosensors based on enzyme field-effect transistors for determination of some substrates and inhibitors. *Anal. Bioanal. Chem.*, 377(3):496–506, 2003.

[28] M.W. Epperly, H.L. Guo, M. Jefferson, S. Nie, J. Gretton, M. Bernarding, D. Bar-Sagi, H. Archer, and J.S. Greenberger. Cell phenotype specific kinetics of expression of intratracheally injected manganese superoxide dismutaseplasmid/liposomes (MnSOD-PL) during lung radioprotective gene therapy. *Gene Ther.*, 10(2):163–171, 2003.

[29] J.M. Fellrath and R.M. du Bois. Idiopathic pulmonary fibrosis/cryptogenic fibrosing alveolitis. *Clin. Exp. Med.*, 3(2):65–83, 2003.

[30] M. Ferrari. Therapeutic Microdevices and Methods of Making and Using Same, U.S. Patent No. 6,107,102, August 22, 2000.

[31] M. Ferrari et al. Particles for Oral Delivery of Peptides and Proteins, US Patent No. 6,355,270, March 12, 2002.

[32] Ferrari. The hallmarks of cancer nanotechnology. *Nature Reviews Cancer*, 3(5):161–171, 2005.

[33] M. Ferrari, and P. Goldschmidt. Personal communication, 2001.

[34] T. Fifis, A. Gamvrellis, B. Crimeen-Irwin, G.A. Pietersz, J. Li, P.L. Mottram, I.F. McKenzie, and M. Plebanski. Size-dependent immunogenicity: therapeutic and protective properties of nano-vaccines against tumors. *J. Immunol.*, 173(5):3148–3154, 2004.

[35] R.M. Fielding, R.O. Lewis, and L. Moon-McDermott. Altered tissue distribution and elimination of amikacin encapsulated in unilamellar, low-clearance liposomes (MiKasome). *Pharm. Res.*, 15:1775–1781, 1998.

[36] R.M. Fielding, L. Moon-McDermott, R.O. Lewis, and M.J. Horner. Pharmacokinetics and urinary excretion of amikacin in low-clearance unilamellar liposomes after a single or repeated intravenous administration in the rhesus monkey. *Antimicrob. Agents Chemother.*, 43:503–509, 1999.

[37] G. Hernandez, P. Rico, E. Diaz, and J. Rello. Nosocomial lung infections in adult intensive care units. *Microbes. Infect.*, 6(11):1004–1014, 2004.

[38] D.H. Geho, N. Lahar, M. Ferrari, E.F. Petricoin, and L.A. Liotta. Opportunities for nanotechnology-based innovation in tissue proteomics. *Biomed. Microdev.*, 6(3):231-239, 2004.

[39] B.E. Gilbert. Liposomal aerosols in the management of pulmonary infections. *J. Aerosol. Med.*, 9(1):111–122, 1996.

[40] R.F. Gillum. Pulmonary embolism and thrombophlebitis in the United States, 1970–1985. *Am. Heart J.*, 114:1262–1264, 1987.

[41] S. Giovagnoli, P. Blasi, C. Vescovi, G. Fardella, I. Chiappini, L. Perioli, M. Ricci, and C. Rossi. Unilamellar vesicles as potential capreomycin sulfate carriers: preparation and physicochemical characterization. *AAPS Pharm. Sci. Tech.*, 4(4):69, 2003.

[42] S.Z. Goldhaber and H. Bounameaux. Thrombolytic therapy in pulmonary embolism. *Semin. Vascul. Med.*, 1(2):213–220, 2001.

[43] J.D. Green. Pharmaco-toxicological expert report Pulmozyme rhDNase Genentech, Inc. *Hum. Exp. Toxicol.*, 13(Suppl 1):S1–S42, 1994.

[44] F. Hafner. Cytosensor® Microphysiometer: technology and recent applications. *Biosens. Bioelectron.*, 15:149–158, 2000.

[45] K. Hamad-Schifferli, J.J. Schwartz, A.T. Santos, S. Zhang, and J.M. Jacobson. Remote electronic control of DNA hybridization through inductive coupling to an attached metal nanocrystal antenna. *Nature*, 415(6868):152–155, 2002.

[46] S. Hashimoto, J.-F. Pittet, K. Hong, H. Folkesson, G. Bagby, L. Kobzik, C. Frevert, K. Watanabe, S. Tsurufuji, and J. Wiener-Kronish. Depletion of alveolar macrophages decreases neutrophil chemotaxis to *Pseudomonas* airspace infections. *Am. J. Physiol. (Lung Cell Mol. Physiol.)*, 270(14):L819–L828, 1996.

[47] G. Hernandez, P. Rico, E. Diaz, and J. Rello. Nosocomial lung infections in adult intensive care units. *Microbes Infect.*, 6(11): 1004–14, 2004.

[48] L. Hood, J.R. Heath, M.E. Phelps, and B. Lin. Systems biology and new technologies enable predictive and preventative medicine. *Science*, 306(5696):640–643, 2004.

[49] O.R. Hung, S.C. Whynot, J.R. Varvel, S.L. Shafer, and M. Mezei. Pharmacokinetics of inhaled liposomeen-capsulated fentanyl. *Anesthesiology*, 83:277–284, 1995.

[50] G. Ilangovan, A. Bratasz, H. Li, P. Schmalbrock, J.L. Zweier, and P. Kuppusamy. In vivo measurement and imaging of tumor oxygenation using coembedded paramagnetic particulates. *Magn. Reson. Med.*, 52(3):650–657, 2004.

[51] M. Iregui, S. Ward, G. Sherman, V.J. Fraser, and M.H. Kollef. Clinical importance of delays in the initiation of appropriate antibiotic treatment for ventilator-associated pneumonia. *Chest*, 122:262–268, 2001.

[52] H. Ishii, K.R. Dumon, A. Vecchione, L.Y. Fong, R. Baffa, K. Huebner, and C.M. Croce. Potential cancer therapy with the fragile histidine triad gene: review of the preclinical studies. *JAMA*, 286(19):2441–2449, 2001.

[53] P.K. Jeffery. Remodeling in asthma and chronic obstructive lung disease. *Am. J. Respir. Crit. Care Med.*, 164(10Pt 2):S28–S38, 2001.

[54] E. Jin, M. Fujiwara, M. Nagashima, H. Shimizu, M. Ghazizadeh, X. Pan, S. Arai, Y. Ohaki, M. Gomibuchi, T. Takemura. and O. Kawanami. Aerogenous spread of primary lung adenocarcinoma induces ultrastructural remodeling of the alveolar capillary endothelium. *Hum. Pathol.*, 32(10):1050–1058, 2001.

[55] M. Joshi and A. Misra. Dry powder inhalation of liposomal ketotifen fumarate: formulation and characterization. *Int. J. Pharm.*, 223(1-2):15–27, 2001.

[56] E. Katz and I. Willner. Biomolecule-functionalized carbon nanotubes: applications in nanobioelectronics. *Chemphyschemistry*, 5(8):1084–1104, 2004.

[57] G. Kersten and H. Hirschberg. Antigen delivery systems. *Expert Rev. Vaccines*, 3(4):453–462, 2004.

[58] L.H. Ketai, B.A. Muggenberg, G.L. McIntire, E.R. Bacon, R. Rosenberg, P.E. Losco, J.L. Toner, K.J. Nikula, and P. Haley. CT imaging of intrathoracic lymph nodes in dogs with bronchoscopically administered iodinated nanoparticles. *Acad. Radiol.*, 6(1):49–54, 1999.

[59] A. Kierkegaard. Incidence and diagnosis of deep vein thrombosis associated with pregnancy. *Acta. Obstet. Gynecol. Scand.*, 62:239–243, 1983.

[60] C. Khanna, P.M. Anderson, D.E. Hasz, E. Katsanis, M. Neville, and J.S. Klausner. Interleukin-2 liposome inhalation therapy is safe and effective for dogs with spontaneous pulmonary metastases. *Cancer*, 79(7):1409–1421, 1997.

[61] W. Kruse, W. Eggert-Kruse, J. Rampmaier, B. Runnebaum, and E. Weber. Dosage frequency and drug-compliance behaviour–a comparative study on compliance with a medication to be taken twice or four times daily. *Eur. J. Clin. Pharmacol.*, 41(6):589–592, 1991.

[62] S.C. Lee, R. Parthasarathy, K. Botwin, D. Kunneman, E. Rowold, G. Lange, J. Klover, A. Abegg, J. Zobel, T. Beck, T. Miller, W. Hood, J. Monahan, J.P. McKearn, R. Jansson, and C.F. Voliva. Biochemical and immunological properties of cytokines conjugated to dendritic polymers. *Biomed. Microdev.*, 6(3):191–202, 2004.

[63] G. Lesinski, S. Sharma, K. Varker, P. Sinha, M. Ferrari, and W. Carson. Release of Biologically Functional Interferon- Alpha from a Nanochannel Delivery System. (in review), 2005.

[64] J. Liu and M. Ferrari. Mechanical spectral signatures of malignant disease? A small-sample, comparative study of continuum vs. nano-biomechanical data analyses. *Dis. Markers*, 18(4):175–183, 2002.

[65] J. Lu and Z. Rosenzweig. Nanoscale fluorescent sensors for intracellular analysis. *Fresenius J. Anal. Chem.*, 366(6-7):569–575, 2000.

[66] J.-F. Marier, J. Lavigne, and M.P. Ducharme. Pharmacokinetics and efficacies of liposomal and conventional formulations of tobramycin after intratracheal administration in rats with pulmonary. *Burkholderia cepacia infection*. *Antimicrob. Agents Chemother.*, 46(12):3776–3781, 2002.

[67] T. Minko, A. Stefanov, and V. Pozharov. Lung edema clearance: 20years of progress selected contribution: lung hypoxia: antioxidant and antiapoptotic effects of liposomal á-tocopherol. *J. Appl. Physiol.*, 93:1550–1560, 2002.

[68] S.M. Moghimi, A.C. Hunter, and J.C. Murray. Long-circulating and target-specific nanoparticles: theory to practice. *Pharmacol. Rev.*, 53(2):283–318, 2001.

[69] M.S. Niederman, L.A. Mandell, A. Anzueto, J.B. Bass, W.A. Broughton, G.D. Campbell, N. Dean, T. File, M.J. Fine, P.A. Gross, F. Martinez, T.J. Marrie, J.F. Plouffe, J. Ramirez, G.A. Sarosi, A. Torres, R. Wilson, and V.L. Yu. American thoracic society guidelines for the management of adults with community-acquired pneumonia: Diagnosis, assessment of severity, antimicrobial therapy, and prevention. *Am. J. Respir. Crit. Care Med.*, 163(7):1730–1754, 2004.

[70] S.S. Nigavekar, L.Y. Sung, M. Llanes, El-Jawahri, T.S. Lawrence, C.W. Becker, L. Balogh, and M.K. Khan. 3H dendrimer nanoparticle organ/tumor distribution. *Pharm. Res.*, 21(3):476–483, 2004.

[71] A. Omri, C. Beaulac, M. Bouhajib, S. Montplaisir, M. Sharkawi, and J. Lagace. Pulmonary retention of free and liposome-encapsulated tobramycin after intratracheal administration in uninfected rats and rats infected with *Pseudomonas aeruginosa*. *Antimicrob. Agents Chemother.*, 38:1090–1095, 1994.

[72] J.J. Oosterheert, M.J. Bonten, E. Hak, M.M. Schneider, and I.M. Hoepelman. How good is the evidence for the recommended empirical antimicrobial treatment of patients hospitalized because of community-acquired pneumonia? A systematic review. *J. Antimicrob. Chemother.*, 52(4):555–563, 2003.

[73] M. Ottey, S.Y. Han, T. Druck, B.L. Barnoski, K.A. McCorkell, C.M. Croce, C. Raventos-Suarez, C.R. Fairchild, Y.Wang, and K. Huebner. Fhit-deficient normal and cancer cells are mitomycin C and UVC resistant. *Br. J. Cancer*, 91(9):1669–1677, 2004.

[74] M.O. Oyewumi, R.A. Yokel, M. Jay, T. Coakley, and R.J. Mumper. Comparison of cell uptake, biodistribution and tumor retention of folate-coated and PEG-coated gadolinium nanoparticles in tumor-bearing mice. *J. Control Rel.*, 95(3):613–626, 2004.

[75] W.A. Palmisano, K.K. Divine, G. Saccomanno, F.D. Gilliland, S.B. Baylin, J.G. Herman, and S.A. Belinsky. Predicting lung cancer by detecting aberrant promoter methylation in sputum. *Cancer Res.*, 60(21):5954–5958, 2000.

[76] R.A. Panettieri Jr. Airway smooth muscle: immunomodulatory cells that modulate airway remodeling? *Respir. Physiol. Neurobiol.*, 137(2–3):277–293, 2003.

[77] S. Prabha and V. Labhasetwar. Critical determinants in PLGA/PLA nanoparticle-mediated gene expression. *Pharm. Res.*, 21(2):354–364, 2004.

[78] L.A.G. Ries, M.P. Eisner, C.L. Kosary, B.F. Hankey, B.A. Miller, L. Clegg, A. Mariotto, E.J. Feuer, and B.K. Edwards. SEER *Cancer Statistics Review*, 1975–2001, National Cancer Institute. Bethesda, MD, http://seer.cancer.gov/csr/1975 2001/, 2004.

[79] M. Roser, D. Fischer, and T. Kissel. Surface-modified biodegradable albumin nano- and microspheres. II: effect of surface charges on in vitro phagocytosis and biodistribution in rats. *Eur. J. Pharm. Biopharm.*, 46(3):255–263, 1998.

[80] E.J. Ruijgrok and E.W.M. VultoAGandVan Etten. Efficacy of aerosolized amphotericin B desoxycholate and liposomal amphotericin B in the treatment of invasive pulmonary aspergillosis in severely immunocompromised rats. *J. Antimicrob. Chemother.*, 48:89–95, 2001.

[81] D.V. Schidlow. Maintaining the horizontal line: early intervention and prevention of CF lung disease. *J. Cyst. Fibros.*, 3(2):63–66, 2004.

[82] M.J. Schoning and A. Poghossian. Recent advances in biologically sensitive field-effect transistors (BioFETs). *Analyst*, 127(9):1137–1151, 2002.

[83] M.D. Silverstein, J.A. Heit, and D.N. Mohr. Trends in the incidence of deep vein thrombosis and pulmonary embolism: a 25-year population-based study. *Arch. Intern. Med.*, 158:585–593, 1998.

[84] P.M. Sinha, G.J. Valco, S. Sharma, X. Liu, and M. Ferrari. Nanoengineered device for drug delivery application. *Nanotechnology*, 15:S585–S589, 2004.

[85] A. Stikeman. *The Programmable Pill*, MIT Technology Review, May, 78–82.

[86] Z.E. Suntres and P.N. Shek. Incorporation of alpha-tocopherol in liposomes promotes the retention of liposomeencapsulated glutathione in the rat lung. *J. Pharm. Pharmacol.*, 46:23–28, 1994.

[87] W. Tan, R. Krishnaraj, and T.A. Desai. Evaluation of nanostructured composite collagen–chitosan matrices for tissue engineering. *Tissue Eng.*, 7(2):203–210, 2001.

[88] G. Viegi, A. Scognamiglio, S. Baldacci, F. Pistelli, and L. Carrozzi. Epidemiology of chronic obstructive pulmonary disease (COPD). *Respiration*, 68(1):4–19, 2001.

[89] J. Wang and Q. Zhang. Uptake of cyclosporineAloaded colloidal drug carriers by mouse peritoneal macrophages in vitro. *Acta Pharmacol. Sin.*, 22(1):57–61, 2001.

[90] Z. Wang J. El-Ali, M. Engelund, T. Gotsaed, I.R. Perch-Nielsen, K.B. Mogensen, D. Snakenborg, J.P. Kutter, and A. Wolff. Measurements of scattered light on a microchip flow cytometer with integrated polymer based optical elements. *Lab. Chip.*, 4(4):372–377, 2004.

[91] K.B. Weiss and D.K.S.O. Wagener. Asthma surveillance in the United States: a review of current trends and knowledge gaps. *Chest*, 98(5 Suppl):179S–184S, 1990.

[92] K.B. Weiss, P.J. Gergen, and E.F.S.O. Crain. Inner-city asthma: the epidemiology of an emerging US public health concern. *Chest*, 101(6 Suppl):362S–367S, 1992.

[93] T. Yokoyama, K. Murai, T. Murozuka, A. Wakisaka, M. Tanifuji, N. Fujii, and T. Tomono. Removal of small non-enveloped viruses by nanofiltration. *Vox Sang.*, 86(4):225–229, 2004.

[94] R. Ziesche, E. Hofbauer, K. Wittmann, V. Petkov, and L.H. Block. A preliminary study of long-term treatment with interferon gamma-1b and low-dose prednisolone in patients with idiopathic pulmonary fibrosis. *N. Engl. J. Med.*, 341(17):1264–1269, 1999.

[95] Y. Zhang, N. Kohler, and M. Zhang. Surface modification of superparamagnetic magnetite nanoparticles and their intracellular uptake. *Biomaterials*, 23(7):1553–1561, 2002.

[96] F. Zhao, Y. Yin, W.W. Lu, J.C. Leong, W. Zhang, J. Zhang, M. Zhang, and K. Yao. Preparation and histological evaluation of biomimetic three-dimensional hydroxyapatite/chitosan-gelatin network composite scaffolds. *Biomaterials*, 23(15):3227–3234, 2002.

12

Nanodesigned Pore-Containing Systems for Biosensing and Controlled Drug Release

Frédérique Cunin,[a] Yang Yang Li,[b] and Michael J. Sailor[b]

[a]Dr. F. Cunin, UMR CNRS/ENSCM 5618, 8 rue de l'école normale,
34296 Montpellier cedex 5, France
[b]Prof. M. J. Sailor, Yang Yang Li, Department of Chemistry and Biochemistry, The University of California, San Diego, 9500 Gillman Drive, La Jolla, CA 92039-0358, USA

For medical treatment of disease, optimal therapeutic efficiency of a drug is governed by both the therapeutic activity of the drug itself and the way in which it is delivered to the patient. Once administered, it is as crucial to control the rate at which a drug is released in the body as it is to control its transport to the desired location. Numerous systems have been introduced over the past three decades to get drugs into the body [1] and there are many more innovative concepts currently in development. The role that nanotechnology plays in this effort is increasing dramatically.

Whereas synthetic chemistry involves the manipulation of matter at the molecular level, nanotechnology can be thought of as a discipline in which the tools of the chemist are applied to problems whose size and level of complexity lie above the molecular level. Biology operates in this domain; a living cell is a complex assembly of interconnected molecular machines and hierarchical structures. Nanotechnology operates in the same size regime, although its tools and its applications are not restricted to biology. In particular, nanostructured porous materials offer a degree of control in both the rate and the location of drug delivery that is just beginning to be recognized. This article will survey the widely accepted methods for controlled drug delivery and then focus on nanostructured materials-in particular silicon-based photonic and templated materials as examples.

12.1. SYSTEM DESIGN CONSIDERATIONS

The most common forms of drug delivery are oral injected tablets, injection, transdermal patches and implantable fixtures. Each of these modes of delivery imposes its own set of requirements. For example, the rate of release of a drug depends on numerous bio-environmental conditions such as pH, circulation of fluid, viscosity, temperature, ionic strength, adsorption of specific or non-specific biomolecules, and local redox potential of the surrounding medium. Some general requirements include:

- Either before or after release, the drug must display some preferential targeting or partitioning to the intended tissue or organ.
- The system must protect the rest of the organism from intoxication, which generally means that it should avoid any release during transport to the target.
- It must load the appropriate amount of active molecule for optimal availability and therapeutic efficacy upon arrival at the target.
- It must protect the incorporated active molecule from degradation before reaching the target. This implies that the drug delivery system is chemically inert to the entrapped drug.
- It must control the rate of delivery, which is specific to each therapeutic case. For example, one might want to obtain a sustained profile of release over an extended period of time for prolonged therapeutic effect. A pulsative release that mimics physiological secretion behaviour is preferred in the case of delivery of insulin for the treatment of diabetes, for example.
- In some cases it is desirable that the system initiates release in response to an external stimulus (magnetic, electrical, ultrasonic, photonic, thermal, etc.).
- In the case of a self-regulated system, release in response to changes in the local bio-environment such as pH, temperature, enzyme activity, antibody concentration, etc. This implies a biosensing feature in addition to release.
- The fixture must disintegrate in the body for biocompatibility and biodegradability purposes.
- Finally, the drug delivery system must satisfy long term toxicological requirements while respecting the patient's compliance with the regimen.

Traditional controlled-release drug delivery systems have used polymer materials as key components because of the particular range of physical and chemical properties they offer such as diffusivity, permeability, biocompatibility, solubility, and their response to pH or temperature changes. The diffusion, dissolution, permeation and swelling characteristics of these materials have been utilized to obtain constant release of entrapped molecules [2, 3]. In most cases, polylactide (PLA), and polyglycolide (PGA) have been used in either homopolymer or copolymer form for *in vivo* clinical applications because of their low toxicity.

12.2. POROUS MATERIAL-BASED SYSTEMS

Porous nanostructured materials (with pore sizes ranging from a few nanometers to several microns) have emerged as a new class of efficient vehicles for drug delivery. Made of various components including polymers [4], lipids (for example liposomes, micelles,

TABLE 12.1. Common "Soft" materials used in controlled drug release.

Material	Comments

Polymer nanocapsules [7]
SIZE: Smaller than 1000 nm
DESCRIPTION: Single polymeric
  membrane enclosing an aqueous
  or oily cavity
	—provides efficient drug protection
	—provides efficient controlled release of entrapped drug
	—biodegradable
	—tunable surface chemistry for targeting
	—suitable for cancer therapy (accommodates multiple drug resistance).

Liposomes [8]
SIZE: 25–2500 nm
DESCRIPTION: vesicles made from
  single or multi bilayered
  phospholipid membranes
  enclosing an aqueous cavity.
	—well tolerated excipient, safe for medical applications
	—provides efficient drug protection
	—Biomimetic artificial system (model for cell membranes when made
	  of natural phospholipids)
	—can be made sensitive to pH changes, light, magnetic fields.
	—can entrap substances in either the membrane or in the cavity
	—flexibility in formulation of chemotherapeutic agents
	—amenable to mass production
	—poor storage stability
	—faster degradation rate in vivo than polymers

Solid lipid nanoparticles [6]
SIZE: 80–1000 nm
DESCRIPTION: particles made
  from lipids that are solid at room
  temperature and body temperature
  (in contrast to emulsions, where
  the lipid phase is liquid).
	—more stable in biological fluid and during storage than liposomes
	—good protection against drug degradation

"Liposil": silicic
  liposome-templated nanocapsules
  [9, 10]
SIZE: 1 to several microns
DESCRIPTION: unilamellar
  phospholipid membrane vesicle
  encapsulated in a non porous
  silica shell
	—very stable formulation
	—silica shell resistant to low pH (gastric conditions)
	—promising for oral administration drug delivery systems
	—possibility of long-term storage and protection of molecules
	—slow release rates-amenable to longer term regimen
	—dry formulation

microemulsions, nanoemulsions, solid lipid nanoparticles) [5, 6], and inorganic materials, reservoir-containing structures in general have the advantage of providing flexibility in preparation. One can control morphological and chemical parameters such as the size and shape of the material, the number of "reservoirs" and their volume, the wall thickness as well as the surface chemistry, permeability and resorption rate. Porous materials allow the design of more sophisticated drug carrier systems for better control of the vectorisation and the release kinetics of the drug. A variety of soft materials are under study, outlined in Table 12.1.

12.3. SILICON-BASED POROUS MATERIALS

A recent approach in the development of controllable reservoir-based drug delivery systems involves the use of more rigid inorganic porous solids as substrate materials for the preparation of functionalized organic/inorganic nanostructured drug carriers [11]. In particular, nanostructured materials based on Si are very promising platforms for pharmaceutical applications. Porous silica [12, 13] and porous Si [14] are good examples of this class. Their more complex architectural and chemical structures provide these hybrid systems with specific functionalities that allow them to respond to a designated stimulus. Widely studied and characterized, mesoporous silica from the M41S family are templated materials made by a chemical route. The synthesis typically involves self assembly of silica using surfactant micelles as the structuring agent [15–21]. These non-toxic materials exhibit well-defined, ordered porosity with a large specific surface area (up to 1000 $m^2 g^{-1}$), a large mesoporous volume, and thermal stability. The pore sizes can be controlled during the synthesis, and typically range from 15 Å to 100 Å. The silanol-terminated pore walls can be functionalized using convenient chemistry to provide specificity for drug absorption and release schemes [13, 22].

12.4. "OBEDIENT" MATERIALS

Like electronically engineered microchips, some hybrid drug delivery systems can be designed to respond to external commands. For example "gates" that will open and close the pores can be installed, releasing the right amount of drug when desired. Tanaka and coworkers have developed a system where coumarin ligands are attached at the entrance of the pores of the mesoporous silica MCM41 [23, 24]. Irradiation with light causes dimerization of the coumarin and closes the pores of the matrix. The process is reversible, allowing the pores to open and release the stored active compound. In a similar manner Lin and coworkers have shown that nanocrystals made of CdS can be functionalized and attached to the surface of the pores of porous silica MCM41 and play the role of caps to trap drugs in the porous matrix. Release of the drug is then triggered by a specific chemical or enzymatic reaction that removes the CdS caps [12].

12.5. POROUS SILICON

The attraction of placing active electronic circuit components into *in-vivo* drug delivery materials led to the exploration of elemental silicon as a biomaterial. In particular a porous form of Si produced by an electrochemical corrosion reaction has been of interest. Since the pioneering work of Canham and others in the late 90's demonstrating the biocompatibility and biodegradability of porous Si *in vitro* and *in vivo* [25–35], this material has been under intensive investigation for controlled drug delivery applications. Like mesoporous silica of the MCM41 class, porous Si offers tuneable structural properties: a large specific surface area, large free volume, and pore sizes that can be controlled from a few nanometers to several hundreds of nanometers depending on the preparation conditions. The surface of freshly prepared porous Si is easily modified via convenient chemistry with a large range of organic or biological molecules (ex: antibody, proteins, etc.) [36]. Recently Swaan and coworkers have performed *in vitro* experiments showing that porous Si particles can be used as efficient delivery vehicles of insulin across intestinal epithelial cells [37]. The drug permeation rate through the membrane is dramatically enhanced when delivered via porous Si particles compared with conventional liquid formulations.

Like other Si-based materials, porous Si offers attractive morphological and chemical properties for biomedical applications but it has one supplementary dimension: its optical properties. Porous Si displays fluorescence deriving from Si quantum dot structures that are produced during the etch [38], and it can also display unique optical reflectivity spectra [39, 40]. Both of these features allow porous Si to exhibit a signal that is affected in a predictable way when exposed to environmental changes [32, 41–46]. This presents new possibilities for the development of more advanced functional systems, referred to as "intelligent systems," that incorporate a sensor for either diagnostic or therapeutic functions. The ease with which porous Si can be integrated into well-established Si microelectronics fabrication techniques should lead to more sophisticated, active devices for medical applications [35, 47–48].

12.6. TEMPLATED NANOMATERIALS

Any porous solid can act as a structural template, and the fabrication of ordered nanostructures using templates has been investigated extensively [49]. Porous alumina membranes [50–51], zeolites [20], and crystalline colloidal arrays [52–54] are commonly used as templates to construct elaborate electronic, mechanical, or optical structures. Recent work has employed other template materials such as porous Si [55–56], and the use of templated materials for biosensing [57] and drug delivery [56] has been demonstrated. For in-vivo applications, templated structures allow one to impart a unique micro- or nanostructure into a recognized biocompatible and therapeutically useful material. One obvious advantage of such an approach is that it can provide an additional degree of control over the dissolution rate of the fixture and the drug release profile. Most recently, the possibility of incorporating a unique optical signature such as a photonic crystal has emerged as an additional feature of the template approach.

12.7. PHOTONIC CRYSTALS AS SELF-REPORTING BIOMATERIALS

Photonic crystals are materials whose index of refraction varies with a periodicity on the order of a few hundred nanometers over length scales of at least a few microns. Such structures can diffract visible light. Familiar examples from Nature include opals, the inside of abalone shells [58], and the carapace of many species of beetle [59]. Photonic crystals constructed from inorganic materials are an active area of research for optical switching, optical computing, and other optoelectronics applications [60], and the capabilities of these materials to act as sensors for chemical or biological compounds has led to a series of developments in the biomedical field. One of the early demonstrations of the potential for photonic crystals in medicine came from the laboratories of Sanford Asher at the University of Pittsburgh. By incorporating a photonic crystal into a biocompatible hydrogel matrix, the Asher group developed contact lenses that change color depending on the concentration of glucose in the wearer's blood [57, 61]. Sensors for proteins, DNA, and small molecules have also been developed based on these and related photonic crystals [44, 62–64].

12.8. USING POROUS SI AS A TEMPLATE FOR OPTICAL NANOSTRUCTURES

As mentioned above, porous Si offers tuneable structural properties, which makes it an interesting candidate material for controlled drug delivery. In addition, with its easily

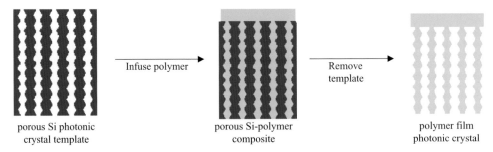

FIGURE 12.1. Template approach for constructing photonic materials for controlled release and monitoring of drugs.

controlled nanostructure, porous Si has been demonstrated to be an excellent template material for construction of elaborate photonic organic and biological polymers [56]. The advantage of such an approach is that it allows one to impart the desirable optical features of the porous Si master to a polymer that possesses the required biocompatibility, resorbability, or drug solubility parameters.

A schematic of the templating method used to produce an optical structure suitable for medical applications is shown in Figure 12.1. This approach was first demonstrated using a rugate dielectric mirror as the template [56]. A rugate dielectric mirror is a structure that contains a sinusoidal refractive index variation, producing a sharp spectral feature in the optical spectrum. The multilayered porous Si template containing nanometer-scale pores is prepared by anodic electrochemical etch of a crystalline Si wafer using a pseudo-sinusoidal current-time waveform [39, 62, 65–69]. The sharp features in the optical reflectivity spectrum are controlled by adjustment of the frequency and amplitude of the sinusoidal current-time waveform [70].

One challenge of the templating approach is to efficiently remove the master from the templated biomaterial, particularly if the biocompatibility of the master is an issue. The porous Si multilayer masters can be converted to SiO_2 by thermal oxidation prior to solution-casting or injection-molding of the daughter material. Removal of the porous SiO_2 template from the polymer or biopolymer imprint can then be achieved by exposure to a dilute solution of HF. Alternatively, the porous Si template dissolves in strong aqueous base. In either case, chemical dissolution of the template provides a freestanding porous polymer film [56]. The polymer replicas inherit an inverse of the optical structure of the template.

It is possible to fabricate more sophisticated optical structures from porous Si films. In 1995, Pavesi and Mazzoleni reported the first microcavity made entirely out of porous Si [40]. These were luminescent structures comprised of planar microcavities with a luminescent porous Si active medium sandwiched between two distributed Bragg reflectors. Recently the superposition of multiple spectral features in one monolithic film has been demonstrated [70–71]. By constructing more elaborate optical structures, one can design into the material characteristic spectral "bar codes" that allow the fixture to be distinguished from tissue, light scattering centers, or highly colored materials in the body [56, 66].

A self-reporting drug delivery matrix is desired for various non-invasive or minimally invasive applications. Implants that change their spectral properties as they degrade or as they release a loaded drug could be imaged using visible light if implanted in transparent media such as the vitreous body of the eye. Alternatively, if near-infrared, tissue-penetrating

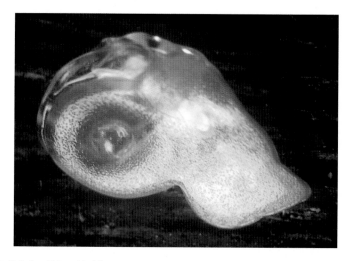

FIGURE 12.2. Poly(lactide) molded from a porous Si photonic crystal template. This polymer contains the drug caffeine. Decay of the optical spectrum (the green color visible in the image) from this fixture provides a surrogate measure of the drug delivery rate.

spectral features are encoded into the material, the fixture conceivably could be probed through the skin or though several millimetres of visibly opaque tissue. This latter concept has been demonstrated with a rugate optical structure made of biodegradable poly-lactide, impregnated with a test drug (Figure 12.2) [56]. Drug release correlates to the decrease in intensity of light reflected by the rugate structure as expected. By placing the spectral feature of the poly-lactide imprint within the low absorbance, near infrared window of human tissue, a drug delivery matrix that could be read through the skin was demonstrated [56].

12.9. OUTLOOK FOR NANOTECHNOLOGY IN PHARMACEUTICAL RESEARCH

The use of porous Si photonic crystals as templates is just one example among many of the application of nanomaterials to the pharmaceutical field. Additional examples include polymer nanoparticles in which the drug is uniformly dispersed, metallic nanoparticles that contain a surface coating of drug, and microporous materials that contain the drug within nanometer-dimension voids. In all of these systems, the high surface to volume ratio of nanoparticles makes them amenable to surface modification for efficient targeting. Advances in nanostructured materials should lead to other approaches for the controlled release of drugs, and they will continue to provide exciting opportunities in medicine.

ACKNOWLEDGEMENTS

The authors gratefully acknowledge helpful discussions with Prof. Sangeeta Bhatia of the Department of Bioengineering at the University of California, San Diego, Dr. Erkki Ruoslahti of the Burnham Institute in La Jolla, California, Dr. Lingyun Cheng and

Dr. William Freeman of the Shiley Eye Center at the University of California, San Diego, and Dr. Jean-Marie Devoisselle and Sylvie Begu of the UMR CNRS/ENSCM in Montpellier, France. The authors thank TREX industries, inc. and the National Cancer Institute of the National Institutes of Health for financial support.

REFERENCES

[1] D.L. Wise. (ed.). *Handbook of Pharmaceutical Controlled Release Technology.* Marcel Dekker Inc., New York, 2000.

[2] W. Amass, A. Amass, and B. Tighe. A review of biodegradable polymers: uses, current developments in the synthesis and characterization of biodegradable polyesters, blends of biodegradable polymers and recent advances in biodegradation studies. *Polym. Int.*, 47:89–144, 1998.

[3] K.E. Uhrich, S.M. Cannizzaro, R.S. Langer, and K.M. Shakesheff. Polymeric systems for controlled drug release. *Chem. Rev.*, 11:3181–3198, 1999.

[4] K.S. Soppimath, T.M. Aminabhavi, A.R. Kulkarni, and W.E. Rudzinski. Biodegradable polymeric nanoparticles as drug delivery devices. *Control. Rel.*, 70:1–20, 2001.

[5] B. Heurtault, P. Saulnier, B. Pech, J.E. Proust, and J.P Benoit. *Biomaterials*, 24:4283, 2003.

[6] R.H. Muller, M. Radke, and S.A. Wissing. Solid lipid nanoparticles (SLN) and nanostructured lipid carriers (NLC) in cosmetics and dermatological preparations. *Adv. Drug Delivery Rev.*, 54 (Suppl. 1):S131–S155, 2002.

[7] I. Brigger, C. Dubernet, and P. Couvreur. Nanoparticles in cancer therapy and diagnosis. *Adv. Drug Del. Rev.*, 54:631–651, 2002.

[8] R.R.C. New (ed.). *Liposomes a practical approach*, Oxford University Press, New York, 1990.

[9] S. Bégu et al. Preparation and characterization of silicious material using liposomes as template. *Chem. Commun.*, 640–641, 2003.

[10] S. Bégu et al. Characterization of a phospholipid bilayer entrapped into non-porous silica nanospheres. *J. Mater. Chem.*, 14:1316–1320, 2004.

[11] E. Ruiz-Hitzky. Functionalizing inorganic solids: towards organic-inorganic nanostructured materials for intelligent and bioinspired systems. *The Chem. Rec.*, 3:88–100, 2003.

[12] C.-Y. Lai et al. A mesoporous silica nanosphere-based carrier system with chemically removable CdS nanoparticle caps for stimuli-responsive controlled release of neurotransmitters and drug molecules. *J. Am. Chem. Soc.*, 125:4451–4459, 2003.

[13] R. Aiello et al. Mesoporous silicate as matrix for drug delivery systems of non-steroidal antiinflamatory drugs. *Stud. Surf. Sci. Catal.*, 14:1165–1172, 2002.

[14] L.T. Canham et al. Derivatized porous silicon mirrors: implantable optical components with slow resorbability. *Phys. Stat. Sol. A*, 182:521–525, 2000.

[15] U. Cieslaa and F. Schüth. Ordered mesoporous materials. *Micropor. Mesopor. Mater.*, 27:131–149, 1999.

[16] J.S. Beck and J.C. Vartuli Recent advances in the synthesis, characterization and applications of mesoporous molecular sieves. *Curr. Opin. Sol. State. Mat. Sci.*, 1:76–87, 1996.

[17] J.Y. Ying, C.P. Mehnert, and M.S. Wong, *Angew. Chem.*, 38:58, 1999.

[18] N. Hüsing and U. Schubert. Aerogels—airy materials: chemistry, structure, and properties. *Angew. Chem.*, 37:22–45, 1998.

[19] A. Sayari. Catalysis by crystalline mesoporous molecular sieves. *Chem. Mater.*, 8:1840–1852, 1996.

[20] K. Moller and T. Bein. Inclusion chemistry in periodic mesoporous hosts. *Chem. Mater.*, 10:2950–2963, 1998.

[21] A. Stein, B.J. Melde, and R.C. Schroden. Hybrid inorganic-organic meso-porous silicates-nanoscopic reactors coming of age. *Adv. Mater.*, 12:1403–1419, 2000.

[22] C. Tourné-Peteilh et al. The potential of ordered mesoporous for the storage of drugs: the example of a pentapeptide encapsulated in a MSU-Tween 80. *Chem. Phys. Chem.*, 3:2003.

[23] N.K. Mal, M. Fujiwara, and Y. Tanaka. Photocontrolled reversible release of guest molecules from coumarin-modified mesoporous silica. *Nature*, 421:350–353, 2003.

[24] N.K. Mal, M. Fujiwara, Y. Tanaka, T. Taguchi, and M. Matsukata. Photo-switched storage and release of guest molecules in the pore void of coumarin-modified MCM-41. *Chem. Mater.*, 15:3385–3394, 2003.

[25] S.H.C. Anderson, H. Elliott, D.J. Wallis, L.T. Canham, and J.J. Powell. Dissolution of different forms of partially porous silicon wafers under simulated physiological conditions. *Phys. Stat. Solidi a-Appl. Res.*, 197:331–335, 2003.

[26] J.M. Ji, X. Li, L.T. Canham, and J.L. Coffer. Use of microcontact printing methods to direct pattern formation of calcified mesoporous silicon. *Adv. Mat.*, 14:41–43, 2002.

[27] T. Jay, L.T. Canham, K. Heald, C.L. Reeves, and R. Downing. Autoclaving of porous silicon within a hospital environment: potential benefits and problems. *Phys. Stat. Solid. A-Appl. Res.*, 182, 555–560, 2000.

[28] L.T. Canham et al. Derivatized mesoporous silicon with dramatically improved stability in simulated human blood plasma. *Adv. Mat.*, 11:1505-+, 1999.

[29] M. Wainwright, L.T. Canham, K. Al-Wajeeh, and C.L. Reeves. Morphological changes (including filamentation) in Escherichia coli grown under starvation conditions on silicon wafers and other surfaces. *Lett. Appl. Microbiol.*, 29:224–227, 1999.

[30] L.T. Canham et al. Calcium phosphate nucleation on porous silicon: Factors influencing kinetics in acellular simulated body fluids. *Thin Solid Films*, 297:304–307, 1997.

[31] S.C. Bayliss et al. Phosphate and cell growth on nanostructured semiconductors. *J. Mat. Sci. Lett.*, 16:737–740, 1997.

[32] S.C. Bayliss, L.D. Buckberry, I. Fletcher, and M.J. Tobin. The culture of neurons on silicon. *Sens. Actu. A-Phys.*, 74:139–142, 1999.

[33] S.C. Bayliss, R. Heald, D.I. Fletcher, and L.D. Buckberry. The culture of mammalian cells on nanostructured silicon. *Adv. Mat.* 11:318–321, 1999.

[34] S.C. Bayliss, L.D. Buckberry, P.J. Harris, and M Tobin. Nature of the silicon-animal cell interface. *J. Por. Mat.*, 7:191–195, 2000.

[35] A.H. Mayne, S.C. Bayliss, P. Barr, M. Tobin, and L. D. Buckberry. Biologically interfaced porous silicon devices. *Phys. Stat. Sol. A*, 182:505–513 2000.

[36] J. M. Buriak. Organometallic chemistry on silicon and germanium surfaces. *Chem. Rev.*, 102:1272–1308, 2002.

[37] A.B. Foraker et al. Microfabricated porous silicon particles enhance paracellular delivery of insulin across intestinal Caco-2 cell monolayers. *Pharma. Res.*, 20:110–116, 2003.

[38] R.T. Collins, P.M. Fauchet, and M.A. Tischler. Porous silicon: from luminescence to LEDs. *Phys. Today*, 50:24–31, 1997.

[39] L. Pavesi and P. Dubos. Random porous silicon multilayers: application to distributed Bragg reflectors and interferential Fabry-Perot filters. *Semicon. Sci. Tech.*, 12:570–575, 1997.

[40] C. Mazzoleni and L. Pavesi. Application to optical components of dielectric porous silicon multilayers. *Appl. Phys. Lett.*, 67:2983–2985, 1995.

[41] V.S. Lin, K. Motesharei, K.S. Dancil, M.J. Sailor, and M.R. Ghadiri. A porous silicon-based optical interferometric biosensor. *Science*, 278:840–843, 1997.

[42] M.J. Sailor. *Properties of Porous Silicon.* Canham, L. (ed.), Short Run Press Ltd., London, pp. 364–370, 1997.

[43] K.-P.S. Dancil, D.P. Greiner, and M.J. Sailor. A porous silicon optical biosensor: detection of reversible binding of IgG to a protein A-modified surface. *J. Am. Chem. Soc.*, 121:7925–7930, 1999.

[44] S. Chan, P.M. Fauchet, Y. Li, L.J. Rothberg, and B.L. Miller. Porous silicon microcavities for biosensing applications. *Phys. Status Solid A*, 182:541–546, 2000.

[45] S. Zangooie, R. Jansson, and H. Arwin. Ellipsometric characterization of anisotropic porous silicon Fabry-Perot filters and investigation of temperature effects on capillary condensation efficiency. *J. Appl. Phys.*, 86, 850–858, 1999.

[46] A.M Tinsley-Bown et al. Tuning the pore size and surface chemistry of porous silicon for immunoassays. *Phys. Status Solid. A*, 182:547–553, 2000.

[47] A.G. Nassiopoulos et al. Sub-micrometre luminescent porous silicon structures using lithographically patterned substrates. *Thin Sol. Films*, 255:329–333, 1995.

[48] M.P. Stewart and J.M. Buriak. Photopatterned hydrosilylation on porous silicon. *Angew. Chem. Int. Ed. Engl.*, 37:3257–3260, 1998.

[49] S. Polarz and M. Antonietti. Porous materials via nanocasting procedures: innovative materials and learning about soft-matter organization. *JCS Chem. Commun.*, 2593–2604, 2002.

[50] M. Wirtz, M. Parker, Y. Kobayashi, and C.R. Martin. Template synthesized nanotubes for chemical separations and analysis. *Chem. Eur. J.*, 16:3572–3578, 2002.

[51] J.C. Hulteen and C.R. Martin. A general template-based method for the preparation of nanomaterials. *J. Mater. Chem.*, 7:1075–1087, 1997.

[52] C.E. Reese, M.E. Baltusavich, J.P. Keim, and S.A. Asher. Development of an intelligent polymerized crystalline colloidal array colorimetric reagent. *Anal. Chem.*, 73:5038–5042, 2001.

[53] X. Xu, S.A. Majetich, and S.A. Asher. Mesoscopic monodisperse ferromagnetic colloids enable magnetically controlled photonic crystals. *J. Am. Chem. Soc.*, 124:13864–13868, 2002.

[54] C. L. Haynes and R. P. Van Duyne. Nanosphere lithography: A versatile nanofabrication tool for studies of size-dependent nanoparticle optics. *J. Phys. Chem. B*, 105:5599–5611, 2001.

[55] S. Matthias et al. Monodisperse diameter-modulated gold microwires. *Adv. Mater.*, 14:1618–1621, 2002.

[56] Y.Y. Li et al. Polymer replicas of photonic porous silicon for sensing and drug delivery applications.*Science*, 299:2045–2047, 2003.

[57] V.L.S. Alexeev, C. Anjal A.V. Goponenko, S Das, I.K. Lednev, C.S. Wilcox, D.N. Finegold, and S.A. Asher. High ionic strength glucose-sensing photonic crystal. *Anal. Chem.*, 75:2316–2323, 2003.

[58] G. Mayer and M. Sarikaya. Rigid biological composite materials: structural examples for biomimetic design. *Exper. Mech.*, 42:395–403, 2002.

[59] A.R. Parker, D.R. Mckenzie, and M.C.J Large. Multilayer reflectors in animals using green and gold beetles as contrasting examples. *J. Exp. Biol.*, 201:1307–1313, 1998.

[60] E. Chomski and G.A. Ozin Panoscopic silicon-a material for all length scales. *Adv. Mater.*, 12:1071–1078, 2000.

[61] A.C.J. Sharma, K. Tushar; S. Rasu, V. Lianjun; A. Mohamed, D.N. Finegold, and S.A. Asher. A general photonic crystal sensing motif: creatinine in bodily fluids. *J. Am. Chem. Soc.*, 126:2971–2977, 2004.

[62] P.A. Snow, E.K. Squire, P.S.J. Russell, and L.T. Canham, Vapor sensing using the optical properties of porous silicon Bragg mirrors. *J. Appl. Phys.*, 86:1781–1784, 1999.

[63] S. Chan, S.R. Horner, B.L. Miller, and P.M.Fauchet. Identification of gram negative bacteria using nanoscale silicon microcavities. *J. Am. Chem. Soc.*, 123:11797–11798, 2001.

[64] S. Chan and P.M. Fauchet. Nanoscale microcavities for biomedical sensor applications. *Proc. SPIE 3912*, 23:2000.

[65] T.A. Schmedake, F. Cunin, J.R. Link, and M.J.S. Sailor. Detection of chemicals using porous silicon "smart dust" particles. *Adv. Mater.*, 14:1270–1272, 2002.

[66] F. Cunin et al. Biomolecular screening with encoded porous silicon photonic crystals. *Nat. Mater.*, 1:39–41 2002.

[67] V. Lehmann, R. Stengl, H. Reisinger, R. Detemple, and W. Theiss. Optical shortpass filters based on macroporous silicon. *Appl. Phys. Lett.*, 78:589–591, 2001.

[68] M. Thonissen and M.G.Berger. *Properties of Porous Silicon.* Canham, L. (ed.) Short Run Press Ltd., London, pp. 30–37 1997.

[69] G. Vincent Optical properties of porous silicon superlattices. *Appl. Phys. Lett.*, 64:2367–2369, 1994.

[70] M.G. Berger et al. Dielectric filters made of porous silicon: advanced performance by oxidation and new layer structures. *Thin Sol. Films*, 297:237–240, 1997.

[71] S.O. Meade, M.S. Yoon, K.H. Ahn, and M.J. Sailor. Porous silicon photonic crystals as encoded microcarriers. *Adv. Mater.*, 2004 (in press).

13

Transdermal Drug Delivery using Low-Frequency Sonophoresis

Samir Mitragotri

*Department of Chemical Engineering, University of California,
Santa Barbara, CA 93106*

13.1. INTRODUCTION

Transdermal drug delivery offers several advantages over traditional drug delivery systems such as oral delivery and injection, especially in regard to protein delivery. These advantages include:

13.1.1. Avoiding Drug Degradation in Gastrointestinal Tract

Orally administered drugs are highly susceptible to gastro-enteric and first pass metabolism. On the other hand, in the case of transdermal drug delivery, the drug diffuses through the skin and is absorbed by the capillary network under the skin. Accordingly, the gastro-intestinal metabolism of the drug is avoided.

13.1.2. Better Patient Compliance

Transdermal drug devices are easier to handle and use than injections. In addition, the pain associated with the injection can also be avoided. These characteristics of transdermal drug delivery offer a significant advantage in cases where frequent drug doses are required, for example, in the case of insulin delivery.

13.1.3. Sustained Release of the Drug can be Obtained

Transdermal drug delivery can provide sustained release of drugs over a sufficiently long time (up to a week). Hence, it is possible to maintain a steady drug concentration in the blood. This is especially important for drugs with a narrow therapeutic window.

Transdermal drug delivery, however, suffers from the severe limitation that the permeability of the skin is very low. Therefore, it is difficult to deliver drugs across the skin at a therapeutically relevant rate. This, in fact, is the main reason why only a handful of low-molecular weight drugs are clinically administered by this route today.

13.2. ULTRASOUND IN MEDICAL APPLICATIONS

Ultrasound is used in various medical therapies including lithotripsy [1], hyperthermia [2], thrombolysis [3], lipoplasty [4], wound healing [5], fracture healing [6], and drug delivery (sonophoresis [7–9], sonoporation [10–13], triggered drug release [14–17], and targeted drug delivery [18, 19]). A majority of these applications have come about in the last decade. Ultrasound has also been used to enhance transdermal transport of various drugs including macromolecules [7–9, 20–69]. This type of enhancement is termed sonophoresis, indicating the enhanced transport of molecules under the influence of ultrasound. Ultrasound at various frequencies in the range of 20 kHz–16 MHz has been used to enhance skin permeability. However, transdermal transport enhancement induced by low-frequency ultrasound ($f < 100$ kHz) has been found to be more significant than that induced by high frequency ultrasound [7, 8, 70].

13.3. SONOPHORESIS: ULTRASOUND-MEDIATED TRANSDERMAL TRANSPORT

The first published report on sonophoresis dates back to 1950's. Fellinger and Schmidt [71] reported successful treatment of polyarthritis of the hand's digital joints using hydrocortisone ointment with sonophoresis. It was subsequently shown that hydrocortisone injection combined with ultrasound "massage" yielded better outcome compared to simple hydrocortisone injections for bursitis treatment [72]. Cameroy [26] reported success using carbocaine sonophoresis for closed Colle's fractures. In a series of publications Griffin et al. showed improved treatment of elbow epicondylitis, bicipital tendonitis, shoulder osteoarthritis, shoulder bursitis and knee osteoarthritis by combined application of hydrocortisone and ultrasound [29–31, 73]. Improved dermal penetration using ultrasound was also reported for local anesthetics [20, 74, 75].

Studies demonstrated that ultrasound enhanced the percutaneous absorption of methyl and ethyl nicotinate by disordering the structured lipids in the stratum corneum. Similar conclusions were reached by Hofman and Moll [76] who studied the percutaneous absorption of benzyl nicotinate. While several investigators reported positive effect of ultrasound on drug permeation, lack of an effect of ultrasound on skin permeation was also reported in certain cases. For example, Williams reported no detectable effect of ultrasound on the rate of penetration of three anesthetic preparations through human skin [63].

Levy et al. [77] showed that 3–5 minutes of ultrasound exposure (1 MHz, 1.5 W/cm^2) increased transdermal permeation of mannitol and physostigmine across hairless rat skin *in vivo* by up to 15-fold. They also reported that the lag time typically associated with transdermal drug delivery was nearly-completely eliminated after exposure to ultrasound. Mitragotri et al. reported *in vitro* permeation enhncement of several low-molecular weight drugs under the same ultrasound conditions [41].

Bommanan et al. [23, 24] hypothesized that since the absorption coefficient of the skin varies directly with the ultrasound frequency, high frequency ultrasound energy would concentrate more in the epidermis, thus leading to higher enhancements. In order to assess this hypothesis, they studied the effect of high-frequency ultrasound (2 MHz-16 MHz) on permeability of salicylic acid (dissolved in a gel) through hairless guinea pig skin *in vivo*. They found that a 20 minute application of ultrasound (0.2 W/cm^2) at a frequency of 2 MHz did not significantly enhance the amount of salicylic acid penetrating the skin. However, 10 MHz ultrasound under otherwise same conditions resulted in a 4-fold increase and 16 MHz ultrasound resulted in about a 2.5-fold increase in transdermal salicylic acid transport [23, 24].

13.4. LOW-FREQUENCY SONOPHORESIS

Low-frequency sonophoresis has been a topic of extensive research only in the last 10 years. Tachibana et al. [60–62] reported that application of low-frequency ultrasound (48 kHz) enhanced transdermal transport of lidocaine and insulin across hairless rat skin *in vivo*. They found that the blood glucose level of a hairless rat immersed in a beaker filled with insulin solution (20 U/ml) and placed in an ultrasound bath (48 kHz, 5000 Pa) decreased by 50 % in 240 minutes [62]. They also showed that application of ultrasound under similar conditions prolonged the anesthetic effect of transdermally administered lidocaine in hairless rats [61] and enhanced transdermal insulin transport in rabbits. Mitragotri et al. [8, 43] showed that application of ultrasound at even lower frequencies (20 kHz) enhances transdermal transport of various low-molecular weight drugs including corticosterone and high-molecular weight proteins such as insulin, γ-interferon, and erythropoeitin across the human skin *in vitro*. Quantitatively, Mitragotri et al. compared the enhancement ratios (ratio of the sonophoretic and passive permeabilities measured *in vitro* across human cadaver skin) induced by therapeutic ultrasound (1 MHz) and low-frequency ultrasound (20 kHz) for four permeants, butanol, corticosterone, salicylic acid, and sucrose. They found that the enhancement induced by low-frequency ultrasound is up to 1000-fold higher than that induced by therapeutic ultrasound [43].

Low-frequency sonophoresis can be classified into two categories; simultaneous sonophoresis and pretreatment sonophoresis. Simultaneous sonophoresis corresponds to a simultaneous application of drug and ultrasound to the skin. This was the first mode in which low-frequency sonophoresis was shown to be effective. This method enhances trans-dermal transport in two ways: i) enhanced diffusion through structural alterations of the skin and ii) convection induced by ultrasound. Transdermal transport enhancement induced by this type of sonophoresis decreases after ultrasound is turned off [49]. Although this method can be used to achieve a temporal control over skin permeability, it requires that the patients use a wearable ultrasound device for drug delivery. In pretreatment sonophoresis, a short application of ultrasound is used to permeabilize skin prior to drug delivery. The skin remains in a state of high permeability for several hours. Drugs can be delivered through permeabilized skin during this period. In this approach, the patient does not need to wear the ultrasound device.

13.5. LOW-FREQUENCY SONOPHORESIS: CHOICE OF PARAMETERS

The enhancement induced by low-frequency sonophoresis is determined by four main ultrasound parameters, frequency, intensity, duty cycle, and application time. A detailed investigation of the dependence of permeability enhancement on frequency and intensity in the low-frequency regime (20 kHz $< f <$ 100 kHz) has been reported by Tezel et al. [7]. At each frequency, there exists an intensity below which no detectable enhancement is observed. This intensity is referred to as the threshold intensity. Once the intensity exceeds this threshold, the enhancement increases strongly with the intensity until another threshold intensity, referred to as the decoupling intensity is reached. Beyond this intensity, the enhancement does not increase with further increase in the intensity due to acoustic decoupling. The threshold intensity for porcine skin increased from about 0.11 W/cm^2 at 19.6 kHz to more than 2 W/cm^2 at 93.4 kHz. At a given intensity, the enhancement decreased with increasing ultrasound frequency.

The dependence of enhancement on intensity, duty cycle, and application time can be combined into a single parameter, total energy density delivered from the transducer, $E = It$ where I is the ultrasound intensity (W/cm^2), t is the net exposure time (seconds). As a general trend, no significant enhancement is observed until a threshold energy dose is reached. The threshold energy doses for various frequencies were found to be 10 J/cm^2 for 19.6 kHz, 63 J/cm^2 at 36.9 kHz, 103 J/cm^2 at 58.9 kHz, 304 J/cm^2 for 76.6 kHz, and 1305 J/cm^2 at 93.4 kHz. Thus, the threshold energy dose increased by about 130-fold as the frequency increased from 19.6 kHz to 93.4 kHz. The dependence of enhancement on energy density after the threshold is different for different frequencies. For extremely high-energy doses (say 10^4 J/cm^2), the enhancement induced by all the frequencies is comparable. However, for lower energy doses, the differences between various different frequencies are significant and the choice of frequency may affect the effectiveness of sonophoresis.

In addition to frequency and energy density, sonophoretic enhancement also depends on additional parameters including the distance between the transducer and the skin, gas concentration in the coupling medium, and the transducer geometry. Detailed dependence of enhancement on these parameters has not been yet studied.

13.6. MACROMOLECULAR DELIVERY

13.6.1. Peptides and Proteins

Low-frequency sonophoresis has been shown to deliver several macromolecular drugs (Figure 13.1). Tachibana and Tachibana demonstrated that a 5 minute exposure to ultrasound (40 kHz 3000–5000 Pa) induced a significant reduction of blood glucose levels in rats

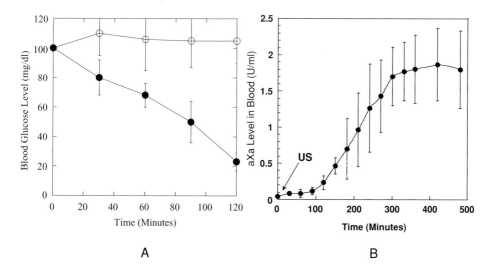

FIGURE 13.1. (**A**) Transdermal insulin delivery using low-frequency sonophoresis (closed circles). Open circles show controls (insulin placed on untreated skin. Skin was permeabilized by a short application of ultrasound (~60-fold enhancement of skin conductivity). Insulin (500 U/ml) was placed on permeabilized skin. (**B**) Transdermal delivery of low-molecular weight heparin (LMWH) with low-frequency sonophoresis. Skin was pretreated with ultrasound for 2 minutes and skin conductance was increased. A solution of LMWH was placed on sonophoretically permeabilized skin. Delivery of LMWH was assessed by measuring plasma aXa activity. Application of LMWH on non-sonicated skin did not increase plasma aXa activity.

exposed to insulin [62]. Specifically, the glucose level decreased to 34% of the initial value at lower pressures and to 22% of the initial value at higher acoustic pressures. Comparable results were obtained in rabbits at somewhat higher frequencies (150 kHz). Mitragotri et al. performed *in vitro* and *in vivo* evaluation of the effect of low-frequency ultrasound on transdermal delivery of proteins [8]. Application of low-frequency ultrasound (20 kHz, 125 mW/cm^2, 100 msec pulses applied every second) enhanced transdermal transport of proteins including insulin, γ-interferon, and erythropoeitin across human cadaver skin *in vitro* [8]. Ultrasound under the same conditions delivered therapeutic doses of insulin across hairless rat skin *in vivo* from a chamber glued on the rat's back and filled with an insulin solution (100 U/ml) [8]. A simultaneous application of insulin and ultrasound (20 kHz, 225 mW/cm^2, 100 msec pulses applied every second) reduced the blood glucose level of diabetic hairless rats from about 400 mg/dL to 200 mg/dL in 30 minutes. A corresponding change in plasma insulin levels was observed during sonophoresis. Boucaud et al. also demonstrated a dose-dependent hypoglycemia in hairless rats exposed to ultrasound and insulin [70]. At an energy dose of 900 J/cm^2 ~75% reduction in glucose levels was reported. Pretreatment of skin by low-frequency ultrasound (20 kHz, ~7 W/cm^2) has also been shown to enhance skin permeability to insulin [78].

13.6.2. Low-molecular Weight Heparin

Low-frequency ultrasound has also been shown to deliver low molecular weight heparin (LMWH) across the skin [48]. Transdermal LMWH delivery was measured by monitoring aXa activity in blood. No significant aXa activity was observed when LMWH was placed on non-treated skin. However, significant amount of LMWH was transported transdermally after ultrasound pretreatment. aXa activity in the blood increased slowly for about 2 hours, after which, it increased rapidly before achieving a steady state after 4 hours at a value of about 2U/ml [48]. Effect of transdermally delivered LMWH was observed well beyond 6 hours in contrast to intravenous or subcutaneous injections, which resulted only in transient biological activity.

13.6.3. Oligonucleotides

Low-frequency ultrasound has also been shown to enhance dermal penetration of oligonucleotides (ODN). A 10-minute application of ultrasound (20 kHz and 2.4 W/cm^2) increased skin ODN permeability to 4.5×10^{-5} cm/hr compared to nearly undetectable values of non-treated skin. A significant amount of ODN was also localized in the skin. Greater enhancements of ODN delivery were obtained by simultaneous application of ultrasound and ODN. Experiments performed with FITC-labeled ODN revealed that ODN is largely localized in the superficial layers of the skin. An estimate of local concentration of ODN in the skin was performed. Assuming a depth of penetration of 100–1000 μm the estimated concentration of ODN in the skin at the end of ultrasound application was about 0.53–5.3 % of the donor concentration. ODN penetration into skin due to LFS was heterogeneous. Heterogeneity of dermal penetration was visualized by monitoring penetration of a dye, sulforhodamine B (SRB) that was incorporated in the coupling medium. SRB penetration clearly indicated 4–5 intensely stained spots ($\sim$1 mm in diameter), which were termed as Localized Transport Pathways, LTPs. To further ensure that ODN penetrated into skin without losing integrity, skin exposed to ISIS 13920 in the presence of ultrasound was assessed using immunohistochemistry. No visible staining was observed in case of passive delivery, however the skin treated with LFS was heavily stained suggesting penetration of oligonucleotide delivery. ODN was localized in the epidermis as well as dermis. Furthermore microscopy studies suggested that ODN penetrated into epidermal cells. This is a particularly appealing feature since viable epidermal cells are an attractive target for ODN delivery.

13.6.4. Vaccines

Recently, low-frequency sonophoresis has also been used to deliver vaccines across the skin (Unpublished data). Transcutaneous immunization promises to be a potent novel vaccination technique since topical immunization elicits both systemic and mucosal immunity [79]. The latter is of great importance, since a significant number of pathogens invade the host via mucosal surfaces [80]. TCI is based on the premise that systemic and mucosal immune responses can be initiated by stimulation of the LCs in the skin.

Ultrasonic delivery of TTx generated a strong IgG response in animals. Delivery of 1.3 μg of TTx generated an immune response comparable to that induced by 10 μg subcutaneous injection. Studies have shown that an IgG antibody response generated by only 5 μg subcutaneous injection is sufficient for protection against a lethal dose of tetanus toxin [81]. Ultrasonic delivery of TTx also generated a strong mucosal immune response. A large number of TT specific plasma cells were found in the intestine. In addition, significant presence of anti-TT antibodies was observed throughout the intestine (unpublished data). Secretory IgA produced by local antibody secreting cells (ASC), i.e. B-lymphocytes, is an important line of defense for mucosal surfaces such as the respiratory or intestinal tract. The polymeric structure of the secretory IgA results in effective cross-linking of large antigens, which are then easily trapped in mucus and eliminated by the action of cilia in the respiratory tract or peristalsis in the gut [80, 82]. Secretory IgA also inhibits infection and colonization by directly preventing the adhesion of bacteria and viruses to mucosal epithelial cells [82, 83].

Two possible mechanisms were proposed to explain why pretreatment of skin with low-frequency ultrasound prior to contact with the antigen vaccine may enhance the immune response. One possible mechanism is that ultrasound pretreatment results in increased delivery of the vaccine compared to control, thus enabling sufficient amount of vaccine to enter the skin in order to activate the skins immune response. However, a comparison of the response obtained by TCI and subcutaneous immunization shows that IgG immune response elicited by TCI is almost 10 -fold more effective per dose compared to subcutaneous injections. The second mechanism involves the involvement of Langerhans and immune cells of the skin that effectively capture the antigen and present it to the immune system. Clear activation of LCs was observed after ultrasonic TTx delivery. LC activation is partly induced by the entry of the antigen and partly by the direct effect of ultrasound on skin. Mechanisms responsible for ultrasound-induced activation of LCs are not clear, although barrier disruption or release of pro-inflammatory signals by the keratinocytes are possible candidates.

13.7. TRANSDERMAL GLUCOSE EXTRACTION USING SONOPHORESIS

Low-frequency ultrasound skin pretreatment has also been used to extract glucose and other analytes from the skin (transport in the opposite direction; from the interstitial compartment through the skin into a reservoir filled with water placed on the top of the pretreated skin). Kost et al. demonstrated that ultrasound pretreatment followed by vacuum application extracts sufficient amounts of interstitial fluid to perform continuous glucose monitoring [37]. The measured glucose flux after application of vacuum (10 in Hg) on ultrasound-pretreated skin was about 52 ± 30 $\mu g/cm^2/hr$ when the average serum glucose concentration of the rat was 183 mg/dL, a flux about 100 time higher compared to passive flux across non-treated skin [84]. This flux corresponds to interstitial fluid (ISF) extraction rate of 25.6 $\mu l/cm^2/hr$.

Correlation between sonophoretically extracted glucose and blood glucose values was assessed in rats [85]. In these experiments, rat skin was exposed to ultrasound. Multiple extractions were performed using vacuum (10 in Hg for 5 minutes applied every 20 minutes) over a period of 2 hours. The first transdermal flux was used to calculate the calibration factor. Blood glucose levels of rats were varied by infusing insulin intravenously at a rate of 10 mU/min for 2 hours. Transdermally extracted glucose flux correlated well with the changes in the blood glucose level in the hypo- and hyperglycemic range. The relationship between the predicted and measured glucose values was linear (r = 0.97). Similar results were reported by Kost et al. in the tests performed in human volunteers [37]. Specifically, ultrasound was used to permeabilize skin of human volunteers. A short application of ultrasound permeabilized skin for about 15 hours. During this period, interstitial fluid was extracted every 30 minutes. Concentration of glucose in the extracted fluid was measured and compared with blood glucose values. The results showed good correlation between glucose in the interstitial fluid and in the blood. Furthermore, patients reported no pain upon ultrasound application. More recently, Kost et al. reported clinical studies performed on diabetic volunteers where ultrasound was used to permeabilize skin and glucose was collected by diffusion instead of vacuum through permeabilized skin. Glucose flux through sonicated site averaged 11 nmol/cm^2/hours [86].

Safety of low-frequency sonophoresis has been evaluated in several studies. Histological studies performed on rat and pig skin indicated no structural changes in the skin on a length scale of $\sim \mu$m [43]. Accordingly, the structural changes in the stratum corneum appear to occur at a sub-micron scale. Singer et al. performed a toxicological analysis of low-frequency sonophoresis. They found a dose-dependent effect of ultrasound on skin. They concluded that low-frequency ultrasound at low intensities appears safe for enhancing the topical delivery of medications, producing only minimal urticarial reactions. Higher-intensity ultrasound produced significant thermal effects. [87] Boucaud et al. also performed a microstructural analysis of skin samples exposed to ultrasound. They reported no detectable changes in the skin structure of human skin at an intensity of 2.5 W/cm^2. Hairless rat skin exposed to the same intensity showed slight and transient erythema and dermal necrosis at 24 hours [88]. Tolerance of low-frequency ultrasound by patients has been reported in a number of studies. Kost et al. reported that low-frequency ultrasound was well-tolerated by patients [37]. More recently, a clinical study on the use of low-frequency ultrasound for lidocaine delivery has also been reported [89]. However, it must be realized that ultrasound, like any other energy source, is likely to exhibit a window of parameters within which safe application can be practiced. Accordingly, a careful selection of parameters must be performed in sonophoresis studies.

13.8. MECHANISMS OF LOW-FREQUENCY SONOPHORESIS

Significant attention has been devoted to understand the mechanisms of low-frequency sonophoresis [23, 41, 64, 90, 91]. A consensus has been reached that acoustic cavitation, the formation and collapse of gaseous cavities, is responsible for low-frequency sonophoresis [9, 41, 57, 91]. Below, we summarize the current conclusions of the mechanistic investigations of sonophoresis.

During low-frequency sonophoresis, cavitation is predominantly induced in the coupling medium (the liquid present between the ultrasound transducer and the skin [57]). The maximum radius of the cavitation bubbles is related to the frequency and acoustic pressure amplitude. Under the conditions used for low-frequency sonophoresis ($f \sim 20$–100 kHz and pressure amplitudes $\sim$1–2.4 bar) the maximum bubble radius is estimated to be between 10–100 μm. Owing to the large bubble size, cavitation is unlikely to occur within the 15 μm thick SC during low-frequency sonophoresis. Accordingly, cavitation in the coupling medium is of primary interest during low-frequency sonophoresis.

Two types of cavitation, stable or inertial have been evaluated for their role in sonophoresis. Stable cavitation corresponds to periodic growth and oscillations of bubbles while inertial cavitation corresponds to violent growth and collapse of cavitation bubbles [92]. Using

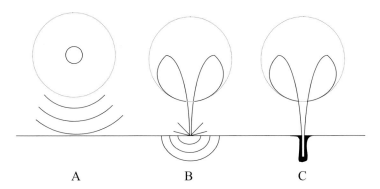

FIGURE 13.2. Three possible modes through which inertial cavitation may enhance SC permeability. **(A)** Spherical collapse near the SC surface emits shock waves, which can potentially disrupt the SC lipid bilayers. **(B)** Impact of an acoustic microjet on the SC surface. The microjet possessing a radius about one tenth of the maximum bubble diameter impacts the SC surface without penetrating into it. The impact pressure of the microjet may enhance SC permeability by disrupting SC lipid bilayers. **(C)** Microjets may physically penetrate into the SC and enhance the SC permeability.

acoustic spectroscopy, stable as well as inertial cavitation has been quantified [9, 57]. The overall dependence of inertial cavitation on ultrasound intensity was found to be similar to that of conductivity enhancement [9, 57]. Specifically, ultrasound intensity above threshold intensity is required before inception of inertial cavitation is observed. This threshold corresponds to minimum pressure amplitude required to induce rapid growth and collapse of cavitation nuclei. Beyond this threshold, white noise (indicator of inertial cavitation) increased linearly with ultrasound intensity, although at any given intensity, inertial cavitation activity decreased rapidly with ultrasound frequency [57]. The threshold intensity for the occurrence of inertial cavitation increased with increasing ultrasound frequency. This dependence reflects the fact that growth of cavitation bubbles becomes increasingly difficult with increasing ultrasound frequency. Tezel et al. showed that regardless of the intensity and frequency, skin conductivity enhancement correlated universally with the cavitation energy density [57]. These data suggested a strong role played by inertial cavitation in low-frequency sonophoresis.

Inertial cavitation occurs in the bulk coupling medium as well as near the skin surface. Inertial cavitation at both locations may potentially be responsible for conductivity enhancement. Three mechanisms by which inertial cavitation events might enhance SC permeability were proposed [93] (Figure 13.2). These include bubbles that collapse symmetrically and emit a shock wave, which can disrupt the SC lipid bilayers and acoustic microjets that might impact the SC without penetration. Impact of microjets may also be responsible for SC lipid bilayer disruption. Microjets resulting from collapsing bubbles near the SC surface may also potentially penetrate into the SC and disrupt the structure.

Inertial cavitation in the vicinity of a surface is fundamentally different from that away from the surface. Specifically, collapse of spherical cavitation bubbles in the bulk solution is symmetric and results in the formation of a shock wave. This shock wave can potentially disrupt the structure of the lipid bilayers. However, the amplitude of the shock wave decreases rapidly with the distance. Collapse of cavitation bubbles near boundaries (especially rigid ones) has been extensively studied in the literature [94]. Specifically, Naude and Ellis showed that cavitation bubbles travel under the influence of ultrasound field towards the boundary and collapse near the boundary depending on its proximity to the surface [95]. The collapse of cavitation bubbles near the boundary is asymmetric due to the difference in the surrounding condition on either side of the bubble. Specifically, the asymmetry in the surroundings leads to the generation of asymmetry in the pressure, which ultimately leads to the formation of a liquid microjet directed towards the surface. The diameter of the microjet is much smaller than that of the maximum bubble radius. There have been several estimates of the speed of the liquid microjet when it strikes the surface (between 50 and 180 m/s [96–98]).

In a recent study, Tezel et al. evaluated the effect of spherical collapses as well as microjets on skin permeability enhancement [91]. They concluded that both types of cavitation events may be responsible for sonophoresis. Regardless of the precise mode of collapse, about 10 collapses/second/cm^2 in the form of spherical collapses or microjets near the surface of the stratum corneum were suggested to explain experimentally observed conductivity enhancements. They also reported that bubble collapses only close to the stratum corneum surface ($\sim$50 μm) contribute to sonophoresis.

Disruption of SC lipid bilayers due to bubble-induced shock waves or microjet impact may enhance skin permeability by at least two mechanisms. First, a moderate level of disruption decreases the structural order of lipid bilayers and increases solute diffusion coefficient [99]. At a higher level of disruption, lipid bilayers may loose structural integrity and facilitate penetration of the coupling medium into the SC. Since many sonophoresis experiments reported in the literature are performed using coupling media comprising aqueous solutions of surfactants, disruption of SC lipid bilayers enhances incorporation of surfactants into lipid bilayers. Incorporation of excessive water and surfactants further promotes bilayer disruption, thereby opening pathways for solute permeation [100, 101]. Recently, Alvarez-Roman reported that lipid extraction also plays a role in low-frequency sonophoresis [64]. They reported that about 30% of the stratum corneum lipids were removed during low-frequency sonophoresis.

13.9. CONCLUSIONS

Low-frequency sonophoresis has been shown to increase skin permeability to a variety of low- as well as high-molecular weight drugs including insulin and low-molecular weight heparin. Ultrasound-mediated enhancement of transdermal transport is mediated by inertial cavitation. Collapse of cavitation bubbles near the stratum corneum is hypothesized to disrupt its structure due to cavitation-generated shock waves or microjets.

REFERENCES

[1] A.J. Coleman and J.E. Saunders. A review of the physical properties and biological effects of the high amplitude acoustic field used in extracorporeal lithotripsy. *Ultrasonics*, 31:75–89, 1993.
[2] C.J. Diederich and K. Hynnen. Ultrasound technology for hyperthemia. *Ultrasound Med. Biol.*, 25:871–887, 1999.

[3] A.V. Alexandrov. Ultrasound-enhanced thrombolysis for stroke: clinical significance. *Eur. J. Ultrasound*, 16:131–140, 2002.

[4] J.C. Goes and A. Landecker. Ultrasound-induced lipoplasty (UAL) in breast surgery. *Aesthetic Plast. Surg.*, 26:1–9, 2002.

[5] C.A. Speed. Therapeutic ultrasound in soft tissue lesions. *Rheumatology*, 40:1331–1336, 2001.

[6] M. Hadjiargyrou, K. McLeod, J.P. Ryaby, and C. Rubin.Enhancement of fracture healing by low intensity *Ultrasound. Clin. Orthop.*, 355(Suppl):S216–S229, 1998.

[7] A. Tezel, A. Sens, J. Tuscherer, and S. Mitragotri. Frequency dependece of sonophoresis. *Pharm. Res.*, 18:1694–1700, 2001.

[8] S. Mitragotri, D. Blankschtein, and R. Langer, Ultrasound-mediated transdermal protein delivery. *Science*, 269:850–853, 1995.

[9] H. Tang, D. Blankschtein, and R. Langer. An investigation of the role of cavitation in low-ferquency ultrasound-mediated transdermal drug transport. *Pharm. Res.*, 19:1160–1169, 2002.

[10] H.R. Guzman, D.X. Nguyen, S. Khan, and M.R. Prausnitz. Ultrasound-mediated disruption of cell membranes. I. Quantification of molecular uptake and viability. *J. Acoust. Soc. Am.*, 110:588–596, 2001.

[11] M.B. Sundaram J. and Mitragotri S. An experimental analysis of ultrasound-induced permeabilization. *Biophy. J.*, 2002.

[12] D. Miller and J. Quddus. Sonoporation of monolayer cells by diagnostic ultrasound activation of contrast-agent gas bodies. *Ultrasound Med. Biol.*, 26:661–667, 2000.

[13] J.Wu, J.P. Ross and J.-F. Chiu. Reparable sonoporation generated by microstreaming. *J. Acoust. Soc. Amer.*, 111:1460–1464, 2002.

[14] J.L. Nelson, B.L. Roeder, J.C. Carmen, F. Roloff, and W.G. Pitt. Ultrasonically activated chemotherapeutic drug delivery in a rat model. *Cancer Res.*, 62:7280–7283, 2002.

[15] R.J. Price and S. Kaul. Contrast ultrasound targeted drug and gene delivery: an update on a new therapeutic modality. *J. Cardiovasc. Pharmacol. Ther.*, 7:171–180, 2002.

[16] C.S. Kwok, P.D. Mourad, L.A. Crum, and B.D. Ratner. Self-assembled molecular structures as ultrasonically-responsive barrier membranes for pulastile delivery. *J. Biomed. Mater. Res.*, 57:151–164, 2001.

[17] J. Kost, K. Leong, and R. Langer. Ultrasound-enhanced polymer degradation and release of incorporated substances. *Proc. Natl. Acad. Sci.*, 86:7663–7666, 1989.

[18] J.R. Linder. Evolving applications of contrast ultrasound. *Am. J. Cardiol.*, 90:72J–80J, 2002.

[19] E.C. Unger, E. Hersh, M. Vannan, T.O. Matsunaga, and T. McCreery. Local drug and genen delivery through microbubbles. *Prog. Cardiovasc.*, 44:45–54, 2001.

[20] H.A.E. Benson, J.C. McElnay, and R. Harland. Phonophoresis of lingocaine and prilocaine from Emla cream. *Int. J. Pharm.*, 44:65–69, 1988.

[21] H.A.E. Benson, J.C. McElnay, and R. Harland. Use of ultrasound to enhance percutaneous absorption of benzydamine. *Phys. Ther.*, 69:113–118, 1989.

[22] H.A.E. Benson, J.C. McElnay, and J. Hadgraft. Influence of ultrasound on the percutaneous absorption of nicotinate esters. *Pharm. Res.*, 9:1279–1283, 1991.

[23] D. Bommannan, G.K. Menon, H. Okuyama, P.M. Elias, and R.H. Guy. Sonophoresis. II. examination of the mechanism(s) of ultrasound-enhanced transdermal drug delivery. *Pharm. Res.*, 9:1043–1047, 1992.

[24] D. Bommannan, H. Okuyama, P. Stauffer, and R.H. Guy. Sonophoresis. I. The use of high-frequency ultrasounnd to enhance transdermal drug delivery. *Pharm. Res.*, 9:559–564, 1992.

[25] N.N. Byl, A. McKenzie, B. Halliday, T. Wong, and J. O'nConell. The effects of phonophoresis with corticosteroids: a controlled pilot study. *J. Orth. Sports Phys. Ther.*, 18:590–600, 1993.

[26] B.M. Cameroy. Ultrasound enhanced local anesthesia. *Am. J. Orthoped.*, 8:47, 1966.

[27] C.D. Ciccone, B.Q. Leggin, and J.J. Callamaro. Effects of ultrasound and trolamine salicylate phonophoresis on delayed-onset muscle soreness. *Phys. Ther.*, 71:666–678, 1991.

[28] J.E. Griffin and J. Touchstone. Ultrasonic movement of cortisol in to pig tissue. *Am. J. Phys. Med.*, 44:20–25.

[29] J.E. Griffin, J.L. Echternach, R.E. Proce, and J.C. Touchstone. Patients treated with ultrasonic driven hydrocortisone and with ultrasound alone. *Phys. Ther.*, 47:600–601, 1967.

[30] J.E. Griffin and J.C. Touchstone. Low-intensity phonophoresis of cortisol in swine. *Phys. Ther.*, 48:1136–1344, 1968.

[31] J. Griffin, E. and J.C. Touchstone. Effects of ultrasonic frequency on phonophoresis of cortisol into swine tissues. *Am. J. Phys. Med.*, 51:62–78, 1972.

[32] M.E. Johnson, S. Mitragotri, A. Patel, D. Blankschtein, and R. Langer. Synergistic effect of ultrasound and chemical enhancers on transdermal drug delivery. *J. Pharm. Sci.*, 85:670–679, 1996.

[33] J.A. Kleinkort and F. Wood. Phonophoresis with 1 percent versus 10 percent hydrocortisone. *Phys. Ther.*, 55:1320–1324, 1975.

[34] J. Kost and R. Langer, Ultrasound-mediated transdermal drug delivery. In V.P. Shah and H.I. Maibach (eds.) *Topical Drug Bioavailability, Bioequivalence, and Penetration.* Plennum: New York. pp. 91–103, 1993,

[35] J. Kost, U. Pliquett, S. Mitragotri, A. Yamamoto, J.Weaver, and R. Langer, Enhanced transdermal delivery: synergistic effect of ultrasound and electroporation. *Pharm. Res.*, 13:633–638, 1996.

[36] J. Kost, S. Mitragotri, and R. Langer. Phonophoresis. In R. Bronaugh and H.I. Maibach (eds.) *Percutaneous Absorption,.* pp. 615–631, 1999.

[37] J. Kost, S. Mitragotri, R. Gabbay, M. Pishko, and L. R. Transdermal extraction of glucose and other analytes using ultrasound. *Nat. Med.*, 6:347–350, 2000.

[38] L. Le, J.Kost, and S. Mitragotri. Combined effect of low-frequency ultrasound and iontophoresis: applications for transdermal heparin delivery. *Pharm. Res.*, 17:1151–1154, 2000.

[39] M. Machluf and J. Kost. Ultrasonically enhanced transdermal drug delivery. Experimental approaches to elucidate the mechanism. *J. Biomat. Sci.*, 5:147–156, 1993.

[40] G. Menon, D. Bommanon, and P. Elias. High-frequency sonophoresis: permeation pathways and structural basis for enhanced permeability. *Skin Pharmacol.*, 7:130–139, 1994.

[41] S. Mitragotri, D. Edwards, D. Blankschtein, and R. Langer. A mechanistic study of ultrasonically enhanced transdermal drug delivery. *J. Pharm. Sci.*, 84:697–706, 1995.

[42] S. Mitragotri, D. Blankschtein, and R. Langer. Sonophoresis: ultrasound mediated transdermal drug delivery. In J. Swarbrick, Boylan, J. (eds.) *Encl. of Pharm. Tech.*, Marcel Dekker, 1995.

[43] S. Mitragotri, D. Blankschtein, and R. Langer. Transdermal drug delivery using low-frequency sonophoresis. *Pharm. Res.*, 13:411–420, 1996.

[44] S. Mitragotri, D. Blankschtein, and R. Langer. Sonophoresis: enhanced transdermal drug delivery by application of ultrasound. In S. J. and J. Boylan, (eds.), *Encyl. Pharm. Tech.,.* pp. 103–122, 1996.

[45] S. Mitragotri, D. Blankschtein, and R. Langer. An explanation for the variation of the sonophoretic transdermal transport enhancement from drug to drug. *J. Pharm. Sci.*, 86:1190–1192, 1997.

[46] S. Mitragotri, J. Farrell, T. H., T. Terahara, J. Kost, and R. Langer. Determination of the threshold energy dose for ultrasound-induced transdermal drug delivery. *J. Control. Rel.*, 63:41–52, 2000.

[47] S. Mitragotri, D. Ray, J. Farrell, H. Tang, B. Yu, J. Kost, D. Blankschtein, and R. Langer. Synergistic effect of ultrasound and sodium lauryl sulfate on transdermal drug delivery. *J. Pharm. Sci.*, 89:892–900, 2000.

[48] S. Mitragotri and J. Kost. Transdermal delivery of heparin and low-molecular weight heparin using low-frequency ultrasound. *Pharm. Res.*, 18:1151–1156, 2000.

[49] S. Mitragotri and J. Kost. Low-frequency sonophoresis: a non-invasive method for drug delivery and diagnostics. *Biotech. Progress*, 16:488–492, 2000.

[50] S. Mitragotri. Synergistic effect of enhancers for transdermal drug delivery. *Pharm. Res.*, 17:1354–1359, 2000.

[51] W.S. Quillen. Phonophoresis: a review of the literature and technique. *Athelet. Train.*, 15:109–110, 1980.

[52] A. Tezel, A. Sanders, J. Tuchscherer, and S. Mitragotri. Synergistic effect of low-frequency ultrasound and surfactant on skin permeability. *J. Pharm. Sci.*, 91:91–100, 2001.

[53] T. Terahara, S. Mitragotri, and R. Langer. Porous resins as a cavitation enhancer for low-frequency sonophoresis. *J. Pharm. Sci.*, 91:753–759, 2002.

[54] T. Terahara, S. Mitragotri, J.Kost, and R. Langer. Dependence of low-frequency sonophoresis on ultrasound parameters; distance of the horn and intensity. *Int. J. Pharm.*, 235:35–42, 2002.

[55] H. Tang, D. Blankschtein, and R. Langer. Effects of low-frequency ultrasound on the transdermal penetration of mannitol: comparative studies with in vivo and in vitro studies. *J. Pharm. Sci.*, 91:1776–1794, 2002.

[56] H. Tang, S. Mitragotri, D. Blankschtein, and R. Langer. Theoretical description of transdermal transport of hydrophilic permeants: application to low-frequency sonophoresis. *J. Pharm. Sci.*, 90:543–566, 2001.

[57] A. Tezel, A. Sens, and S. Mitragotri. Investigations of the role of cavitation in low-frequency sonophoresis using acoustic spectroscopy. *J. Pharm. Sci.*, 91:444–453, 2002.

[58] A. Tezel, A. Sens, and S. Mitragotri. A theoretical analysis of low-frequency sonophoresis: dependence of transdermal transport pathways on frequency and energy density. *Pharm. Res.*, 19:1841–1846, 2002.

[59] A. Tezel, A. Sens, and S. Mitragotri. A theoretical description of transdermal transport of hydrophilic solutes induced by low-frequency sonophoresis. *J. Pharm. Sci.*, 92:381–393, 2003.

[60] K. Tachibana. Transdermal delivery of insulin to alloxan-diabetc rabits by ultrasound exposure. *Pharm. Res.*, 9:952–954, 1992.

[61] K. Tachibana and S. Tachibana. Use of ultrasound to enhance the local anesthetic effect of topically applied aqueous lidocaine. *Anestheiology*, 78:1091–1096, 1993.

[62] K. Tachibana and S. Tachibana. Transdermal delivery of insulin by ultrasonic vibration. *J. Pharm. Pharmacol.*, 43:270–271, 1991.

[63] A.R.Williams. Phonophoresis: an in vivo evaluation using three topical anaesthetic preparations. *Ultrasonics*, 28:137–141, 1990.

[64] R. Alvarez-Roman, G. Merriono, Y.N. Kalia, A. Naik, and R. Guy. Skin permeability enhancement by low-frequency sonophoresis-lipid extraction and transport pathways. *J. Pharm. Sci.*, 92:1138–1146, 2003.

[65] G. Merrino, Y.N. Kalia, and R.H. Guy. Ultrasound-enhanced transdermal transport. *J. Pharm. Sci.*, 92:1125–1137, 2003.

[66] G. Merriono, Y.N. Kalia, M.B. Delgado-Charro, R. Potts, and R.H. Guy. Frequency and thermal effects on the enhancement of transdermal transport by sonophoresis. *J. Control. Rel.*, 88:85–94, 2003.

[67] L.J. Weimann and J. Wu. Transdermal delivery of poly-l-lysine by sonomacroporation. *Ultrasound Med. Biol.*, 28:1173–1180, 2002.

[68] A. Joshi and J. Raje. Sonicated transdermal drug transport. *J. Control. Rel.*, 83:13–22, 2002.

[69] L. Machet and A. Boucaud. Phonophoresis: efficiency, mechanisms, and skin tolerance. *Int. J. Pharm.*, 243:1–15, 2002.

[70] A. Boucaud, M.A. Garrigue, L. Machet, L. Vaillant, and F. Patat. Effect of sonication parameters on transdemral delivery of insulin to hairless rats. *J. Pharm. Sci.*, 91:113–119, 2002.

[71] K. Fellinger and J. Schmidt. Klinik and Therapies des Chromischen Gelenkreumatismus. Maudrich Vienna, Austria, pp. 549–552, 1954.

[72] G.L. Coodley. Bursitis and post-traumatic lesions. *Am. Pract.*, 11:181–187, 1960.

[73] J.E. Griffin. Physiological effects of ultrasonica energy as it is used clinically. *J. Am. Phys. Ther. Assoc.*, 46:18–26, 1966.

[74] J.C. McElnay, M.P. Matthews, R. Harland, and D.F. McCafferty. The effect of ultrasound on the percutaneous absorption of lingocaine. *Br. J. Clin. Pharmacol.*, 20:421–424, 1985.

[75] M.A. Moll. New approaches to pain. *US Armed Forces Med. Serv. DIg.*, 30:8–11, 1979.

[76] D. Hofman and F. Moll. The effect of ultrasound on in vitro liberation and in vivo penetration of benzyl nicotinate. *J. Control. Rel.*, 27:187–192, 1993.

[77] D. Levy, J. Kost, Y. Meshulam, and R. Langer. Effect of ultrasound on transdermal drug delivery to rats and guinea pigs. *J. Clin. Invest.*, 83:2974–2078, 1989.

[78] S. Mitragotri and J. Kost. Low-frequency sonophoresis: a review. *Adv. Drug Deliv. Rev.*, 56:589–601, 2004.

[79] C.M. Gockel, S. Bao, and K.W. Beagley. Transcutaneous immunization induces mucosal and systemic immunity:a potent method for targeting immunity to the female reproductive tract. *Mol. Immunol.*, 37:37–544, 2000.

[80] P.L. Ogra, J. Mestecky, M.E. Lamm, W. Strober, J.R. McGhee, and J. Bienenstock. *Handbook of Mucosal Immunology*. New York: Acedemic Press, 1994.

[81] T. Scharton-Kersten, J.Yu, R.Vassell, D. O'Hagan, C.R. Alving, and G.M. Glenn. Transcutaneous immunization with bacterial ADP-ribosylating exotoxins, subunits, and unrelated adjuvants. *Infect. Immun.*, 68:5306–5313, 2000.

[82] I. Roitt, J. Brostoff, and D. Male. *Immunology,*(5th Ed.). London: Mosby, 1998.

[83] Tana, S.Watarai, E. Isogai, and K. Oguma. Induction of intestinal IgA and IgG antibodies preventing adhesion of verotoxin-producing Escherichia coli to Caco-2 cells by oral immunization with liposomes. *Lett. Appl. Microbiol.*, 36:135–139, 2003.

[84] S. Mitragotri, M. Coleman, J. Kost, and R. Langer. Transdermal extraction of analyets using low-frequency ultrasound. *Pharm. Res.*, 17:466–470, 2000.

[85] S. Mitragotri, M. Coleman, J. Kost, and R. Langer. Analysis of ultrasonically extracted interstitial fluid as a predictor of blood glucose values. *J. Appl. Physiol.*, 89:961–966, 2000.

[86] J. Kost, S. Mitragotri, T. Elstrom, N. Warner, and S. Kellogg. Ultrasonic Skin Permeabilizing Device for Transdermal Glucose Monitoring. *2nd La Jolla Conference on Diabetes*, San Diego, 2003.

[87] A.J. Singer, C.S. Homan, A.L. Church, and S.A. McClain. Low-frequency sonophoresis: pathologic and thermal effects in dogs. *Acad. Emerg. Med.*, 5:35–40, 1998.

[88] A. Boucaud, J. Montharu, L. Machet, B. Arbeille, M.C. Machet, F. Patat, and L. Vaillant. Clinical, histologic, and electron microscopy study of skin exposed to low-frequency ultrasound. *Anatom. Rec.*, 264:114–119, 2001

[89] N.P. Katz, D.E. Shapiro, T.E. Hermann, J. Kost, and L. Custer. Rapid onset of cutaneous anesthesia with EMLA cream after pretreatment with a new ultrasound emitting device. *Anesth. Analg.*, 98:371–376, 2004.

[90] Simmonin, On the mechanisms of in vitro and in vivo phonophoresis. *J. Control. Rel.*, 33:125–141, 1995.

[91] A. Tezel and S. Mitragotri. Interactions of inertial cavitation collapses with stratum corneum lipid bilayers. *Biophys. J.* (in press), 2003.

[92] K.S. Suslick. *Ultrasound:Its Chemical, Physical and Biological Effects.* VCH Publishers, 1989.

[93] A. Tezel and S. Mitragotri. Interactions of inertial cavitation bubbles with stratum corneum lipid bilayers during low-frequency sonophoresis. *Biophys. J.*, 85:3502–3512, 2003.

[94] J. Blake and D. Gibson. Cavitation bubbles near boundaries. In J. Lumley, M. Van Dyke and H. Reed (eds.), *Annual Reviews of Fluid Mechanics*, pp. 99–123, 1985.

[95] Naude and A. Ellis. On the mechanisms of cavitation damage by non-hemispherical cavities in contact with solid boundary. *Trans. ASME J. Basic Eng.*, 83:648–556, 1961.

[96] T. Benjamin and A. Ellis. The collapse of cavitation bubbles and the pressures thereby produced against solid boundaries. *Philos. Trans. R. Soc. London Ser A*, 260:221–240, 1966.

[97] W. Lauterborn and H. Bolle. Experimental investigations of cavitation bubble collapse in the neighbourhood of a solid boundary. *J. Fluid. Mech.*, 72:391–399, 1975.

[98] M. Plesset and R. Chapman. Collapse of an initially spherical vapour cavity in the neighbourhood of a solid boundary. *J. Fluid. Mech.*, 47:283–290, 1971.

[99] S. Mitragotri. Effect of bilayer disruption on transdermal transport of low-molecularweight hydrophobic solutes. *Pharm. Res.*, 18:1022–1028, 2001.

[100] G. Black. Interaction between anionic surfactants and skin. In K.Walters and J. Hadgraft, (eds.), *Pharmaceutical Skin Penetration Enhancement*, Marcel Dekker, New York, Basel, Hong Kong, pp. 145–174, 1993.

[101] K.A. Walters, Surfactants and percutaneous absorption. In R.C. Scott, R.H. Guy, and J. Hadgraft (eds.), *Predictions of Percutaneous Penetration*, IBC Technical Services: London. pp. 148–162, 1990.

14

Microdevices for Oral Drug Delivery

Sarah L. Tao[1] and Tejal A. Desai[2]

[1] Department of Biomedical Engineering, Boston University Boston, MA 02215

[2] Department of Bioengineering and Physiology, University of California, San Francisco, CA

14.1. INTRODUCTION

The application of micro- and nanotechnology to the biomedical arena has tremendous potential in terms of developing new diagnostic and therapeutic modalities. Recent advances in micromachining and microelectriomechanical systems (MEMS) technology have provided the opportunity to fabricate miniature biomedical devices for a variety of applications. While the majority of research has focused on the development of miniaturized *diagnostic* tools, researchers have more recently concentrated on the development of microdevices for *therapeutic* applications. The progression of microfabrication technology has enabled the creation of entirely new classes of drug delivery devices which can possess a combination of structural, mechanical, chemical, and electronic features to surmount the challenges associated with conventional systems. Some therapeutic applications of microtechnology include microneedles for transdermal delivery [1, 2], as well as various implantable systems, such as immunoisolating biocapules [3, 4] and microchips for controlled release [5, 6]. However in this research, the strengths of microfabrication and micromachining were capitalized to create a completely novel microdevice from silicon dioxide, porous silicon, and poly(methyl methacrylate) for cytoadhesive oral drug delivery with inherent features providing controlled release.

14.1.1. Current Challenges in Drug Delivery

The ultimate aim of pharmacy and medicine is the delivery of any drug at the right time in a safe and reproducible manner to a specific target at the required level [7]. Conventional dosage forms, such as oral delivery and injection, are the predominant routes for drug administration. However, these types of dosages are not easily able to control the rate of drug delivery or the target area of the drug and are associated with an immediate, rapid release of drug. Consequently, the initial concentration of the drug in the body peaks above the level of toxicity and then gradually diminishes over time to an ineffective level. The duration of therapeutic efficacy then becomes dependent on the frequency of administration and the half-life of the drug. High dosages of non-targeted drugs are often administered in order to achieve an effective blood concentration. In recent years, increasingly sophisticated and potent drugs have been developed by the biotech industry. For many of these new protein-based and DNA-based compounds, the therapeutic concentration range is small and varies with time, and toxicity is observed for concentration spikes, which renders traditional methods of drug delivery ineffective [8]. An immense amount of interest has been increasingly placed on controlled release drug delivery systems to maintain the therapeutic efficacy of these drugs. There are a number of mechanisms that can provide such controlled release of drugs, including transdermal patches, implants, bioadhesive systems, and microencapsulation. In addition, attractive drug delivery technologies such as inhaled and sustained release injectable peptide/protein drugs are currently under intensive study.

14.1.2. Oral Drug Delivery

Oral drug delivery is one of the most preferred methods of drug administration due to its non-invasive nature. However, it is generally not a viable method for peptide and protein delivery. This is, in part, a consequence of the first-pass effect of the liver, but primarily due to the acidic environment of the stomach and the resistance exerted by the intestine. The human gastrointestinal (GI) tract by nature resists absorption of peptides, proteins, and other macromolecules until they are broken down into smaller molecules. The acidic environment of the stomach combined with an array of enzymes and physical barriers in the intestines either destroy or prevent absorption of nearly all macromolecules. This problem leads to unacceptably low oral bioavailability of the drug, which then fails to induce a clinical response. Several approaches are currently being developed to enhance the oral delivery of macromolecular drugs. Protective coatings, such as lipids and polymers, have been used to protect peptides during transport through the acidic environment of the stomach and improve transport across the intestinal wall [9, 10]. Bioadhesive agents have been used to induce contact of the peptide to the intestinal wall [11–13]. This allows for local delivery of the peptide to sites in the GI tract generating greater levels of absorption and stability. In addition, the use of permeation enhancers to augment uptake and transport through the intestinal wall, and protease inhibitors to protect peptides from enzymatic degradation is being explored [14, 15]. While all these approaches have been shown to improve upon the oral bioavailability of large molecules, none of these approaches offers a complete solution for adequate and safe oral administration.

14.1.3. Bioadhesion in the Gastrointestinal Tract

Bioadhesive drug delivery systems have been given considerable interest due to their potential for delaying transit and prolonging the residence time at the site of drug action or absorption. Bioadhesion refers to the adherence of molecules (bioadhesives) to biological

surfaces [16]. "Targeting" refers to strategies of effectively directing molecules to a particular tissue, cell type, or subcellular compartment [17]. Localization of the delivery system
at a given target site prone to drug action could increase the drug concentration gradient
due to intense contact [11,13].

The gastrointestinal tract is the most important body surface for drug administration
due to its massive surface area. Two main targets exist for anchoring a delivery system in the
gastrointestinal tract: the mucus gel layer and mucosal tissue. Intestinal mucus is composed
of high molecular weight glycoproteins. The mucin gel layer covers the mucosal tissue with
a continuous adherent blanket. Its thickness varies regionally along the GI tract, decreasing
from 50–500 μm in the stomach to 15–150 μm in the colon [11]. Adhesive interactions
with the complex structure of mucus gel layer are generally entered through non-specific
Van der Waals and/or hydrophobic interactions. The non-specificity of the interaction limits
the maximum duration of mucoadhesion to the turnover time of the mucus gel layer. In the
intestine, mucus turnover has been estimated to be between 47 and 270 minutes, which is
on the same scale as normal intestinal transit (three to four hours) [13, 17, 18]. In this case,
drug release and absorption must be completed during this limited adhesion time.

An alternative to mucoadhesion is "cytoadhesion," where adhesion occurs directly to
the cell surface [13]. The mucosal epithelium of the small intestine is comprised of a single
layer of polarized epithelial cells, predominantly consisting of enterocytes. The enterocytes
are connected through tight intercellular junctions with pore diameters reported to be less
than 10 angstroms [17]. Follicle associated epithelium (FAE) is differentiated from other
intestinal epithelial sites by the presence of specialized antigen sampling membranous
epithelial (M) cells. The transcytotic capacity of these cells is much greater than other
intestinal epithelial cell types, and therefore, a potentially important site for drug delivery
[19]. Molecules can be transported across the epithelial barrier by passive diffusion through
transcellular or paracellular pathways, or actively transported across by membrane-derived
vesicles or membrane bound carriers. Uptake can also occur due to adsorptive endocytosis
via clathrin-coated pits and vesicles, fluid phase endocytosis, and phagocytosis induced by
M cell antigen sampling. Figure 14.1 shows a schematic of the epithelium and the various
pathways for molecular transport across the barrier.

Cytoadhesion of a drug delivery system would involve grafting a ligand that shows
affinity for a receptor in the gastrointestinal cavity to mediate an adhesive interaction between the particle and the biological surface. Enterocytes and M cells are both potential
target sites for cytoadhesion. Although particle adsorption by enterocytes is relatively inefficient when compared to that of M cells, the greater surface area that they occupy offsets
their inefficiency.

Lectins, deemed "second generation" bioadhesives, are one group of molecules that
are able to recognize surface structures of intestinal cells. Lectins are a class of carbohydrate binding proteins or glycoproteins of non-immune origin. Most lectins contain two
or more sugar binding sites and can agglutinate cells and precipitate glycoconjugates with
considerable specificity without altering the covalent structure of the glycosyl moiety [11].
Lectins are an ideal possibility for targeted drug delivery because of their relative resistance
to low pH and enzymatic degradation. And, because the intestinal epithelial cells maintain
a cell surface glycocalyx consisting of membrane-anchored glycoconjugates, a multitude
of lectin binding sites exist along the gastrointestinal tract. The structural diversity of cell-
surface glycoconjugates encodes unique signals recognized by lectins in a complementary

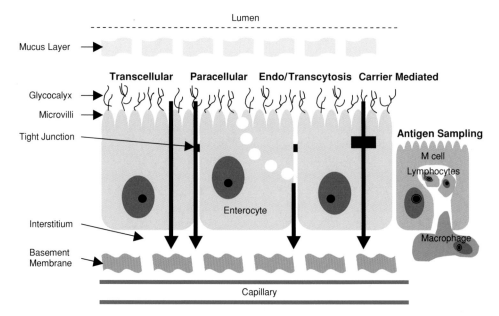

FIGURE 14.1. Schematic of the structure of the epithelium. Molecules can be transported across the epithelial barrier by passive diffusion through transcellular or paracellular pathways, or actively transported across by membrane-derived vesicles or membrane bound carriers. Uptake can also occur due to adsorptive endocytosis via clathrin-coated pits and vesicles, fluid phase endocytosis, and phagocytosis induced by M cell antigen sampling.

way analogous to ligand-receptor interactions [20]. These interactions are even thought to be associated with cell internalization signaling pathways. A number of orally administered plant lectins have been found to bind to the luminal surface of the gut with high affinity (K_d values of 10^{-4} to 10^{-6} M) [21]. Their binding is primarily dependent on the structure and accessibility of carbohydrate moieties on the epithelial cells and the sugar specificity of the lectins. Table 14.1 shows examples of commonly found lectins and their sugar binding specificities.

14.1.4. Microdevice Technology

Micromachined platforms, when combined with complementary approaches, may address some of the shortcomings of current oral delivery systems for peptides and proteins by

TABLE 14.1. Examples of common lectins.

Lectin	Abbreviation	MW (Da)	Sugar Specificity*
Triticum vulgaris (Wheat germ)	WGA	36,000	N-AcGlu;SA
Lycopersicon esculentum (Tomato)	TL	71,000	N-AcGlu
Canavalia ensiformis (concavalin A)	Con A	102,000	Man;Glu
Phaseolus vulgaris (Red kidney bean)	PHA	115,000	Gal;N-AcGal
Arachis hypogea (Peanut)	PNA	120,000	β-Gal

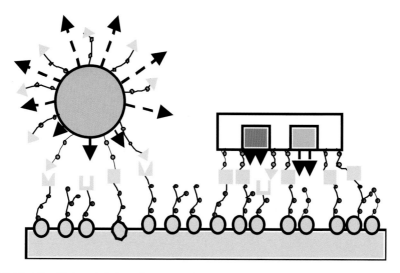

FIGURE 14.2. Microfabricated platforms have many advantages over traditional microsphere delivery systems. In contrast to microspheres, microfabricated devices may be specifically designed flat and thin to maximize contact area with the intestinal lining. This flat design minimizes the side areas exposed to constant flow and can be microfabricated to incorporate single or multiple drug reservoirs to contain a number of drugs/biomolecules of interest. These reservoirs also allow for unidirectional release of the drug. Furthermore, specific areas of the device can be targeted for surface modification.

combining several features into a single drug delivery vehicle [22]. First, one can achieve control over the size and shape of the delivery device [23–25]. Unlike traditional delivery particles, such as microspheres, microfabricated devices may be specifically designed flat, thin, and disc-shaped to maximize contact area with the intestinal lining and minimize the side areas exposed to the constant flow of liquids through the intestines (Figure 14.2). The size of the microdevices can be selected small enough to have good contact with the undulations of the intestinal wall and large enough to avoid endocytosis of the entire particle. Secondly, one can selectively attach bioadhesive agents targeting specific cell in the GI tract onto the device surface using relatively simple surface chemical modification strategies [23,24]. Finally, micromachining presents the opportunity to create multiple reservoirs of a desired size to contain any number of biomolecules of interest which can then be released the shortest distance towards the intestinal epithelium [23–25].

14.2. MATERIALS

The use of silicon as a starting substrate material in the integrated circuit industry was based on its intrinsic mechanical stability and the feasibility of integrating sensing and electronics in self-contained microsystems capable of performing intelligent tasks, such as sensing and signal processing, at a very low manufacturing cost [26, 27]. From its roots in electrical engineering, the discipline of miniaturization has become widely accepted in fields such as mechanical engineering and biomedical engineering. This in turn has led to a wider array of materials used in microfabrication.

In the realm of microfabrication, even for biomedical applications, silicon remains the starting substrate of choice namely due to the fact that a great deal of equipment has been built to accommodate silicon wafers. As other substrates are harder to accommodate, silicon intrinsically appears to have a convenient advantage. Other advantages of silicon include its superior mechanical stability and well-established chemical techniques for modification utilizing silane chemistry. Although preliminary medical evidence is thought to suggest that silicon remains benign in the body [28], it is not recognized as a Food and Drug Administration (FDA) approved biomaterial. Therefore considerable efforts have been made towards increasing silicon biocompatibility.

14.2.1. Silicon Dioxide

Silicon readily oxidizes to form a stable oxide layer. Silicon oxides usually function as insulation between conducting layers, diffusion and ion implementation masks, capping doped oxides to prevent loss of dopants, for gettering impurities, and for passivation to protect devices from impurities moisture and scratches. In terms of biological applications, this oxide layer may completely isolate underlying layers of a device thereby interacting more favorably with the cellular environment.

14.2.2. Porous Silicon

The formation of porous silicon was first discovered by Uhlir in 1956 [29] and has led to creation of many devices from quantum structures and permeable membranes to photoluminescent and electroluminescent devices. Porous silicon has numerous important features. Pore size can range in diameter from 20 Å to 10 μm and follow crystallographic orientation. It is also highly reactive, and therefore oxidizes and etches at very high rate [27]. Furthermore, recent interest has been gathered in the use of porous silicon in biological applications from biosensors to mammalian cell culture [30–32].

For biomedical applications, polymers have become a favorable alternative to the silicon substrate. Although surface modification protocols for modifying polymers are not as established, polymer substrates still have advantages comparable to silicon. First, various methods can be used to construct the devices inexpensively and in large number, such as injection molding, laser ablation, imprinting, and hot embossing [33]. For this reason, polymers are ideal for the fabrication of disposable devices. Second, high-aspect-ratio microstructure devices can be produced with polymer substrates that are often very difficult to achieve with silicon [33]. Third, polymers can exist either in a hard, glassy state, or a soft rubbery state, an option typically unassociated with structures based on silicon. Fourth, in contrast to silicon, the FDA has already accepted a variety of polymers for internal use. Poly methylmethacrylate (PMMA), polylactic acid (PLA), polyglycolic acid (PGA), and polyethylene (PE) are among such polymers. Devices microfabricated from these polymer alternatives may provide a more suitable interface with biological tissue.

14.2.3. Poly(methyl methacrylate)

PMMA is a synthetic non-biodegradable polymer first discovered in Germany in the early 1900s. In terms of microfabrication, PMMA is used in short wavelength photolithographic systems such as deep ultraviolet, electron beam, ion beam, and x-ray lithography. PMMA is a model substrate for microfabricated systems geared towards biomedical application. Besides being used extensively as a resist in photolithographic applications, PMMA is known to be biocompatible and is FDA approved. It is already used in many medical applications such as bone cement [34, 35], contact lenses [36], and intraocular lenses [37].

14.3. MICROFABRICATION

The overall concept of constructing a micromachined platform for drug delivery remains the same across all three substrates (silicon dioxide, porous silicon and PMMA). As a prototype, rectangular-shaped microparticles were fabricated with dimensions that would potentially allow for *in vivo* transit. The rectangular dimensions were chosen for simplicity of design and fabrication, the rectangle being the simplest shape which photomask generating equipment can create. However, by simply changing the photomask features, any shape or size particle can be easily fabricated. The size of the particles may greatly affect the response generated in the body [38]. Although smaller particles, less than five microns, have shown an increased adherence in the whole gut, they are more likely to induce a localized inflammatory response followed by easy uptake by macrophages [39]. In this case, there is an increased risk that the carrier system might be degraded after internalization, followed by loss of its activity [39, 20]. Particles with a much larger size are taken up less effectively by macrophages, however, carrier systems greater than 200 microns may illicit a foreign body response if retained in the GI tract. For these reasons, the devices were fabricated with several dimensions ranging from lengths and widths of 50 to 200 microns. The thickness of the devices is limited by the processing capability of the microfabrication techniques used. Efforts here were concentrated towards maximizing drug loading volume while also maximizing the surface area of the face to side wall ratio. Microdevices between two and 25 microns thick were fabricated. The fabrication processes for creating such a system in the three substrates remains very different. However, these processes were all built upon standard MEMS fabrication techniques including photolithography, etching, and thin film deposition.

14.3.1. Silicon Dioxide [23]

Silicon dioxide microdevices were fabricated on silicon <111> p-type wafers cleaned by the RCA standard three step procedure. An etch stop layer comprised of thermal silicon dioxide was first formed by wet thermal oxidation. Low-pressure chemical vapor deposition (LPCVD) was then used to deposit a sacrificial layer of polysilicon atop the thermal oxide. Next, a layer of low temperature oxide (LTO) was deposited by LPCVD to form the device layer. The processed wafers were then spin-coated with a positive resist and photolithography was performed by exposing the wafers to ultraviolet light through a photomask defining the shape of the device reservoir. Following photoresist development, the exposed LTO device layer was carved by performing either a wet or dry etching procedure. A wet etch created a rounded reservoir and was performed by timed submersion of the masked wafer in a buffered oxide solution followed by excess photoresist removal. A dry etch created a

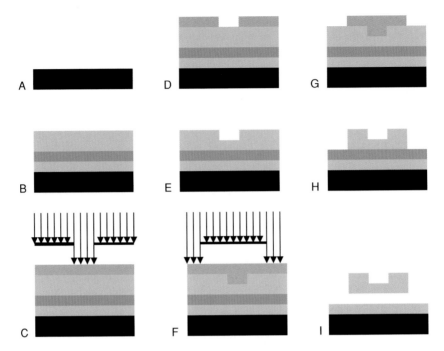

FIGURE 14.3. Process for creating silicon dioxide microdevices. A) Clean silicon wafer. B) Etch stop, sacrificial and device layer deposited respectively on the wafer. C) Positive lithography to define reservoirs in resist. D) Pattern developed. E) Reservoir carved from device layer and remaining photoresist stripped. F) Positive lithography to define microdevice body in resist. G) Pattern developed. H) Device body etched and remaining photoresist stripped. I) Microdevice released from wafer.

straight-walled reservoir. In this case, a reactive ion etch (RIE) with SF_6 and O_2 was used to carve the reservoirs into the LTO device layer and any remaining photoresist was then removed.

A second positive lithography step was performed to define the microdevice body. Reservoir features on the wafer were aligned to the photomask defining the device body using front side alignment on the mask aligner. After exposure and development of the photoresist, a reactive ion etch with SF_6 and O_2 was used to etch completely through the unmasked areas of the LTO layer and any remaining photoresist was then removed. The welled microdevices were released into solution by performing a wet etch of the sacrificial polysilicon layer with dilute potassium hydroxide. The potassium hydroxide solution was neutralized and filtered to isolate the microdevices. The microfabrication process for producing silicon dioxide microdevices is shown in Figure 14.3. Figure 14.4 provides examples of the various features which can be implemented in the device utilizing these processes.

14.3.2. Porous Silicon [25]

Porous silicon microdevices were fabricated on a single-side polished silicon <100> p^+-type wafer. First, a layer of silicon nitride was deposited by LVCD. The processed

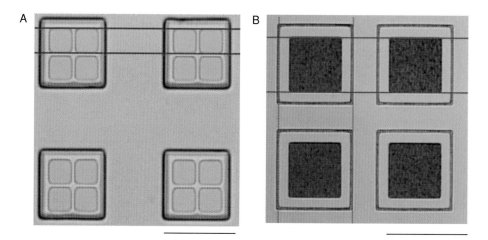

FIGURE 14.4. Silicon dioxide microdevices with different sized reservoirs and bodies. A) 35 μm wells inside a 100 μm body. Bar represents 100 μm. B) 100 μm well inside a 150 μm body. Bar represents 150 μm.

wafers were then spin-coated with a positive resist and photolithography was performed by exposing the frontside of the wafers to ultraviolet light through a photomask defining the area of intended porosity while protecting all other areas of the wafer. The backside of the wafer was then etched with SF_6 to strip the layer of silicon nitride. Exposed silicon nitride on the frontside of the wafer was etched with SF_6 and the remaining photoresist stripped. The patterned wafers were then anodized and electropolished in a custom built anodization tank with platinum electrodes. An ethanol-hydrofluoric acid (1:1) solution was placed in the cell and a current was applied upon connection of the electrodes to a power supply. Porosification exclusively took place along the anodic side of the silicon wafers. Anodization of the silicon wafer can be described through the following reactions:

$$Si + 2HF + \lambda h^+ \rightarrow SiF_2 + 2H^+ + (2 - \lambda)e^- \qquad (14.1)$$

$$SiF_2 + 2HF \rightarrow SiF_4 + H_2 \qquad (14.2)$$

$$SiH_4 + 2HF \rightarrow H_2SiF_6 \qquad (14.3)$$

Where h^+ and e^- are the exchanged hole and electron, respectively, and λ is the number of charges exchanged during the elementary step. Pore size and shape depend on the type of silicon used (n/p/p$^+$), the resistivity of the silicon, the current density, and the concentration of hydrofluoric acid solution. A schematic diagram of the double sided Teflon anodization tank in Figure 14.5A and Figure 14.5B shows a scanning electron micrograph of the pores which can be created utilizing this method. The porous particles were separated from the wafer by an electropolishing step in a dilute ethanol-hydrofluoric acid (1:4) solution. The microdevices were then capped by sputtering a thin layer of gold and removed from the wafer utilizing water soluble tape. The entire process for fabricating the porous silicon microdevices is shown in Figure 14.6. A scanning electron micrograph of a porous silicon microdevice is shown in Figure 14.7.

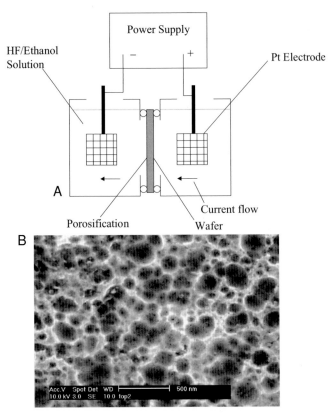

FIGURE 14.5. A) Schematic of the double-sided Teflon anodization tank. B) Scanning electron micrograph demonstrating pore size distribution between 20–200nm. Courtesy of IMEDD, Inc.

14.3.3. Pol(methyl methacrylate) [24]

PMMA microdevices were fabricated on RCA cleaned silicon <111> p-type wafers. The wafers were first spin-coated with multilayers of PMMA resist film, followed by a layer of positive photoresist. Lithography was used to pattern the device body. After exposure and development of the photoresist, the unmasked area of PMMA was reactive ion etched completely through using an O_2 plasma and any remaining resist was subsequently removed.

A second positive lithography step was performed to expose the intended reservoir areas in the PMMA. Device features on the wafer were aligned to the photomask defining the reservoir by front side alignment on the mask aligner. After exposure and development, the exposed PMMA area was then etched with an O_2 plasma to carve the reservoirs into the PMMA layer and any remaining photoresist was subsequently removed. The microdevices were then either left on the wafer for further chemical processing or released from the wafer. Since the PMMA adheres to the silicon by linkage to the native oxide layer, the microdevices were released by soaking the wafer in basic solution to break this bonding. The microfabrication process for the PMMA microdevices is depicted in Figure 14.8. A scanning electron micrograph of an array of PMMA microdevices still attached to the silicon wafer is shown in Figure 14.9.

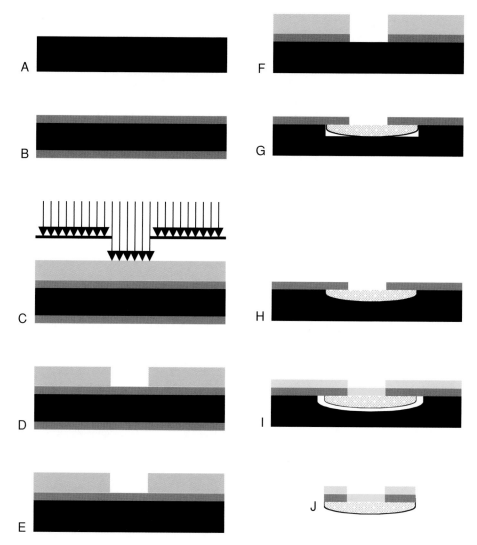

FIGURE 14.6. Process for creating porous silicon microdevices. A) Bare silicon wafer. B) Nitride layer is deposited on surface. C) Photolithography used to pattern protective area. D) Resist developed. E) Backside nitride layer stripped. F) Frontside nitride layer patterned and remaining photoresist stripped. G) Pores formed by anodization of exposed silicon. H) Wafer electropolished to separate devices. I) Devices capped by sputtering gold. J) Devices removed from wafer by lifting off with water soluble tape.

14.4. SURFACE CHEMISTRY

Techniques used to associate proteins and polymers include covalent linkage and passive adsorption. Although covalent linkage is often more stable than simple adsorption, it is possible that the stability and function of the conjugates may be impaired. Covalent linkage requires the presence of functional groups on the surface of the substrate. Protein-carrier conjugates have been prepared through activation of hydroxyl, carboxyl, and amino groups

FIGURE 14.7. Scanning electron micrograph of a porous silicon microdevice. Courtesy 8 IMEDD. Inc.

by cyanogenbromide [41], carbodiimide [42], and glutaraldehyde [43, 44], respectively. However, it is uncertain whether these types of crosslinking agents reduce the lectin's carbohydrate binding abilities. In addition, the amino acid composition, molecular weight, and tertiary structure are so diverse amongst the various types of lectins that direct conjugation to a carrier system would ultimately require a separate modification protocol for

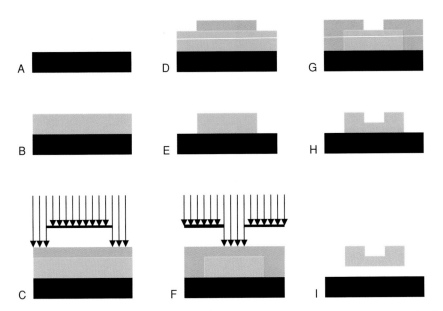

FIGURE 14.8. Process for PMMA microdevices. A) Silicon wafer cleaned by RCA. B) Multilayers of PMMA spun on wafer. C) Photolithography used to pattern body in resist. D) Pattern developed. E) PMMA etched and remaining photoresist stripped. F) Photolithography used to expose reservoir area. G) Pattern developed. H) Reservoir etched from body and remaining photoresist stripped. I) Devices released from wafer.

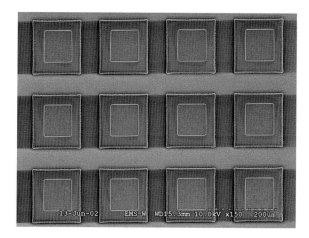

FIGURE 14.9. Scanning electron micrograph of a PMMA microdevice array.

each type of lectin. It is instead desirable to have a single method for attaching various lectins to a surface. One such method is to modify the lectin and surface each with one or another of a complementary pair of materials that have a high affinity for each another. A plausible pair of linking materials is avidin and biotin [45]. The bond between avidin and biotin is established very rapidly and is the strongest non-covalent biological interaction between a protein and a ligand. Once the bond is formed it is undisturbed by low pH, exposure to organic solvents and other denaturing agents [45]. By first conjugating avidin to the surface of the carrier, commercially available biotinylated lectins can be attached via avidin-biotin interaction. The basic chemistry for incorporating cytoadhesive properties into this microfabricated platform was performed in three major steps: (1) amine groups were formed on the device; (2) avidin was coupled with convential carbodiimide coupling reagents to form an avidin carboxylate which in turn was reacted with the amine groups of the substrate to form an avidin conjugate; (3) biotinylated lectins were attached to the surface utilizing the strong interaction and affinity between avidin and biotin [23,46] (Figure 14.10).

14.4.1. Aimine Functionalization

Amine groups were formed on the surface of *silicon dioxide and porous silicon* microdevices through a two-step reaction. First, hydroxyl groups were formed on the silicon surfaces (Figure 14.11A), followed by timed silanization in a (3-aminopropyl)triethoxysilane (APTES) solution to form amine terminated surfaces (Figure 14.11). Heterogeneous modification of *PMMA* with N-lithioethylenediamine as an aminolyzing agent leads to amination of its ester groups [47, 48]. This reaction renders a layer of amine sites tethered to the PMMA backbone by stable amide groups. First, *N*-lithioethylenediamine was synthesized by purging reacting ethylenediamine with *n*-butyllithium under nitrogen atmosphere. Amine groups were placed on the surface of the reservoir side of the microdevices by reacting the wafer bound devices with *N*-lithioethylenediamine. (Figure 14.12).

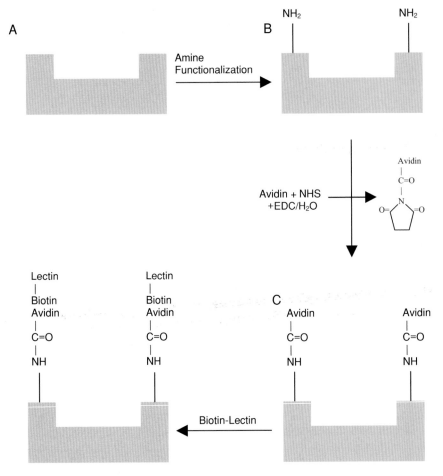

FIGURE 14.10. Surface chemistry used to attach lectins to microdevice surfaces. A) Amine functionalization. B) Avidin immobilization using a hydroxy-succinimide catalyzed carbodiimide reagent. C) Lectin attachment utilizing avidin-biotin attraction and interaction.

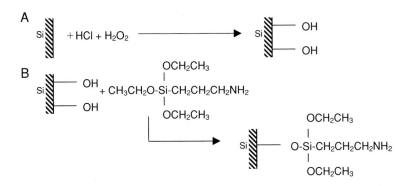

FIGURE 14.11. Surface chemistry for generating amine terminated silicon surfaces. A) Oxidation. B) Silanization.

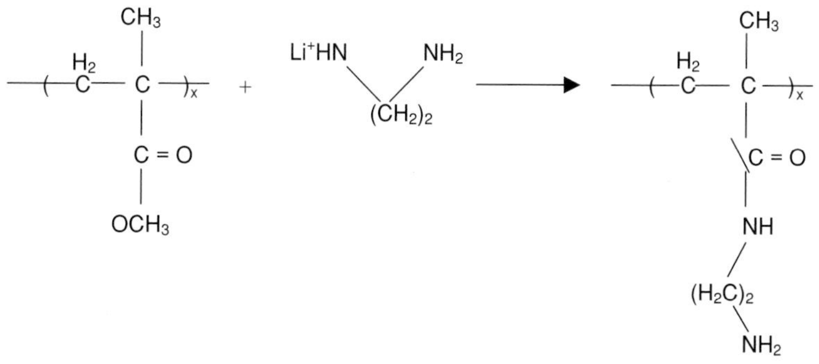

FIGURE 14.12. Chemical reaction for generating aminated PMMA.

14.4.2. Avidin Immobilization

Biological molecules containing carboxyl groups can be reacted with traditional carbodiimide coupling agents to form a reactive intermediate ester that is susceptible to attack by amines. This reaction can be applied to self-assemble a layer of covalently coupled biological molecules to the surface of the silicon dioxide, porous silicon, and PMMA microdevices which have been modified to contain amine sites. Here, molecules of avidin were attached to the microdevice surfaces by timed incubation in a solution of 1-ethyl-3-(3-dimethylaminopropyl) carbodiimide (EDC), N-Hydroxysuccinimide (NHS), and avidin.

14.4.3. Lectin Conjugation

Lectins were attached to the avidin-modified surfaces by means of an avidin-biotin conjugation method. The avidin-immobilized surfaces were time incubated in a solution of biotinylated lectin optimized for a 1:1 binding ratio of avidin to lectin.

14.5. SURFACE CHARACTERIZATION

Surfaces were characterized after modification utilizing several procedures including chemical assay, fluorescent labeling as well as x-ray photoelectron spectroscopy (XPS) and atomic force microscopy (AFM). A simple, qualitative method using sulfo-succinimidyl-4-O-(4,4-dimethoxytrityl)-butyrate (s-SDTB) was employed to determine the coverage of amine groups on the substrate surfaces [49]. The reaction of s-SDTB with amine groups on the substrate surface results in stable bonds at alkaline conditions. 4,4'-dimethoxytrityl cation is liberated as a result of this reaction. Since it has a high absorbance at 498 nm and an extinction coefficient of 70,000 M^{-1} cm^{-1}, concentration can easily be determined by UV/VIS spectrophotometry. As an example, Figure 14.13 shows the concentration of PMMA surface amine groups plotted as a function of reaction time with N-lithioethylenediamine. It was found that a ten minute incubation time rendered approximately 4 nmol cm^{-2}.

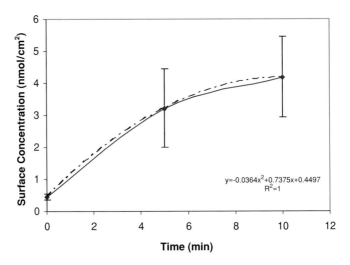

FIGURE 14.13. Coverage of immobilized amine groups on the PMMA surface as a function of reaction time.

To determine the maximum amount of avidin able to bind successfully to the amine-terminated surfaces, the microdevices were conjugated with fluorescein isothyocyanate (FITC) labeled avidin. Figure 14.14 A and B respectively show aminated porous silicon and PMMA microdevices labeled with avidin-FITC.

XPS analysis was performed to ensure the presence of amine groups, avidin, and lectin. Survey scans were taken of substrate samples following amine functionalization, avidin immobilization and lectin conjugation. The changes in elemental surface composition suggest successful modification (Table 14.2). After amine functionalization, an increase of nitrogen was expected. It was shown that this was the case among all three substrates. Analysis of APTES modified surfaces revealed an increase in nitrogen suggesting desirable APTES binding. Furthermore, the amount of carbon also increased after APTES modification. This was expected due to the long hydrocarbon chains of APTES. Analysis of the aminated

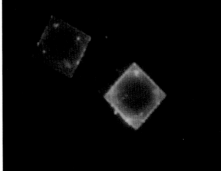

FIGURE 14.14. Fluorescently labeled avidin immobilized onto A) aminated porous silicon and B) aminated PMMA.

TABLE 14.2. Relative elemental surface composition of silicon dioxide, porous silicon and PMMA surfaces.

Sample	%O	%N	%C	%Si
Silicon Dioxide	31.6	1.2	26.9	40.3
Amine-Functionalized Silicon Dioxide	29.0	10.6	47.6	12.1
Avidin-lmmobilized Silicon Dioxide	24.2	11.2	50.5	12.9
Lectin-Conjugated Silicon Dioxide	20.2	15.9	59.6	0.5
Porous Silicon	59.6	0.0	8.7	31.7
Amine-Functionalized Porous Silicon	42.7	2.4	35.3	19.5
Avidin-Immobilized Porous Silicon	38.9	2.5	42.4	16.3
Lectin-Conjugated Porous Silicon	37.3	2.6	47.0	13.3
PMMA	26.2	0.0	73.8	0.0
AmineFunctionalized PMMA	12.5	10.8	76.8	0.0
Avidin-Immoblized PMMA	19.8	12.6	67.7	0.0
Lectin-Conjugated PMMA	18.7	15.1	66.2	0.0

PMMA surface also showed a significant increase in nitrogen. In addition, the peak area ratio of oxygen to carbon in the amine functionalized PMMA decreased by 12.3% in comparison with that of pristine PMMA. This suggested a successful loss of ester groups and the addition of amide and amine groups on the surface. For silicon substrates, the addition of avidin resulted in an increase in nitrogen composition and a significant increase in carbon. Attachment of lectin resulted in a further increase in nitrogen as well as carbon content on the surface. The indication of protein on the surfaces is supported further by the large increase in the ratio of nitrogen to silicon after protein conjugation. For the PMMA substrate, the addition of avidin and lectin also resulted in a considerable increase of nitrogen on the surface. Successful protein immobilization was confirmed by the increase in the nitrogen to carbon ratio following each procedure.

After chemical modification procedures, subsequent changes in topography were examined using AFM. The surface of pristine, amine-functionalized, avidin-immobilized and lectin-conjugated samples were visualized and characterized by AFM in order to determine the impact of modification on surface coverage. Distinct changes in surface topography were observed on all three substrates. For example, Figure 14.15A shows PMMA surfaces before and after incorporation of amine sites on the surface. Distinct peaks, equally distributed and of relatively uniform height can be seen after amine functionalization. Figure 14.15B shows a comparison of the porous silicon microdevice surfaces before modification and after lectin conjugation. From the AFM images it was determined that the modification procedures did not clog the pores of the microdevice, thereby showing no damage to the releasing mechanism.

14.6. MIOCRODEVICE LOADING AND RELEASE MECHANISMS

Like the microfabrication processes used to create the welled and porous devices, the mechanisms used to load the devices remain very different. These approaches are described below.

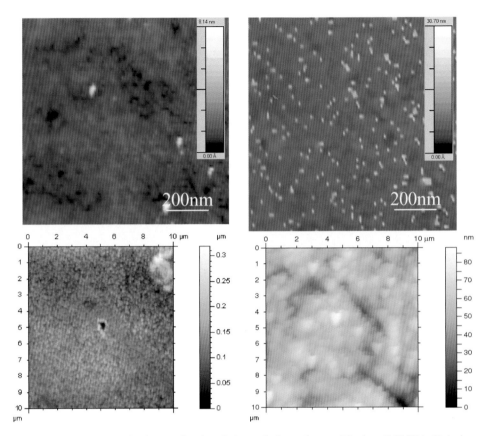

FIGURE 14.15. AFM scanning images of surfaces before and after surface modification. A) PMMA. B) Amine functionalized PMMA. C) Porous silicon. B) Lectin-modified porous silicon.

14.6.1. Welled Silicon Dioxide and PMMA Microdevices

These reservoirs were filled with pico- to nanoliters of a polymeric solution using microinjectors (Figure 14.16). Water quickly evaporates from these reservoirs leaving behind the drug contained in polymer which acts as a timed-release plug. Using a specific type of polymer predetermines the time and rate of release of drug from the reservoir; for example, a hydrogel that swells in response to a specific pH, solvent or temperature or a polymer with a known dissolution rate. Different polymers with various dissolution rates can then be used in separate reservoirs to obtain controlled release of several compounds compounds.

14.6.2. Porous Silicon Microdevices

Capillary action was exploited in order to load the porous silicon microdevices. Homogeneous liquid formulations were prepared and slowly loaded onto the porous silicon particles using a micropipettor. The solution was then taken into the porous particles by

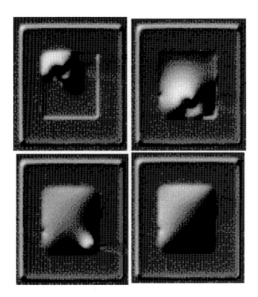

FIGURE 14.16. Time sequence of a microdevice reservoir filled by microinjection.

capillary action. The microdevices were then dried by applying vacuum pressure until any air trapped within the pores was released.

14.6.3. Caco-2 In Vitro Studies

The Caco-2 cell line, derived from human colorectal carcinoma cells, is one of the most useful in vitro models for the study of intestinal epithelial permeability and function as well as a model for drug delivery. The Caco-2 cell line is known to differentiate spontaneously into enterocytes to form polarized monolayers in culture [50]. When grown to confluency these cells highly resemble the small intestinal epithelium both, structurally and functionally. The microvilliated cells in the monolayer maintain tight intercellular junctions and expose only their apical, brush border membranes [51]. The cells also maintain transport systems, enzymes, and ion channels similar to those of the intestinal epithelium. The glycosylation pattern of the Caco-2 cells is characterized by high amounts of N-acetylglucosamine containing oligosaccharides followed by fucosyl- and mannosyl- and minor amounts of galactosamine- and N-acetyl-galactosamine-residues [52].

14.6.4. Cell Culture Conditions

Caco-2 cells were grown in culture medium consisting of Eagle's minimum essential media supplemented with Earle's Balanced Salt Solution, 2.0 mM L-glutamine, 1.0 mM sodium pyruvate, 0.1 mM nonessential amino acids, 1.5 g/L sodium bicarbonate, 20% FBS and 20 μg/ml gentamycin sulfate antibiotic. Cells were maintained in a humidified 5% CO_2/95% air atmosphere at 37°C while renewing media every second day. Cells were subcultured by trypsinization with 0.25% trypsin-EDTA. For experiments, cells were seeded into either six-well plates or Costar Transwell inserts (0.4 μm pore size, and 4.7 cm^2 surface area). Before seeding, culture surfaces were coated with rat collagen in order to promote cell adhesion. Caco-2 monolayers grown between 14 and 30 days were used in all experiments.

14.6.5. Assessing Confluency and Tight Junction Formation

Cell confluency and tight junction formation, as a function of time, can be assessed using a number of techniques. All techniques rely on the measurement of selected substance from one tissue compartment to the opposing compartment. The most frequently used method is by measuring transepithelial electrical resistance. A square current pulse (ΔI) is passed across the epithelium and the resulting voltage response (ΔV) is measured. Using Ohm's law, the conductance can be calculated ($\Delta I/\Delta V$). Here, cell monolayers exhibiting TEER values of at least 220 Ωcm^2 were used in transport studies. Although this method is rapid and simple to implement, it can often be inaccurate due to nonuniform current fields caused by the point source nature of the current passing electrodes and can increase infection of the culture system [53]. A simple, inexpensive spectroscopic method can instead be used to assess cell confluency, tight junction formation, and fluid flow [54]. Using this method, the flux of phenol red across a cell monolayer when grown on a transwell permeable support is observed. Phenol red is a pH indicator used in commercially available media. It is nontoxic to epithelial cells and is not metabolized or synthesized by the cells. Because phenol red is a pH indicator, its absorbance properties are a function of pH; however at a wavelength of 479 nm, its isosbestic point, it is not pH sensitive. The concentration of phenol red in a solution can be calculated by applying the Beer-Lambert Law using the extinction coefficient for phenol red ($8450 \ M^{-1} \ cm^{-1}$). The flux of phenol red and the permeability of the monolayer can be calculated using the following equations:

$$J = (A_{479}^{b} \times Vol_b)/(t \times A \times EC \times Pl) \qquad (14.4)$$

$$P = J/[PR] \qquad (14.5)$$

Where J is the phenol red flux, A_{479}^{b} is the absorbance of the basolateral solution at 479 nm, Vol_b is the volume of the basolateral solution at the time of measurement, t is time, A is the surface area of the epithelium, EC is the extinction coefficient, and Pl is the light path. P denotes the permeability to phenol red and [PR] is the phenol red concentration in the apical compartment. *Days 1–4* represent cell confluence, *days 4–8* represents complete formation of tight junction and inhibition of cell proliferation, and *days 8–21* represent a small but marginally significant further decrease in permeability (Figure 14.17).

14.6.6. Adhesion of Lectin-Modified Microdevices

In vitro studies were performed using the Caco-2 cell line to measure the bioadhesive properties of lectin-conjugated microdevices (Figure 14.18). The binding characteristics of microdevices modified with two types of lectin (tomato, known to agglutinate Caco-2 cells, and peanut, an unrelated lectin) were observed as a function of time. Although both lectin conjugates produce a higher degree of binding than the pristine microdevices, a marked difference still remains between the peanut and tomato lectin conjugates. The amount of binding associated with tomato lectin-modified microdevices is thought to be a direct result of the high amounts of N-acetylglucosamine-containing oligosaccharides which are

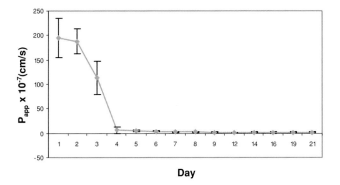

FIGURE 14.17. Change in apparent permeability as the Caco-2 cells become confluent and form tight junctions.

associated with Caco-2 cells. These oligosaccharides provide specific sites for the tomato lectins to bind. In contrast, only minor amounts of galactosamine residues are present on Caco-2 cells, limiting the number of peanut lectin specific binding sites. Figure 14.19 shows the binding trend of unmodified, peanut lectin-modfied, and tomato lectin-modified PMMA microdevices.

14.6.7. Bioavailibility Studies

The Caco-2 model was used to test drug transport through paracellular tight junctions, similar to those found in the intestinal lining, as a model for systemic bioavailability. Upon microdevice application, the rate of permeation of a model drug, FITC-labeled insulin, through the monolayer was measured using fluorescence spectroscopy. Concurrent measurement of the total resistance across the membrane was also measured to determine the magnitude of the physiologic response of tight junction opening. The effect of transport by microdevice delivery was compared to that of a liquid formulation. It was found that drug transport efficiency was augmented when drug formulations were delivered from

FIGURE 14.18. Scanning electron micrograph of adherent PMMA microdevices on a Caco-2 monolayer.

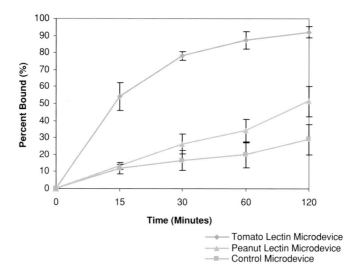

FIGURE 14.19. Binding of PMMA microdevices to Caco-2 cell monolayers as a function of incubation time.

microdevices as compared to liquid formulations. This efficiency increased further with the additional use of a permeation enhancer (Figure 14.20). This dramatically augmented flux when utilizing microdevice delivery is a result of a very high concentration of both permeation enhancer (sodium laurate) and FITC-insulin directly at tight junctions.

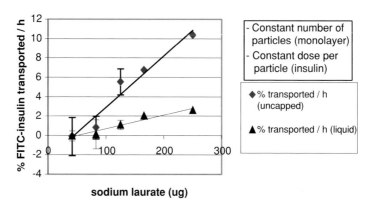

FIGURE 14.20. Transport rate of FITC-insulin across Caco-2 monolayers when delivered from porous silicon microdevices and in liquid formulation in the presence of encorporated sodium laurate permeation enhancer.

ACKNOWLEDGEMENTS

Funding is gratefully acknowledged from iMEDD, Inc., The Whitaker Foundation, The National Science Foundation, The Center for Integrated Medicine and Technology, and The National Aeronautic and Space Administration. Also, special thanks to those who have contributed to this work: Christopher Bonner, Aamer Ahmed, and Michael Lubeley or the University of Illinois-Chicago; Ketul Popat and Simon Su of Boston University; colleagues from iMEDD, Inc.

REFERENCES

[1] S. Henry, D.V. McAllister, M.G. Allen, and M.R. Prausnitz. Microfabricated microneedles: a novel approach to transdermal drug delivery. *J. Pharm. Sci.*, 87:922–925, 1998.

[2] D.V. McAllister, M.G. Allen, and M.R. Prausnitz. Microfabricated microneedles for gene and drug delivery. *Annu. Rev. Biomed.*, 2:289–313, 2000.

[3] L. Leoni and T.A. Desai. Nanoporous biocapsules for the encapsulation of insulinoma cells: biotransport and biocompatibility considers. *IEEE Trans. Biomed. Eng.*, 48:1335–1341, 2001.

[4] L. Leoni, T. Boriarski, and T.A. Desai. Characterization of nonporous membranes for immunoisolation: diffusionproperties and tissue effects. *J. Biomed. Microdev.*, 4:131–139, 2001.

[5] J.T. Santini, M.J. Cima, and R. Langer. A controlled-release microchip. *Nature*, 397:335–338, 2000.

[6] J.T. Santini, A.C. Richards, R. Scheidt, M.J. Cima, and R. Langer. Microchips as controlled drug-delivery devices. *Agnew. Chem. Int. Ed. Engl.*, 39:2396–2407, 2000.

[7] G. Orive, R.M. Hernandez, A.R. Gascon, A. Dominguez-Gil, and J.L. Pedraz. Drug delivery in biotechnology: present and future. *Curr. Opin. Biotechnol.*, 14:659–664, 2003.

[8] S.S. Davis and L. Illum. Drug delivery systems for challenging molecules. *Int. J. Pharm.*, 176:1–8, 1998.

[9] M. Saffran, G.S. Kumar, D.C. Neckers, J. Pena, R.H. Jones, and J.B. Field. Biodegradable azopolymer coating for oral delivery of peptide drugs. *Biochem. Soc. Trans.*, 18:752–754, 1990.

[10] K. Iwanaga, S. Ono, K. Narioka, M. Kakemi, K. Morimoto, S. Yamashita, Y. Namba, and N. Oku. Application of surface-coated liposomes for oral delivery of peptide: effect of coating the liposome's surface on the GI transit of insulin. *J. Pharm. Sci.*, 88:248–252, 1999.

[11] G. Ponchel and J. Irache. Specific and non-specific bioadhesive particulate systems for oral delivery to the gastrointestinal tract. *Adv. Drug Deliv. Rev.*, 34:191–219, 1998.

[12] M.A. Arrangoa, G. Ponchel, A.M. Orecchioni, M.J. Renedo, D. Duchene, and J.M. Irache. Bioadhesive potential of gliadin nanoparticulate systems. *Eur. J. Pharm. Sci.*, 11:333–341, 2000.

[13] C.M. Lehr. Lectin-mediated drug delivery: the second generation of bioadhesives. *J. Control. Rel.*, 65:19–29, 2000.

[14] A. Fasano and S. Uzzau. Modulation of intestinal tight junctions by *Zonula occludens* toxin permits enteral administration of insulin and other macromolecules in an animal model. *J. Clin. Invest.*, 99:1158–1164, 1997.

[15] U.I. Schwarz, T. Gramatte, J. Krappweis, R. Oertel, and W. Kirch. P-glycoprotein inhibitor erythromycin increasesoral bioavailability of talinolol in humans. *Int. J. Clin. Pharmacol. Ther.*, 38:161–167, 2000.

[16] J. Woodley. Bioadhesion: new possibilities for drug administration? *Clin. Pharmacokinet.*, 40:77–84, 2001.

[17] E.C. Lavelle. Targeted delivery of drugs to the gastrointestinal tract. *Crit. Rev. Ther. Drug Carrier Syst.*, 18:341–386, 2001.

[18] M. Montisci, A. Dembri, G. Giovannuci, H. Chacun, D. Duchene, and G. Ponchel. Gastrointestinal transit and mucoadhesion of colloidal suspensions of Lycopersicon esculentum L. and Lotus tetragonolobus lectin-PLA microsphere conjugates in rats. *Pharm. Res.*, 18:829–837, 2001.

[19] M.A. Clark, B.H. Hirst, and M.A. Jepson. Lectin-mediated mucosal delivery of drugs and microparticles. *Adv. Drug Deliv. Rev.*, 43:207–223, 2000.

[20] R. Mody, S. Joshi, and W. Chaney. Use of lectins as diagnostic and therapeutic tools for cancer. *J. J. Pharm. Toxic. Methods.*, 33:1–10, 1995.

[21] A. Putzai. Lectin-targeting of microparticles to different parts of the gastrointestinal tract. *Proc. Intern. Symp. Control. Rel Bioact. Mater.*, 22:161–162, 1995.

[22] F.J. Martin and C. Grove. Microfabricated drug delivery systems: concepts to improve clinical benefit. *Biomed. Microdev.*, 3:97–108, 2001.

[23] A. Ahmed, C. Bonner, and T.A. Desai. Bioadhesive microdevices with multiple reservoirs: a new platform fororal drug delivery. *J Control Rel.*, 3:291–306, 2002.

[24] S.L. Tao, M.W. Lubeley, and T.A. Desai. Bioadhesive poly(methyl methacrylate) microdevices for controlled drug delivery. *J. Control Rel.*, 2215–2228, 2003.

[25] A.B. Foraker, R.J. Walczak, M.H. Cohen, T.A. Boiarski, C.F. Grove, and P. Swann. Microfabricated porous silicon particles enhance paracellular delivery of insulin across intestinal Caco-2 cell monolayers. *Pharmaceut. Res.*, 20:110–116, 2003.

[26] D.L. Polla, A.G. Erdman, W.P. Robbins, D.T. Markus, J. Diaz-Diaz, R. Rizq, Y. Nam, and H.T. Brickner. Microdevices in medicine. *Annu. Rev. Biomed. Eng.*, 2:551–576, 2000.

[27] M. Madou. *Fundamentals of Microfabrication*. Boca Raton, CRC 2003.

[28] N. Maluf. *An Introduction to Microelectromechanical Systems Engineering*. Boston, Artech House, 2000.

[29] A. Uhlir. Electrolytic shaping of germanium and silicon. *Bell. Syst. Tech. J.*, 35:333–347, 1956.

[30] A. Angelescu, I. Kleps, M. Mihaela, M. Simion, T. Neghina, S. Petrescu, N. Moldovan, C. Paduraru, and A. Raducanu. Porous silicon matrix for applications in biology. *Rev. Adv. Mater. Sci.*, 5:240–449, 2003.

[31] L.T. Canham. Properties of porous silicon. *EMIS Datare View Series*. London, INSPEC, 1997.

[32] F. Cunin, T.A. Schmedake, J.R. Link, Y.Y. Li, J. Koh, S.N. Bhatia, and M.J. Sailor. Biomolecular screening with encoded porous-silicon photonic crystals. *Nat. Mater.*, 1:39–41, 2002.

[33] A. Henry, T. Tutt, M. Gallowaty, Y. Davidson, S. McWhorter, S. Soper, and R. McCarley. Surface modification of poly(methyl methacrylate) used in the fabrication of microanalytical devices. *Anal. Chem.*, 72:5331–5337, 2000.

[34] M. Jager and A. Wilke. Comprehensive biocompatibility testing of a new PMMA-hA bone cement versus conventional PMMA cement in vitro. *J. Biomater. Sci. Polym. Ed.*, 11:1283–1298, 2003.

[35] K.T. Chu, Y. Oshida, E.B. Hancock, M.J. Kowolik, T. Barco, and S.L. Zunt. Hydroxyapatite/PMMA composites as bone cements. *Biomed. Mater. Eng.*, 1:87–105, 2004.

[36] M.G. Harris, N.H. Wong, and A.W. Low. Patient response to PMMA contact lenses. *J. Am. Optom. Assoc.*, 46:1184–1187, 1975.

[37] G.N. Papaliodis, Q.D. Nguyen, C.M. Samson, and C.S. Foster. Intraocular lens tolerance in surgery for cataracta complicata: assessment of four implant materials. *Semin. Ophthalmol.*, 17(3–4):120–123, Sep-Dec, 2002.

[38] V.J. Tomazic-Jezic, K. Merritt, and T.H. Umbreit. Significance of the type and the size of biomaterial particles on phagocytosis and tissue distribution. *J. Biomed. Mater. Res.*, 55:523–529, 2001.

[39] Y. Tabata, Y. Inoue, and Y. Ikada. Size effect on systemic and mucosal immune responses induced by oral administration of biodegradable microspheres. *Vaccine*, 14:1677–1685, 1996.

[40] Y. Tabata and Y. Ikada. Phagocytosis of polymer microspheres by macrophages. *Adv. Poly. Sci.*, 94:107–141, 1990.

[41] I.H. Al-Abdulla, G.W. Mellor, M.S. Childerstons, A.M. Sidki, and D.S. Smith. Comparison of three different activation methods for coupling antibodies to magnetisable cellulose particles. *J. Immunol. Meth.*, 122:253–258, 1989.

[42] A. Rembaum, S.P. Yen, and E. Cheong. Functional polymeric microspheres based on 2-hydroxyethyl methacrylate for immunocheical studies. *Macromolecules*, 9:328–336, 1976.

[43] J.H. Bowes and C.W. Cater. The reaction of gluteraldehyde with proteins and other biological materials. *J. Roy. Microsc. Soc.*, 85:193–200, 1966.

[44] H. Otto, H.Takamiya and A.Vogt. A two stage method for crosslinking antibody globulin to ferritin by glutaraldehyde. Comparison between the one-stage and the two-stage method. *J. Immuno. Meth.*, 2:127–146, 1973.

[45] K.J. Sultzbaugh and T.J. Speaker. A method to attach lectins to the surface of spermine alginate microcapsules based on the avidin biotin interaction. *J. Microencapsul.*, 13:363–375, 1996.

[46] S.L. Tao, M.W. Lubeley, and T.A. Desai. Synthesis of cytoadhesive poly(methylmethacrylate) for applications in targeted drug delivery. *J. Biomed. Mater. Res.*, 67:369–75, 2003.

[47] B. Karandikar, J. Puschett, and K. Matyjaszewski. Homogeneous and heterogeneous modificatin of poly(methyl methacrylate) with ethylene diamine. *Polym. Prepr. (Am. Chem. Soc., Div. Polym. Chem.)*, 30:250–251, 1989.

[48] A.C. Henry, T.J. Tutt, M. Galloway, Y. Davidson, C.S. McWhorter, S. Soper, and R. McCarley. Surface odificatin of poly(methyl methacrylate) used in the fabrication of microanalytical devices. *Anal. Chem.*, 72:5331–5337, 2000.

[49] R.K. Gaur and K.C. Gupta. A spectrophotometric method for the estimation of amino groups on polymer supports. *Anal. Biochem.*, 180:253–258, 1989.

[50] C.M. Lehr, J. Bouwstra, W. Kok, A. Noach, A. Boer, and H. Junginger. Bioadhesion by means of specific binding to tomato lectin. *Pharm. Res.*, 9:547–553, 1992.

[51] G.J. Russell-Jones, H. Veitch, and L. Arthur. Lectin-mediated transport of nanoparticles across Caco-2 and OK cells. *Int. J. Pharm.*, 190:165–174, 1999.

[52] F. Gabor, M. Stangl, and M. Wirth. Lectin-mediated bioadhesion: binding characteristics of plant lectins on the enterocyte-like cell lines Caco-2, HT-29 and HCT-8. *J. Control. Rel.*, 55:131–142, 1998.

[53] S.A. Lewis. Assessing epithelial cell confluence by spectroscopy. *Methods Mol. Biol.*, 188:329–336, 2002.

[54] B. Jovov, N.K. Wills, and S.A. Lewis. A spectroscopic method for assessing confluence of epithelial cell cultures. *Am. J. Physiol.*, 261:C1196–C1203, 1991.

15

Nanoporous Implants for Controlled Drug Delivery

Tejal A. Desai[1], Sadhana Sharma[2], Robbie J. Walczak[3], Anthony Boiarski[3], Michael Cohen[3], John Shapiro[2], Teri West[3], Kristie Melnik[3], Carlo Cosentino[4], Piyush M. Sinha, and Mauro Ferrari[5]

[1]*Department of Bioengineering and Physiology, University of California, San Francisco, CA*
[2]*Department of Physiology and Biophysics, University of Illinois at Chicago, Chicago, IL*
[3]*IMEDD Inc., Foster City, CA*
[4]*Department of Experimental and Clinical Medicine, University of Catanzaro "Magna Græcia", Catanzaro, Italy*
[5]*Professor, Brown Institute of Molecular Medicine Chairman, Department of Biomedical Engineering, University of Texas Health Science Center, Houston, TX; Professor of Experimental Therapeutics, University of Texas M.D. Anderson Cancer Center, Houston, TX; Professor of Bioengineering, Rice University, Houston, TX; Professor of Biochemistry and Molecular Biology, University of Texas Medical Branch, Galveston, TX; President, the Texas Alliance for NanoHealth, Houston, TX*

15.1. INTRODUCTION

15.1.1. Concept of Controlled Drug Delivery

Over the last three decades considerable advances have marked the field of drug delivery technology, resulting in many breakthroughs in clinical medicine. However, major unmet needs remain. Among these are broad categories of: 1) Continuous release of therapeutic agents over extended time periods and in accordance to a pre-determined temporal profile [38, 60]; 2) Local delivery at a constant rate to the tumor microenvironment, to overcome much of the systemic toxicity and improve anti-tumor efficacy; 3) Improved ease of administration, increasing patient compliance, while minimizing the needed intervention of healthcare personnel and decreasing the length of hospital stays [6, 46]. Success in addressing some or all of these challenges would potentially lead to improvements in efficacy and patient compliance, as well as minimization of side effects [66].

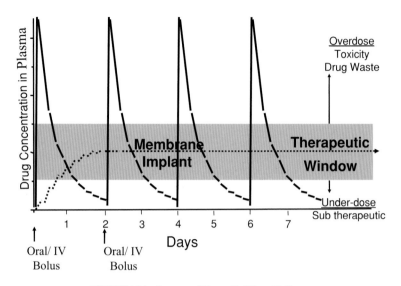

FIGURE 15.1. Concept of Controlled Drug Delivery.

The importance of controlled-release drug delivery systems may be argued with reference to the goal of achieving a continuous drug release profile consistent with zero-order kinetics, wherein blood levels of drugs would remain constant throughout the delivery period. By contrast, injected drugs follow first-order kinetics, with initial high blood levels of the drug after initial administration, followed by an exponential fall in blood concentration [65]. Toxicity often occurs when blood levels peak, while efficacy of the drug diminishes as the drug levels fall below the therapeutic range (Figure 15.1). The potential therapeutic advantages of continuous-release drug delivery systems are thus significant, and encompass: In vivo predictability of release rates on the basis of in vitro data, minimized peak plasma levels and thereby reduced risk of adverse reactions, predictable and extended duration of action, reduced inconvenience of frequent dosing and thereby improved patient compliance [7, 39].

15.1.2. Nanopore Technology

Nanopore technology provides excellent opportunities for controlled drug delivery through the fabrication of microfabricated nanoporous membranes. Unlike conventional polymeric membranes, silicon membranes are biologically, thermally, chemically, and mechanically stable *in vivo*-like environments. Furthermore, compared to alternate nanopore fabrication technology such as ion track-etched polycarbonate (e.g. Osmonics, Minnetonka, MN) and porous alumina (e.g. Whatmann Ann Arbor, Michigan), microfabricated nanoporous membranes have uniform pore size and very low thickness. These membranes are ideally suited for drug delivery applications, including controlled diffusion and sustained release. It is possible to design nanopore membranes, which achieve almost constant rate of drug delivery, avoiding the "burst effect". By precisely controlling pore size, pore length and pore density, the nanopore membrane fitted with a drug reservoir suitable for subcutaneous implantation can serve as a diffusion barrier for a variety of biological drugs [21].

Our group introduced the surface and bulk silicon nanomachining protocols, required for the fabrication of nanopore with exquisite control over pore dimensions and surface composition. Our work has resulted in over 15 issued US and International patents, with claims over the methods of fabrication, the resulting devices, and their methods of use, and a large number of publications [9, 10, 12, 14–15, 19, 34–37, 68, 69]. A variety of progressively improved designs of silicon-based nanochannel systems were developed and investigated for a range of applications, including bioseparation, immunoisolation and controlled drug delivery [9, 25].

The nanoporous silicon membranes, with highly uniform pores in the nanometer range, were first fabricated using standard microfabrication techniques of photolithography, thin film deposition, and selective etching [25]. Nanopores were generated by a key process step, based on the use of thermally grown sacrificial silicon oxide layer, sandwiched between two structural layers- a process termed "sacrificial oxide nanopore formation" [10, 22, 25, 34, 68]. Over the years, nanopore technology has undergone continued improvements. Nevertheless, the basic structure and fabrication protocol for the nanopores has remained the same. The membrane area is made of thin layers of polysilicon, silicon dioxide, and/or single crystalline silicon depending on the design employed. The other main part of the membrane is the anisotropically backside etched wafer. Since photolithography in general has a lower limit of resolution of 0.25-1 µm, strategies using sacrificial layers were utilized to achieve desired pore size down to the tens of nanometers. The strategies were initially based on the use of a sacrificial oxide layer, sandwiched between two structural layers, for the definition of the pore pathways. However, all designs of the microfabricated membrane consisted of a surface micromachined membrane on top of an anisotropically etched silicon wafer, which provides mechanical support. Changes in pore size, density, and geometry as well as path length were the main features changed while optimizing the membrane design.

The first design of nanoporous membranes consisted of a bilayer of polysilicon with L-shaped pore paths. The flow path of fluids and particles through the membrane is shown in (Figure 15.2A) [20]. As shown, fluid enters the pores through openings in the top polysilicon layer, travel laterally through the pores, make a 90° turn, and exit the pores through the bottom of the pore where both the top and bottom polysilicon layers lay on the etch stop layer). While this design performed well for preventing the diffusion of the larger, unwanted immune system molecules, its L-shaped path slowed down and, in some cases, prevented the diffusion of the smaller molecules of interest. The pores in this design were fairly long, which led to the slow diffusion of the desired molecules. Also, because of the large area per pore, it was difficult to increase the pore density and thus the diffusion rate. The next design had an improvement in the production of short, straight, vertical pores through a single crystal base layer. This design had the advantage of direct flow paths (Figure 15.2B) [20]. This direct path allows the smaller molecules of interest to diffuse much quicker through the membrane, while still size-separating the larger molecules. To further improve the reliability of the nanoporous membranes, several basic changes were made in the fabrication protocol from the previous membrane design to eliminate problems with the diffused etch stop layer [43]. This design also incorporated a shorter diffusion path length, based on the thicknesses of the two structural layers. The design of a new membrane fabrication protocol

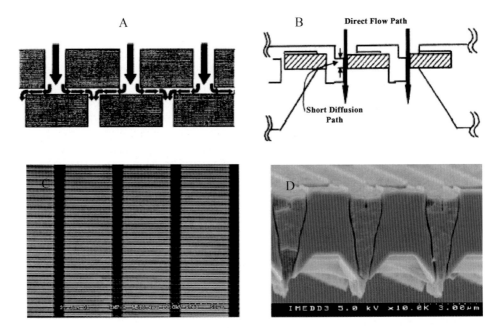

FIGURE 15.2. (A) Flow path through M1 filters, with lateral diffusion through the nanopores defined by sacrificial oxide. (B) Cross-section of M2 design showing dirext flow path. Scanning electron micrographs of microfabricated membrane: (C) top view detail; (D) side view detail.

incorporated several desired improvements: a well defined etch stop layer, precise control of pore dimensions, and a lower stress state in the membrane (Figure 15.2C & 15.2D). The new protocol also increased the exposed pore area of the membranes.

Most recently, we have developed nanochannel delivery systems (called nDS) with improved mechanical stability [62]. These devices are based on bulk micromachining and sandwich encapsulated filter design [68]. The nDS device consists of two bonded silicon wafers: the micromachined filtration structural wafer and the cap wafer (Figure 15.3). Fluids enter through the hole etched in the cap wafer, then flow horizontally through the filtration channel as defined by the gap between the cap wafer and the machined features on the structural wafer, and then out the hole etched in the structural wafer. Since the flow is between two directly bonded wafers, the filter has more mechanical support and is thus structurally stronger. Another advantage to the encapsulated-channel filter lies in the fabrication process: a bulk micromachined structure requires less complicated fabrication steps than a surface-micromachined one and thus leads to a simpler and faster process flow. The features on the structural wafer are shown in Figure 15.3 and are fabricated using bulk micromachining. To achieve high throughputs, interdigitated finger geometry was used. The first devce in our proposed sequence of delivery systems is the passive release device called nDS1. Future embodiments will have the capability of integration of electronics on board, and are being developed for preprogrammed- (electroosmoticaly driven, nDS2) and remote-activated (drug on demand, nDS3) delivery of drugs.

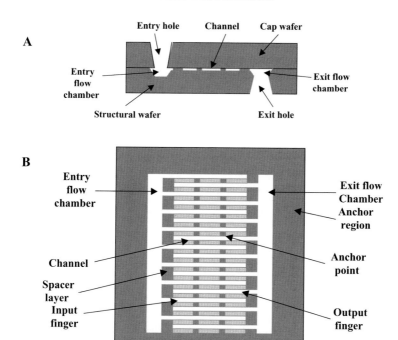

FIGURE 15.3. (A) Cross-sectional view of the nanochannel delivery system (nDS). (B) Top and cross-sectional view of nDS structure on the structural wafer.

15.1.3. Comparison of Nanopore Technology with Existing Drug Delivery Technologies

The nanopore delivery technology discussed here is based on the bulk and surface processing of single crystal silicon wafers, and thin layers of polysilicon, silicon oxide, and silicon nitride. It offers distinct advantages in the scaleability of the manufacture (obviously demonstrated by the world-changing successes of the microelectronics industry and, to a lesser extent, of micro-electro-mechanical-systems (MEMS)), exquisite device replicability, and the possible integration of complex electronic functionalities.

The most important competitor to our proposed technology is the IV administration of bioacive agents, which is associated with obvious difficulties in terms of patient inconvenience, discomfort, required hospital stay, and adverse affects such as phlebitis and infections. While it is impossible to review here in any detail the very broad field of drug-delivery technology, it is noted that development of drug delivery systems encompasses broad categories, such as implantable devices with percutaneous components, fully implantable devices, polymer-based systems, microchips, and osmotic pumps. Krulevitch and Wang [41] described a microfabricated, fully integrated drug delivery system capable of secreting controlled dosages of drugs over long periods of time, while Cao, Lai, and Lee [29] describe a self-regulated drug delivery device that integrates both mechanical and chemical methodologies. Numerous polymer systems have been employed [60, 74, 52] with varying degrees of success. Infection is a major concern with clinically available implantable drug delivery pumps, as are catheter-related complications, such as kinking, dislodgements, disconnections, tears and occlusions [26]. Furthermore, catheter-tip inflammatory masses continue to be a problem with current devices [30]. Poor patient compliance is a significant obstacle that often leads to suboptimal treatment and inferior outcomes [4, 18, 25]. In fact, poor compliance is the most common cause of medication failure [6, 56]. For patients with terminal illness in particular, discouragement and lack of conviction regarding the effectiveness of treatments result in poor compliance [64]. These problems could potentially be ameliorated though the use of appropriate implantable drug delivery systems. Implants with degradable polymers suffer from two major drawbacks. Polymer depots exhibit an initial "burst effect" prior to sustained drug release, and typically are not as efficient in controlling release rates of small molecules [53]. The use of implantable devices with percutaneous components such as ambulatory peritoneal dialysis catheters, intravenous catheters and orthopedic implants, is complicated by such occurrences as infection, marsupialization, permigration, and avulsion [31, 32].

In spite of recent developments in the drug delivery technology, there is no clinically available device that has been shown to be able to perform the controlled, long-term diffusion of the agents on interest here. Potential alternative approaches that might be employed are A: Osmotic pumps [67]; and B: MicroChips developed by Santini, Cima, and Langer [57–59]. The former employs an osmotic piston to provide the zero-order release of drugs, and has been clinically demonstrated for the constant-rate administration of leuprolide in the management of prostate cancer [27, 74]. The device developed by Santini et al. employs the electrochemical dissolution of the cover of a number of reservoirs to obtain the controlled release of their contents. Ideally, this methodology may yield a desired, and potentially variable, release profile. Neither one of these two potentially alternative release devices have been proven capable of providing the zero-order release of the immunomodulating molecules of interest in the research protocol presented herein. Expected problems that might present themselves, were one to attempt the employ osmotic pumps or the MicroChip device by Santini et al. to perform the experimental protocols described in this proposal, include: concerns with formulation and mass transport dynamics, leading to expectedly high and potentially unsafe pressures in the ALZET chamber; and loss of functionality of the therapeutic moieties. Furthermore, it remains to be proven that the desired release profiles, especially for what concerns the variable rates, could be met with the MicroChips. The ALZET pump is, by its very design, incapable of performing variable rates, or arresting the release once it is started. A further disadvantage is the difficulty of developing an effective dosage solution to deliver lipophilic compounds [5, 40]. By contrast, the devices fabricated by our group are based on constrained diffusion and controllable electrokinetic transport, and therefore will not lead to any build up of pressure during its use. It will not require the development of novel formulations, since the release of functionally active molecules has been demonstrated [44]. Future embodiments will have the capabilities for preprogrammed- and remote-activated delivery of drugs.

In this review we will mainly focus on the fabrication, release characteristics, and biocompatibility issues of a small subcutaneous nanoporous implant named NanoGATE incorporating a microfabricated silicon nanoporous membrane engineered to the exact size and requirements of the individual molecule. The NanoGATE device is anticipated to slowly release the encapsulated drug at an optimal rate to mimic a slow infusion, so that the patient will have therapeutic levels of the drug in his/her body for the entire course of therapy. The drug reservoir of the NanoGATE device will contain a highly concentrated form of the drug either as a dry powder or concentrated suspension to minimize the size of the device required to hold the cumulative dose required for an extended period of treatment (e.g., 3 to 6 months). At present NanoGATE technology is being developed to deliver Interferon-α for the treatment of chronic hepatitis C. Interferon α delivered through a NanoGATE subcutaneous implant is expected to provide significant medical advantages over the current treatment modalities.

15.2. FABRICATION OF NANOPOROUS MEMBRANES

We have used top-down microfabrication methods to create nanopore membranes consisting of arrays of parallel rectangular channels which, in their smallest aspect, range from 7 nm to 50 nm (Figure 15.4). The original method pioneered by Ferrari and colleagues [12] consists of two basic steps: (1) Surface micromachining of nanochannels in a thin film on the top of a silicon wafer, and (2) Forming the nanopore membrane by etching away the

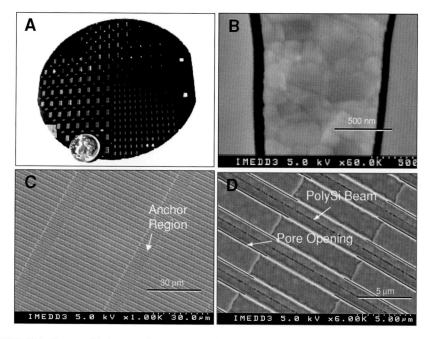

FIGURE 15.4. Photographic images of nanopore membranes. (A) Appearance of 4 in silicon wafer showing 120 small and 100 large membrane dies before being cut into individual units. (B) SEM cross-sectional view of membrane with 50 nm pores separated by silicon and poly-silicon material. (C) SEM top view of membrane with pores at 1000X magnification showing 45 μm long pores separated by 10 μm long anchor regions. (D) 6000X top SEM view of membrane showing details of pore and anchor structures.

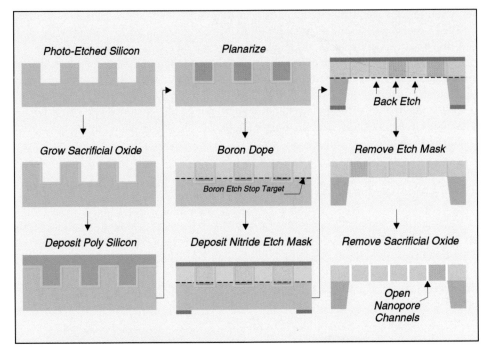

FIGURE 15.5. Schematic of key steps in silicon nanopore membrane fabrication process.

bulk of the silicon wafer underneath the thin-film structure. The overall fabrication process is shown schematically in Figure 15.5.

The first major process step involves etching channels in the silicon substrate over its entire surface to define the overall shape of the pores (Figure 15.5A). The etched channels are 2 μm wide separated by 2 μm and they are 5 μm deep. These channels are formed using a plasma-etch procedure with a thermally grown oxide layer as an etch-mask.

The next major step involves growing a sacrificial oxide-layer over the entire wafer surface including the surface area of the channels (Figure 15.5B). The sacrificial-oxide layer thickness defines the channel width in the final membrane. Proper selection of the time and temperature of this thermal oxidation step allows for control of the sacrificial layer thickness to tolerances of less than 0.5 nm across the entire wafer.

Anchor points are created within the channels to define the pore length (45 μm distance). These anchor points are 10 μm long and they are formed between each 45 μm -long pore region. The anchors provide rigidity to the membrane structure because there are no pores in the anchor regions. The anchor points are formed across the entire wafer by selectively etching through the sacrificial layer in a series of 10 μm-wide strips perpendicular to the channel direction.

The next fabrication step involves filling the channels using a polysilicon deposition process (Figure 15.5C). This filling step forms the membrane structure by providing silicon material on each side of the sacrificial oxide layers. The deposition also provides solid silicon—polysilicon areas in the anchor regions to stabilize the membrane. The deposited polysilicon layer is then "planarized" (see Figure 15.5D) using a plasma-etch process leaving a smooth surface structure with the pores exposed. An SEM photograph image of the planarized membrane surface is shown in Figure 15.1.

Following planarization, a boron-doping step is performed where boron ions are diffused into the surface of the silicon-polysilicon material to a depth of 3 μm (see dashed line in Figure 15.5E). Boron-doped silicon etches in KOH at a much lower rate than undoped silicon so the boron doping provides an etch-stop that will later define the membrane thickness.

A protective nitride mask layer is then deposited on the wafer completely covering both sides. This layer is impervious to the KOH chemical etch that will be used to form the membranes out of the silicon substrate. On the backside of the wafer, windows are plasma-etched in the nitride layer to define the membrane dimensions (i.e., a series of 1 × 2 mm windows for the small membrane dies and 2 × 3.5 mm windows for the large-membrane dies in Figure 15.4A). These windows in the protective nitride layer expose the silicon wafer to KOH etchant in the desired regions (Figure 15.5F).

The wafer is then placed in a 55°C KOH bath to etch the unprotected silicon through the windows in the nitride and up to the boron etch-stop to form the membrane (Figure 15.5G). After the silicon is removed, the protective nitride etch mask (Figure 15.2H) and sacrificial oxide (Figure 15.2I) are removed using an HF etchant. This final etching step opens the pores and provides the desired nanopore membrane structure.

15.3. IMPLANT ASSEMBLY AND LOADING

All NanoGATE implant housing components were obtained from Manufacturing Technical Solutions (Carroll, OH). A 2 × 3 mm nanopore membrane die was affixed over a small-bore opening within a cylindrical methacrylate insert carrier fitted with two rubber O-rings using general purpose silicone adhesive and allowed to cure 3 h at 55°C. The completed carrier was inserted into the titanium encasement until the nanopore membrane region was fully aligned under the grate opening. Methacrylate end caps containing resealable rubber septa were sealed at each end of the titanium encasement using silicone adhesive and allowed to cure. Figure 15.6 shows a drawing the NanoGATE implant fitted with nanopore membrane (top) and a photograph of prototype implant illustrating its size in relation to a US 1 cent piece (bottom). For filling, the implant was oriented vertically and a 27 gauge luer-lock needle was inserted into the upper septa for use as an air vent. The liquid suspension was slowly injected into the implant via the lower septa until all the air within the implant was removed, as indicated by the presence of liquid exuding from the upper needle. The needles were removed under gentle liquid injection pressure to avoid any concomitant influx of air upon withdrawal. The implants were rinsed by immersion in appropriate buffer prior to either placement into a testing vessel or surgical implantation.

15.4. NANOPOROUS IMPLANT DIFFUSION STUDIES

Fick's laws are usually adequate to describe diffusion kinetics of solutes from a region of higher concentration to a region of lower concentration through a thin, semi-permeable

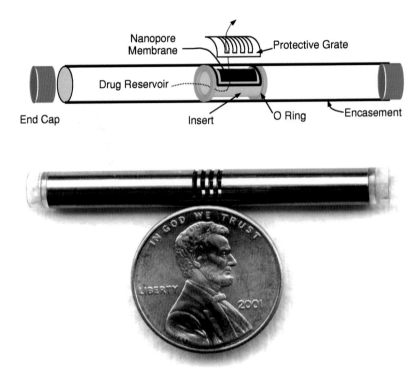

FIGURE 15.6. Implant device fitted with nanopore membrane. (Top) Drawing illustrating key features of the device. The dashed arrow represents a possible diffusion path of a drug molecule held within the device reservoir. (Bottom) Photograph of prototype implant device illustrating its size in relation to a US 1 cent piece.

membrane. As the size of the membrane pores approaches that of the solute, however, unexpected effects can occur, which deviate substantially from those predicted by Fick's laws. Diffusion of molecules in microporous media such as zeolites has led to experimental evidence of such unusual phenomena as molecular traffic control and single file diffusion (SFD) (1-6) [16, 28, 29, 42, 51, 71]. Theoretical treatments and simulations suggest that in the case of SFD solute molecules of equal size cannot pass each other in pores which approximate the dimensions of the molecule itself, regardless of the influence of concentration gradient, and thus their initial rate of movement (or flux) is underestimated by Fick's law [1, 2, 44, 48, 49, 54].

During development of the immunoisolating biocapsule, it was noted that diffusion through nanopore membranes, when using the smaller pore sizes, was slower than predicted from Fick's law. To explore this phenomenon, the relationship between diffusion rates of various solutes and the width of nanopore membranes was investigated [50].

15.4.1. Interferon Release Data

The interferon release profile is shown in Figure 15.7, the membrane used in the experiment has 20 nm pore size, the initial concentration in the donor well is 4.68 mg/ml, and the Stokes Radius of interferon is about 2.3 nm. The solid line has been obtained by

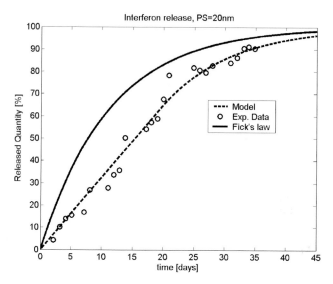

FIGURE 15.7. In vitro interferon diffusion through nanopore membrane (20 nm pore size): experimental data (○), Fick's law prediction (-), model based simulation (- -).

simply combining the first Fick's law with the mass conservation principle, and multiplying the mass flux times the total nominal pores area: the resulting diffusion profile, as well known, results to be exponential. The steady state value is close to 100% of the amount of drug loaded into the donor well, because the acceptor well volume is much larger than the donor well, therefore, in order to make the concentration homogeneous on both sides of the membrane, only a little amount of drug has to be retained in the donor well.

Looking at the experimental measurements (circle markers), it is fairly clear that, even changing the parameters values with respect to the nominal ones, the experimental diffusion profile could not be explained by Fick's law, because it is not exponential. On the contrary, the release rate keeps constant (zero order kinetics) for a long period (about 25 days), until 75% of the total amount has been released. Later on we will discuss how to interpret this result and to devise a suitable theoretical diffusion model fitting the data (dashed line).

15.4.2. Bovine Serum Albumin Release Data

Experiments, like the one described in the previous section, indicate that nanopore membranes can be engineered to control diffusion rates and kinetic order by "fine-tuning" channel width in relation to the size of solutes. Moreover, when the proper balance is struck, zero-order diffusion kinetics is possible. Implantable zero-order output devices are useful to deliver drugs which are not orally bioavailable, particularly in clinical settings where maintenance of a steady state level in the blood stream for long periods is desirable [74]. The nanopore membrane offers a non-mechanical means to achieve such release kinetics for delivering small molecular weight organic drugs as well as larger peptide- and protein-based biopharmaceuticals.

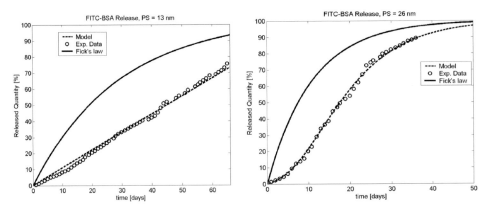

FIGURE 15.8. In vitro diffusion kinetics of fluorescein isothiocyanate (FITC) labeled-BSA through (A) 13 nm pore size and (B) 26 nm membrane under sink conditions: experimental data (o), Fick's law prediction (-), model based simulation (- -).

As a point of reference, the hydrodynamic diameter of a typical biopharmaceutical ranges between about 4 nm and 15 nm and a typical small organic drug may be 3 nm to 5 nm in size [33]. Thus, the widths of these nanopore channels can be made to approximate molecular dimensions. To test the system we selected bovine serum albumin (BSA, MW 66,000 Da, Stokes radius 3.88 nm) as a surrogate of a fairly large protein biopharmaceutical. Virtually no BSA diffusion is seen through nanopore membranes of 7 nm (data not shown), which would be expected since the hydrodynamic diameter of BSA is about 8 nm. In vitro release of fluorescent-labeled BSA loaded into implants fitted with nanopore membranes of two sizes, 13 nm and 26 nm, under sink conditions is shown in Figure 15.8. Fick's expected behavior is depicted in the figure and compared with an interpolation, obtained from a dynamical diffusion model, which will be discussed later on.

The mass flux is constant in the 13 nm case (Figure 15.8A), and zero order kinetics is maintained for the whole experiment duration, while the 26 nm test data (Figure 15.8B) show an exponential release profile. Note that, in the latter case, there is a second order effect, which creates a kind of inertia in the initial days: the release rate starts from a very low value, then increases till the curve assumes a slope close to that predicted by Fick's law. This phenomenon occurs in some experiments, and is caused by an unsuccessful pre-wetting of the pores, thus it has to be neglected on a par with experimental errors.

As one would intuitively expect, even with pore sizes greater than 26 nm, BSA diffusion remains Fickian (data not shown). These data indicate that flux can be controlled by porosity, and kinetic rate by channel width.

In order to capitalize on the constrained diffusion feature offered by these nanopore membranes, we tested their utility as a means of controlling the bioavailability of a protein in vivo. Devices fitted with 13 nm membranes and filled with 300 μl of a solution containing 0.15 mg of ^{125}I-BSA were implanted subcutaneously in the backs of 3 rats. The devices were adjusted to have an in vitro output rate of 15 μg/day and thus were designed to release the albumin for about 100 days. The pharmacokinetics of the labeled BSA in blood was measured and compared to a dose of BSA delivered by a standard subcutaneous bolus injection. Figure 15.9 shows the mean values for blood levels over a period of 45 days

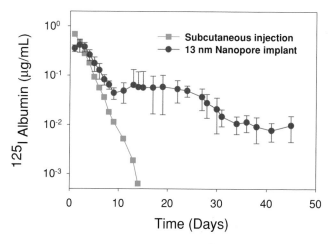

FIGURE 15.9. Pharmacokinetics of ^{125}I-BSA in rats given as a single bolus subcutaneous injection (■) or following implantation of nanopore devices with an in vitro output rate of 15 □g/day (▲). Mean values for three implants and standard deviation are plotted.

after implantation. In the case of the nanopore implant group, following an initial period of rapid decline (during the first 9 days), the rate of clearance of BSA from the central compartment slowed, maintaining measurable levels for the ensuing 4 weeks. The initial decline is attributed to the equilibration of the radiolabeled BSA appearing in blood with the albumin pool in the interstitial fluid volume. This equilibration has been reported to have a half-life of about 3–7 days, which is in line with our results [63]. In comparison to the standard subcutaneous injection, BSA delivered in the nanopore device was detectable for a substantially longer period. As expected, when the devices were recovered from experimental animals they were encapsulated in a fibrous capsule, but upon visual inspection the nanopore membrane itself was free of any tissue intrusion. The encapsulation response did not appear to retard the bioavailability of the albumin released from the device; about half the labeled albumin was recovered from the nanopore device, which conforms to the expectation that half the drug was released during the 7-week implantation period.

15.4.3. Results Interpretation

The situation presented here differs from SFD observed for adsorbate molecules such as methane or CF_4 in crystalline zeolites. The microfabricated nanopore channels used here are of molecular size in only one dimension and the solutes themselves do not tend to adsorb to the silicon surface. Our observations are consistent with the diffusion reported for colloidal particles confined in closed one-dimensional channels of micrometer scale where particle self-diffusion is non-Fickian for long time periods and the distribution of particle displacements is a Gaussian function [71]. In our situation zero-order flux is observed when a chamber filled with a solute is separated from a solute-free external medium by channels that are only several times wider than the hydrodynamic diameter of the individual molecules. The basic principle of diffusion as a mixing process with solutes free to undergo Brownian motion in three dimensions does not apply since in at least one dimension solute movement within the nanopore is physically constrained by the channel walls. Experimental observations of colloidal particles in a density matched fluid confined between two flat plates reveal that particle diffusion becomes anisotropic near the interface; in this case leading to hindered diffusion as a consequence of constrained Brownian motion and hydrodynamic drag effects at distances close to the walls [47]. In our case, it is not entirely certain that the ordering of solutes imposed by the nanopore geometry will be as strict as true cylindrical pores, nor that the sequence of particles passing through the nanopores under the influence of the concentration gradient will remain unchanged over the time required to travel the 4 μm length of the channel; particles could conceivably pass each other laterally. Whether a consequence of a SFD-like phenomenon or drag effects (or a combination of both), the nanopore membrane used here is rate-limiting and, if properly tuned, restricts solute diffusion to a point that flux rates across the membrane are entirely independent of concentration gradient.

15.4.4. Modeling and Data Fitting

In order to achieve a further insight in the mechanisms involved in nano-channel diffusion, we describe the experimental phenomena in mathematical terms, thus yielding to the creation of a dynamical model, which makes it possible to simulate the diffusion experiments and fit the related data.

A detailed description of such model is presented elsewhere [17]. Here, we will limit our description to the main concepts only. Basically, the main core of the model is constituted by first Fick's law combined with the mass conservation principle. The main hypothesis the model relies upon is that the membrane effect can be mathematically described by means of a saturation on the mass flux, where the threshold is intuitively depending on the nano-channel width and the molecular dimensions.

This assumption can be better understood if we look at the results from the interferon diffusion test. Referring to Figure 15.7, we can obtain the release profile depicted by the dashed line by simply simulating the experiment with nominal parameters values and optimizing the mass flux saturation level (which is, therefore, the only free parameter) on the basis of the available data. The simulated mass flux is depicted in Figure 15.10, along with the same quantity obtained from a simulation of the free (Fickian) diffusion case, with the same parameters values.

Clearly, the flux at the beginning assumes the highest value, because the concentration gradient is maximum. Assuming a saturation level below this maximum value results in a constant flux for a certain time interval, in this case about 20 days. The switch from constant to exponentially decreasing profile occurs when the concentration gradient becomes so low that the flux value is less or equal than the fixed threshold.

In the light of the good data fitting obtained, we can argue that the assumption about the flux saturation effect is fairly suitable, although deeper understanding has to be achieved, concerning the basic molecular mechanisms causing it.

Besides the theoretical value, this interpretation of the phenomenon, along with the computational model, proves very useful as a tool for the tuning of the release device, allowing us to substitute long and expensive experimental tests by simulation with different parameters values.

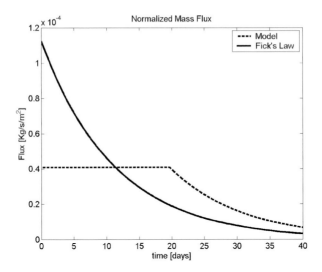

FIGURE 15.10. Simulated mass flux through a 20 nm pore size membrane: Fick's law prediction (-), model based simulation (- -).

The results presented here lead us to believe that devices outfitted with such nanopore membranes can regulate delivery kinetics of a wide range of drugs. Moreover, since the mechanism of release is attributable to a novel constrained diffusion mechanism provided by the precise geometry of the nanopore membrane itself, and no moving parts such as pistons are required, we believe that drugs can be loaded into the device reservoir in a range of physical states including solutions, and crystalline or micronized suspensions. Flexibility with respect to the physical form of encapsulated drugs provides options to substantially increase the loaded dose and duration of therapy as well as approaches to increase the stability of proteins, which are intrinsically unstable in aqueous solution at body temperature.

15.5. BIOCOMPATIBILITY OF NANOPOROUS IMPLANTS

The material-tissue interaction that results from the device implantation is one of the major obstacles in developing viable, long-term implantable drug delivery systems. Membrane biofouling is a process that starts immediately upon contact of a device with the body when cells, proteins and other biological components adhere to the surface, and in some cases, impregnate the pores of the material [72]. Not only does biofouling of the membrane impede drug diffusion (i.e. release) from the implant causing reduced therapeutic levels of drug in the patient's blood stream, it is believed that the adhering proteins are one of the main factors that modulate the longer term cellular and/or fibrous encapsulation response [55].

The site and the extent of injury created in the implantation, biomaterial chemical composition, surface free energy, surface charge, porosity, roughness, and implant size and shape, all govern the degree of fibrosis and vascularization [3]. A thin fibrous tissue reaction may have a negligible diffusion resistance relative to the membrane itself. In contrast, a granular tissue reaction would include vascular structures to facilitate the delivery of therapeutic products. A thin tissue of high vascularity can be induced with membranes of particular porosities or architectures or with membranes coated with biocompatible polymers.

Polymeric membranes commonly used for drug delivery applications do not have all the desired 'ideal' membrane properties such as stability, biocompatibility, and well-controlled permselectivity. Moreover, these membranes do not allow passage of desired biomolecules without biological fouling over time. Recent advances in microfabrication technology have enabled the fabrication of silicon membranes with precisely controlled pore sizes. The straight pore architecture of micromachined membranes as opposed to tortuous-path associated with polymeric membranes offers better antifouling behavior [19]. In addition, the silicon surface chemistry itself does not promote mineralization associated with other membrane materials [24]. Furthermore, the silicon membranes can be coupled with protein-resistant molecules to improve biocompatibility [23, 61].

As mentioned earlier, a key component of NanoGATE implant is a microfabricated silicon nanopore membrane engineered to the exact size and requirements of the individual molecule. We examined the long-term biocompatibility of NanoGATE implant in terms of the fouling of the nanopore membrane and formation of a fibrotic tissue capsule around the implant, and evaluated how these effects influence diffusion of the model drug such as lysozyme from the implant to the patient's vascular compartment.

15.5.1. In Vivo Biocompatibility Evaluation

In vivo membrane biocompatibility was evaluated using glucose as a model molecule. Glucose being relatively small molecule (180 Da) can be used for a broad range of pore sizes (7 to 50 nanometers) and analyzed using very quick, already established, and easy to perform assay procedures. Figure 15.11 shows the ratio of post explantation glucose diffusion rate compared to its initial value. As evident from Figure 15.1, there was no noticeable change in glucose diffusion rates pre- and post-implantation illustrating that the silicon membranes did not foul over a six-month implantation period. The implants placed subcutaneously in mice were removed after seven days and examined visually. There was

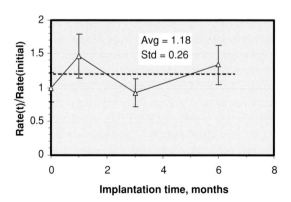

FIGURE 15.11. Ratio of post to pre-implantation glucose diffusion rates.

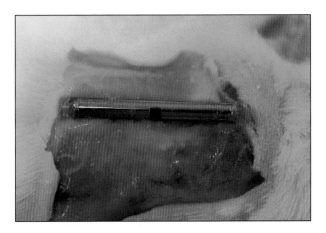

FIGURE 15.12. Photograph of implantation site after thirty days *in vivo*.

no visible evidence of tissue binding to the surface. Figure 15.12 shows a photograph of the implant site after thirty days of implantation. As we can see in Figure 15.12, only a thin vascular capsule forms around the implant as opposed to the avascular fibrous capsule. This minimal tissue response is supposed to be responsible for the comparable pre- and post-implantation glucose diffusion rates observed in this investigation.

15.5.2. Long-Term Lysozyme Diffusion Studies

We examined both the *in vivo* and *in vitro* release of a model drug lysozyme (MW = 14.4 kDa) from a NanoGATE implant. The goal was to determine whether a correlation existed between *in vivo* and *in vitro* drug diffusion kinetics from the implants. Establishing such a correlation would be useful for determining the effects of fibrous tissue capsule on drug release kinetics. Choice of lysozyme as a model drug was prompted by the fact that its molecular weight is very close to the molecular weight to interferon-α (MW = 19.2 kDa), a drug of interest for NanoGATE implant.

For in vivo studies, ^{125}I-labeled lysozyme solution was loaded into several prototype implants at a concentration of 5 mg/mL and investigated *in vivo* using a rat model. This concentration resulted in an initial delivery of 20 μg per day of ^{125}I-labeled lysozyme using a 13 nm silicon nanopore-membrane. The implants were checked after loading the drug to make sure that the release rates were within the expected range prior to surgical implantation for a period of 40–50 days [70].

The blood concentration level of lysozyme released (in μg per mL) from each of the two NanoGATE implants is shown in Figure 15.13. In each case, a single-phase pattern existed for lysozyme levels as represented by a distinct zone of slowly decreasing plasma concentration with time after implantation. In contrast, blood samples from control rats administered with a bolus subcutaneous dose (80 μg) showed radio-labeled lysozyme levels declining rapidly with time (Figure 15.13).

The initial lysozyme plasma concentration of 100 ng/mL agreed well with the predicted value of 60 ng/mL calculated using a clearance, CL, value of 0.23 mL/min. This CL

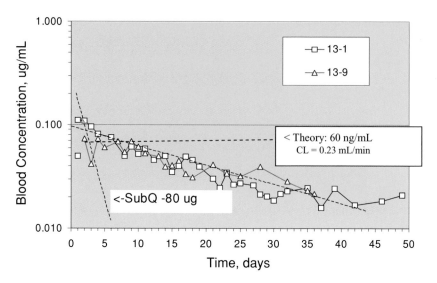

FIGURE 15.13. ^{125}I-lysozyme concentration in plasma of rats for two NanoGATE implants initially releasing 20 μg/day (*in vitro*).

value was calculated from the subcutaneous injection data using standard pharmacokinetic modeling techniques. Ideally, the plasma levels of the drug should remain constant at the predicted level for the entire duration of the experiment. Nevertheless, as shown in Figure 15.13, the plasma levels did not stay at the predicted level, and fell to ~20 ng/mL after 40 days *in vivo*. This slow decrease in lysozyme level in the bloodstream may be an artifact of the radio-labeled lysozyme material used in these experiments and demands further investigation to have a better understanding of the release characteristics of the NanoGATE implant.

A long-term *in vitro* diffusion study was carried out using ^{125}I-labeled and unlabeled lysozyme in order to compare the *in vivo* lysozyme plasma levels to *in vitro* release data (Figure 15.14). Data in Figure 15.5 show that the initial rate of *in vitro* diffusion of ^{125}I-labeled lysozyme was 20 μg/day, but the rate decreased with time suggesting that ^{125}I-labeled lysozyme did not diffuse from the implant with zero-order kinetics. In contrast, unlabeled lysozyme diffuse from NanoGATE implant (with 13nm pore membrane) at a constant rate of 29 μg/day consistent with zero-order kinetics. These *in vitro* results suggested that the radiolabeled lysozyme undergoes some sort of structural rearrangements during the course of experiment, forming aggregates such as dimers, trimers, ... etc with higher molecular weight than the monomer lysozyme. This inference was further supported by size-exclusion chromatography. The chromatogram of radiolabeled implant retentate revealed two major components not present in the unlabeled lysozyme standard. A peak eluting directly at the void volume of the column represented components of very large molecular weight (>2,000 kDa) and suggested major protein aggregation or approximately 5% of the total radioactivity present in the sample. The radiolabels eluting between fractions 20 and

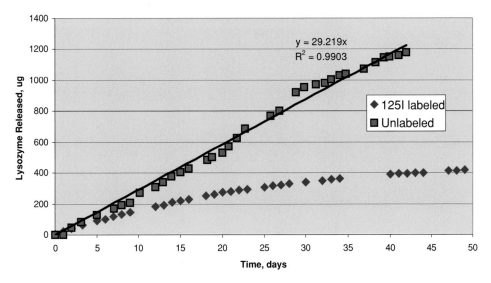

FIGURE 15.14. *In vitro* diffusion of ^{125}I-lysozyme and unlabeled lysozyme.

30 corresponded to material of smaller molecular weight, approximately 30–100 kDa, suggested presence of smaller lysozyme aggregates that are, however, larger than the 14.4 kDa standard material (data not shown). This establishes that essentially the time-dependent aggregation of radiolabeled lysozyme reduced the diffusion rate over time resulting in lower than expected levels of ^{125}I-lysozyme detected in the rat plasma.

15.5.3. In Vivo/In Vitro Correlation

In order to derive a correlation between the *in vivo* plasma levels of ^{125}I-lysozyme (Figure 15.13) and the *in vitro* release rate of this radiolabeled material (Figure 15.14), calculations were performed to determine how the slope of the two plots changed with time. The average plasma concentration, C, from the two implanted devices (Figure 15.13) was divided by the initial plasma concentration, Co, at time zero, and these normalized concentrations are shown in Figure 15.15(A). A least-square fit of the data showed that the slope of C/Co as a function of time is −0.0235.

The daily *in vitro* ^{125}I-lysozyme release rates (µg/day) were derived from total released amounts shown in Figure 15.14. In Figure 15.15(B), the profile shows the normalized release rates i.e. the ratio of the daily rate, Rate, divided by the initial rate, Rate(0) of 20 µg/day. Again, a least square fit of this release rate ratio data shows that the slope of Rate/Rate(0) as a function of time is −0.0210. This value is in very good agreement with the slope of −0.0235 determined from the C/Co plot (Figure 15.15B). The fact that these slopes are nearly identical indicates that there exists a good *in vitro/in vivo* correlation for the ^{125}I-labeled lysozyme diffusion from a NanoGATE implant. In other words, the monotonic drop in lysozyme release from the implant (Figure 15.15B) results in the monotonic drop in plasma levels shown in Figure 15.14. This also indicates that the tissue capsule surrounding

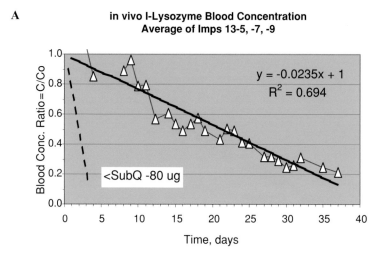

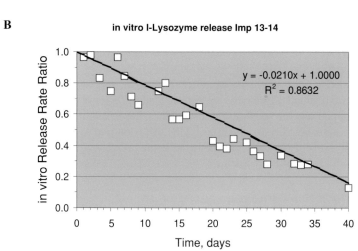

FIGURE 15.15. (A) [125]I lysozyme *in vivo* plasma concentration ratio as a function of time. (B) [125]I lysozyme in vitro release rate ratio as a function of time.

the implant surface does not have any deleterious effects on lysozyme diffusion from the implant for a period of 40 days.

15.5.4. *Post-Implant Diffusion Data*

The effect of the tissue capsule surrounding the implant on drug diffusion was also investigated by measuring the biomolecular release rates from the implant before implantation and after the implant retrieval. The post-implantation diffusion studies were performed using two model molecules, viz., albumin (MW: 66 kDa) and lysozyme (MW: 14.4 kDa). The results of these experiments are shown in Figure 15.16. Data in Figure 15.16

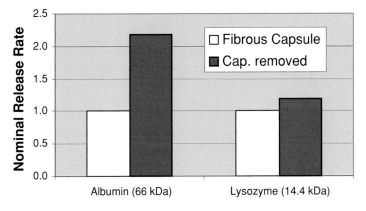

FIGURE 15.16. Post-implantation diffusion testing (with and without tissue capsule).

indicated that for a large molecule like albumin, the diffusion rate doubles once the tissue capsule is removed indicating that the presence of fibrous capsule resulted in lower diffusion rates for albumin. However, for a smaller molecule like lysozyme, there was only a 10% increase in diffusion rate from the implant once the tissue capsule is removed suggesting that the presence of fibrous capsule has only marginal effects on lysozyme diffusion. These results clearly indicated that the fibrous tissue capsule surrounding the implant should not impede the diffusion of interferon-α (MW = 19.2 kDa) from the NanoGATE implant.

15.6. CONCLUSIONS

In this chapter, we discussed about the fabrication, release characteristics, and biocompatibility issues of a small subcutaneous nanoporous implant named NanoGATE designed for the controlled delivery of bioactive molecules. This implant incorporates a microfabricated silicon nanoporous membrane engineered to the exact size and requirements of the individual molecule and offers to slowly release the encapsulated drug at an optimal rate to mimic a slow infusion. At present this implant is being developed to deliver Interferon-α for the treatment of chronic hepatitis C. Nevertheless, this design has the capabilities to deliver a variety of drug molecules and is expected to provide significant medical advantages over the current treatment modalities.

REFERENCES

[1] S.K. Aityan and V.I. Portnov. *Gen. Phys. Biophys.*, 5:351–364, 1986.
[2] S.M. Auerbach. Theory and simulation of jump dynamics, diffusion and phase equilibrium in nanopores. *Int.Rev. Phys. Chem.*, 19:155–198, 2000.
[3] J.E. Babensee, J.M. Anderson, L.V. McIntire, and A.G. Mikos. Host response to tissue engineered devices. *Adv. Drug Del. Rev.*, 33:111–139, 1998.
[4] L.S. Baines, J.T. Joseph, and R.M. Jindal. Compliance and late acute rejection after kidney transplantation: a psychomedical perspective. *Clin. Transplant.*, 16:69–73, 2002.

[5] B. Bittner, T.H. Thelly, H. Isel, and R.J. Mountfield. The impact of co-solvents and the composition of experimental formulations on the pump rate of the ALZET□ osmotic pump. *Intl. J. Pharmaceut.*, 205:195–198, 2000.

[6] P. Boudes. Drug compliance in therapeutic trials: a review. *Control. Clin. Trials.*, 19:257–268, 1998.

[7] D.D. Breimer. Future challenges for drug delivery. *J. Control. Rel.*, 62(1–2):3–6, 1999.

[8] X. Cao, S. Lai, and L.J. Lee. Design of a self-regulated drug delivery device. *Biomed. Microdev.*, 3(2):109–118, 2001.

[9] W. Chu, T. Huen, J. Tu, and M. Ferrari. Silicon-micromachined direct pore filters for ultrafiltration. *Proc. SPIE*, 2978:111–122, 1996.

[10] W.H. Chu and M. Ferrari. Micromachined Capsules having Porous Membranes and Bulk Supports. US Patent No. 5,770,076, 1996.

[11] W.H. Chu and M. Ferrari. Microfabricated Filter with Specially Constructed ChannelWalls, and Containment Well and Capsule Constructed with Such Filters, U.S. Patent No. 6,044,981, April 4, 2000.

[12] W.H. Chu and M. Ferrari. Micromachined Porous Membranes with Bulk Support, US Patent No. 5,985,328, November 16, 1999b.

[13] W.H. Chu, R. Chin, T. Huen, and M. Ferrari. Silicon membrane nanofilters from sacrificial oxide removal. *J. Microelectromechan. Syst.*, 8:34–42, 1999.

[14] W.H. Chu and M. Ferrari. Method for Forming a Filter, U.S. Patent No. 5,985,164, November 16, 1999a.

[15] W.H. Chu and M. Ferrari. Microfabricated Filter with Specially Constructed ChannelWalls, and Containment Well and Capsule Constructed with Such Filters, US Patent No. 5,798,042, August 25, 1998.

[16] L.A. Clark, G.T. Ye, and R.Q. Snurr. Molecular traffic control in a nanoscale system. *Phys. Rev. Lett.*, 84:2893–2896, 2000.

[17] C.,Cosentino, F. Amato, A. Boiarski, R. Walczak, and M. Ferrari.Adynamic model of biomolecules diffusion through two-dimensional nanochannels. *J. Phys. Chem. B*, (submitted), 2004.

[18] D. Debray, V. Furlan, V. Baudouin, L. Houyel, F. Lacaille, and C. Chardot. Therapy for acute rejection in pediatric organ transplant recipients. *Paediatr Drugs*, 5(2):81–93, 2003.

[19] T.A. Desai, M. Ferrari, and D. Hansford. Implantable Analyte Sensor, International Patent No. PCT/EP01/03027, March 16, 2001.

[20] T.A. Desai, D.J. Hansford, and M. Ferrari. Micromachined interfaces: new approaches in cell immunoisolation and biomolecular separation. *Biomolec. Eng.*, 17:23–36, 2000b.

[21] T.A. Desai, D.J. Hansford, L.Kulinsky, A.H. Nashat, G. Rasi, J. Tu, Y. Wang, M. Zhang, and M. Ferrari. Nanopore technology for biomedical applications. *Biomed. Microdev.*, 2(1):11–40, 1999.

[22] T.A. Desai, D.J. Hansford, L. Leoni, M. Essenpreis, and M. Ferrari. Nanoporous anti-fouling silicon membranes for biosensor applications. *Biosens. Bioelectron.*,15:453–, 2000a.

[23] T.A. Desai, K.C. Popat, and S.Sharma. Engineered silicon surfaces for biomimetic interfaces. In *Business Briefing: Medical Device Manufacturing and Technology, World Markets Research Center*. London, pp. 80–82, 2002.

[24] T.A. Desai. *Materials and Design Technology*. ASME, 97–103, 1995.

[25] M.A. Dew, R.L. Kormos, L.H. Roth, S. Murali, A. DiMartini, and B.P. Griffith. Early Post-transplant medical compliance and mental health predict physical morbidity and mortality one to three years after heart transplantation. *J. Heart Lung Transpl.*, 18(6):549–562. 1999.

[26] K.A. Follet and C.P. Naumann.A prospective study of catheter-related complications of intrathecal drug delivery systems. *J. Pain Symp. Manag.*, 19(3):209–215, 2000.

[27] J.E. Fowler, M. Flanagan , D.M. Gleason, I.W. Klimberg, J.E. Gottesman, and R. Sharifi. Evaluation of an implant that delivers leuprolide for one year for the palliative treatment of prostate cancer. *Urology*, 55(5):639–642, 2000.

[28] V. Gupta, S.S. Nivarthi, D. Keffer, A.V. McCormick, D.H.T. Evidence of single-file diffusion in zeolites. *Science*, 274:164–, 1996.

[29] K. Hahn, J. Karger, and V.V. Kukla. Single file-diffusion observation. *Phys. Rev. Lett.*, 76:2762–2765, 1996.

[30] S. Hassenbusch, K. Burchiel, R.J. Coffey, M.J. Cousins, T. Deer, M.B. Hahn, S. Du Pen, K.A. Follett, E. Krames, J.N. Rogers, O. Sagher, P.S. Staats, M. Wallace, and K.D. Willis. Management of intrathecal catheter-tip inflammatory masses: a consensus statement. *Pain Medicine.*, 3(4):313–323, 2002.

[31] K.M. Holgers and A.Ljungh. Cell surface characteristics of microbiological isolates from human percutaneousimplants in the head and neck. *Biomaterials.*, 20:1319–1326, 1999.

[32] N. Jeandidier and S. Boivin. Current status and future prospects of parental insulin regimens, strategies and delivery systems for diabetes treatment. *Advan. Drug Deliv. Rev.*, 35:179–198, 1999.

[33] E.M. Johnson, D.A. Berk, R.K. Jain, and W.M. Deen. Hindered diffusion in agarose gels: test of effective medium model. *Biophy. J.*, 70:1017–23, 1996.

[34] C.G. Keller and M. Ferrari. Microfabricated particle filter. US Patent No. 5,651,900, 1997.

[35] C.G. Keller and M. Ferrari. High Vertical Aspect Ratio thin Film Structures, U.S. Patent No. 6,015,559, January18, 2000.

[36] C.G. Keller and M. Ferrari. Microfabricated Capsules for Immunological Isolation of Cell Transplants, US PatentNo. 5,893,974, April 13, 1999a.

[37] C.G. Keller and M. Ferrari. Microfabricated Particle Thin Film Filter and method of making it, U.S. Patent No.5948255, September 7, 1999b.

[38] A. Kikuchi and T. Okano. Pulsatile drug release using hydrogels. *Advan. Drug Deliv. Rev.*, 54:53–77,2002.

[39] E.A. Klausner, S. Eyal, E. Lavy, M. Friedman, and A. Hoffman. Novel levodopa gastroretentive dosage form: in-vivoevaluation in dogs. *J. Cont. Release*, 88:117–126, 2003.

[40] V.M. Knepp, A. Muchnik, S. Oldmark, and L. Kalashnikova. Stability of nonaqueous suspension formulations of plasma derived factor IX and recombinant human alpha interferon at elevated temperatures. *Pharm. Res.*, 15(7):1090–1095, 1998.

[41] P.A. Krulevitch and A.W. Wang. Microfabricated Injectable Drug Delivery System. US Patent Appl. Publ. No. 2001044620, 2001.

[42] V. Kukla, *et al.* NMR studies of single-file diffusion in unidimensional channel zeolites. *Science.*, 272:702–704, 1996.

[43] L. Leoni, A. Boiarski, and T.A. Desai. Characterization of nanoporous membranes for immunoisolation: diffusion properties and tissue effects. *Biomed. Microdevices.*, 4(2):131–, 2002.

[44] G.B. Lesinski, S. Sharma, K. Varker, P. Sinha, M. Ferrari, and W.E. Carson. Release of biologically functional interferonalpha from a nanochannel delivery system. *Biomed. Microdev.* (in press), 2004.

[45] D.G. Levitt. Dynamics of a single-file pore: non-fickian behavior. *Phys. Rev. A (General Physics)*, 8:3050–3054, 1973.

[46] J.R. Lewis and M. Ferrari. BioMEMS for drug delivery applications. In A. van den Berg (ed.) (Invited Chapter), *Lab-on-a-Chip: Chemistry in Miniaturized Synthesis and Analysis Systems.* pp. 101–115, 2003.

[47] B. Lin, J. Yu, and S. Rice. Direct measurements of constrained Brownian motion of an isolated sphere between two walls. *Phys. Rev. E.*, 62:3909–19, 2000.

[48] J.M.D. MacElroy and S.H. Suh. Self-diffusion in single-file pores of finite length. *J. Chem. Phys.*, 106:8595–97, 1997.

[49] Z. Mao and S.B. Sinnott. A Computational study of molecular diffusion and dynamic flow through carbon nanotubes. *J. Phys. Chem. B.*, 104:4618–4624, 2000.

[50] F. Martin, R. Walczak, A. Boiarski, M. Cohen, T. West, C. Cosentino, and M. Ferrari. Tailoring width of microfabricated nano-channels to solute size can be used to control diffusion kinetics. *J. Control. Rel.* (in press), 2004.

[51] T. Meersmann, *et al.* Exploring single-file diffusion in one-dimensional nanochannels by laser-polarized ^{129}Xe NMR spectroscopy. *J. Phys. Chem. A*, 104:11665–11670, 2000.

[52] Z. Megeed, J. Cappello, J. Ghandehari. Genetically engineered silk-elastin-like protein polymers for controlled drug delivery. *Adv. Drug Del. Rev.*, 54:1075–1091, 2002.

[53] B. Narasimhan and R. Langer. Zero-order release of micro- and macromolecules from polymeric devices: the role of the burst effect. *J. Control. Rel.*, 47(1):13–20, July 7, 1997.

[54] P. Nelson and S. Auerbach. Self-diffusion in single-file zeolite membranes is Fickian at long times. *J. Chem. Phys.*, 110:9235–9244, 1999.

[55] B.D. Ratner, A.S. Hoffman, F.J. Schoen, and J.E. Lemons (eds.). Biomaterials science. *A Introduction to Materials in Medicine.* Academic Press, San Diego, CA, 1996.

[56] J.H. Samet, H. Libman, K.A. Steger, et al. Compliance with zidovudine therapy in patients infected with humanimmunodeficiency virus, type 1: a cross-sectional study in a municipal hospital clinic. *Am. J. Med.*, 92:495–502, 1992.

[57] J.T. Santini, M.J. Cima, and R. Langer. A controlled-release microchip. *Nature*, 397:335–338, 1999.

[58] J.T. Santini, M.J. Cima, and S.A. Uhland. Thermally-activated microchip chemical delivery devices. *PCT Int. Appl.*, 2001.

[59] J.T. Santini, A.C. Richards, R. Scheidt, MJ. Cima, and R. Langer. Microchips as Controlled Drug-delivery Devices Angewandte Chemie, International Edition. 39:2396–2407, 2000.

[60] S. Sershen and J. West. Implantable, polymeric systems for modulated drug delivery. *Adv. Drug Del. Rev.*, 54:1225–1235, 2002.

[61] S. Sharma, K.C. Popat, and T.A. Desai. Controlling non-specific protein interactions in silicon bio-microsystems withpoly(ethylene glycol) films. *Langmuir*, 18(23):8728–8731, 2002.

[62] P. Sinha, G. Valco, S. Sharma, X. Liu, and M. Ferrari. Nanoengineered device for drug delivery application. *Nanotechnology*, 15:S585–S589, 2004.

[63] A. Spiess, V. Mikalunas, V. Carlson, M. Zimmer, and R.M. Craig. Albumin kinetics in hypoalbuminemic patients receiving total parenteral nutrition. *J. Parenter. Enter. Nut.*, 20:424–428, 1996.

[64] B. Spilker. Methods of assessing and improving patient compliance in clinical trails. In B. Spilker (ed.), *Guide toClinical Trials*. New York, Raven Press, pp. 102–114, 1991.

[65] R.K. Stoelting. Pharmacokinetics and pharmacodynamics of injected and inhaled drugs. In *Pharmacology andPhysiology in Anesthetic Practice*. Lippincott-Raven, New York, pp. 3–35, 1999.

[66] S.L. Tao, and T.A. Desai. Microfabricated drug delivery systems: from particles to pores. *Adv. Drug Del.Rev.*, 55:315–328, 2003.

[67] F. Theeuwes and S.I. Yum. Principles of the design and operation of generic osmotic pumps for the delivery of semisolid or liquid drug formulations. *Ann. Biomed. Eng.*, 4:343–353, 1976.

[68] J.K. Tu and M. Ferrari. Microfabricated Filter and Capsule Using a Substrate Sandwich. U.S. Patent No. 5,938,923, Aug. 17th, 1999.

[69] J.K. Tu, T. Huen, R. Szema, and M. Ferrari. Filtration of sub-100nm particles using a bulk-micromachined, direct-bonded silicon filter. *Biom. Microdev.*, 1(2):113–120, 1999.

[70] R. Walczak, A. Boiarski, T. West, J. Shapiro, S. Sharma, M. Ferrari. Long-term biocompatibility of NanoGATE drug delivery implant. *Nanobiotechnology* (invited paper, submitted) 2004.

[71] Q. Wei, C. Bechinger, P. Leiderer. Single-file diffusion of colloids in one-dimensional channels. *Science*, 287:625–27, 2000.

[72] N. Wisniewski, B. Klitzman, B. Miller, and W.M. Reichert. Decreased analyte transport through implanted membranes: Differentiation of biofouling from tissue effects. *J. Biomed. Mater. Res.*, 57:513–521, 2001.

[73] R.C. Wood, E.L. LeCluyse, and J.A. Fix Assessment of a model for measuring drug diffusion through implantgenerated fibrous capsule membranes. *Biomaterial.*, 16(12):957–959, 1995.

[74] J.C. Wright, S.T. Leonard, C.L. Stevenson, J.C. Beck, G. Chen, R.M. Jao, P.A. Johnson, J. Leonard, and R.J. Skowronski. Anin vivo/in vitro comparison with a leuprolide osmotic implant for the treatment of prostate cancer. *J. Control. Rel.*, 75:1–10, 2001.

[75] T. Yasukawa, H. Kimura, Y. Tabata, and Y. Ogura. Biodegradable scleral plugs for vitreoretinal drug delivery. *Adv. Drug Del. Rev.*, 52(1):25–36, Oct. 31, 2001.

III

Molecular Surface Engineering for the Biological Interface

16

Micro and Nanoscale Smart Polymer Technologies in Biomedicine

Samarth Kulkarni, Noah Malmstadt, Allan S. Hoffman, and Patrick S. Stayton

Department of Bioengineering, University of Washington, Seattle, WA 98195

Polymers that exhibit a sharp hydrophilic to hydrophobic phase transition on the application of an environmental stimulus such as pH or temperature are called smart polymers. These smart polymers, or stimuli responsive polymers, have been used to develop several drug delivery technologies. The emergence of the fields of nanotechnology and microfluidics has created new opportunities for smart polymers. We have recently developed two nano- and microscale technologies for diagnostic applications. The first is a reversible particle system using stimuli-responsive polymer-protein conjugates. We have found that conjugates of streptavidin and the temperature-responsive poly(N-isopropylacrylamide) (PNIPAAm), rapidly form stable and uniformly sized mesoscale particles above the lower critical solution temperature (LCST) of the polymer. The size of these particles is dependent on concentration, molecular weight of the polymer used and formulation parameters such as the heating rate. The second is a stimuli-responsive bioanalytical bead system. Latex beads were dual-conjugated with PNIPAAm and an affinity ligand to confer temperature-responsiveness to the beads. Above the LCST of the PNIPAAm, the bead surface becomes hydrophobic and the modified beads aggregate and adhere to the walls of microfluidic channels. They have been used to develop a reversible microfluidic affinity chromatography matrix for the upstream processing of complex fluids and for immunoassays. Both technologies can be used in a wide variety of formats, including microfluidic-based micro-total analytical systems (μTAS) devices and simple, rapid field tests.

16.1. SMART POLYMERS

The 'smartness' of smart polymers is derived from their rapid, reversible and highly cooperative response to simple physical stimuli, such as temperature, pH, or light. Much like highly cooperative biological reactions such as protein folding, smart polymers respond with a property change of large magnitude in response to very small changes in environmental conditions. Upon application of the stimulus, the stimuli-responsive polymers 'switch' from a hydrophilic, expanded state to a hydrophobic collapsed state, a change that can be manifested in one of many forms. Free polymers might switch from a solvated state in aqueous solution to a phase separated state. Cross-linked hydrogels display swelling or shrinking in response to stimuli, and surfaces with smart polymers exhibit a dramatic change in wettability. All these changes are reversible upon restoration of the original conditions.

Since the publication of the first work on the 'smart', temperature sensitive polymer PNIPAAm in 1968 [1], there have been more than 15000 publications in the area of smart polymers. The biggest impact of smart polymers has been in the biomedical field, where the versatility and usefulness of these polymers as molecular engineering tools has been demonstrated in the development of biological sensors, drug delivery vehicles and tissue engineering [2–4]. More recently, the unique properties have been employed in the nano- and microscale space, to develop useful biosensors and bioanalytical technology. This chapter will primarily focus on the use of smart polymers in micro/nanotechnology based biosensors, diagnostics and therapeutics. For a broader review of smart polymers, readers are directed to other reviews of the literature [5, 6].

16.1.1. Mechanism of Aggregation

The molecular engineering of "biohybrid" materials composed of stimuli-responsive polymers and biomolecules can take several forms. Polymer chains can be covalently end-attached or grafted to individual protein, lipid or DNA molecules, or the polymer could be used to modify the surface of a synthetic nanoparticle that is also modified with biomolecules. These materials exhibit a smart response to stimuli, similar to that the original isolated polymer, though some interesting alterations in the polymer response can also be generated and exploited. We have demonstrated the conjugation of smart polymers to proteins, antibodies, oligonucleotides, plasmid DNA, lipids, ligands such as biotin and several synthetic drug molecules for use in drug delivery, molecular separations, and diagnostics [2, 7–10].

PNIPAAm is one of the most extensively studied and utilized smart polymers, with a rich 35 year history after first being reported in 1968 [1]. PNIPAAm exhibits a lower critical solution (LCST) behavior. Below a certain temperature designated as the LCST, it is hydrophilic and highly solvated, while above the LCST, it is aggregated and phase separated. On a macroscopic level, this change is observed as the development of turbidity in a solution of PNIPAAm, which can be measured by a change in its transmittance. Of great interest is the sharp transition from individual chains to the aggregated state over a very narrow temperature range of a few degrees. The change is completely reversible and reversal of the stimulus results in the PNIPAAm going back into solution rapidly. Various techniques have been used to determine the LCST of PNIPAAm including UV turbidimetry, IR spectroscopy, light scattering, calorimetry and fluorescence techniques

[11–14]. Recently, a microfluidic system with a temperature-gradient has been developed by the Cremer and coworkers to measure the LCST of thermosensitive polymers [15].

The mechanism and energetics of the PNIPAAm phase transition have been extensively characterized [16–18]. PNIPAAm is fully solvated below the LCST, with water ordering around the isopropyl surface, along with a complex hydrogen bonding network between water molecules and the polymer hydrogen bond acceptor and donor atoms. Below the LCST, the favorable enthalpy associated with the hydrogen bonding network is sufficient to drive the highly solvated state of the polymer. Chain collapse above the LCST is entropically driven by water release to the higher entropy state of bulk solvent, where the $T\Delta S$ entropy term becomes favorable compared to the ΔH enthalpic contribution to the free energy. The energetics of the PNIPAAm phase separation have been previously reviewed [19].

On a molecular level, the mechanistic events following the elevation of the temperature above the LCST have also been characterized using light scattering techniques. Using high molecular weight chains ($> 10^6$ g/mol) at dilute concentrations, collapse of individual chains of PNIPAAm above the LCST to form thermodynamically stable globules was observed. The water released from the fully solvated polymer chain results in the rapid collapse of the polymer, determined by measuring the radius of gyration by Wu *et al.* [20]. A schematic model for the post-stimulus phase separation behavior of PNIPAAm is shown in Figure 16.1. At these dilute concentrations (picomolar), chain collapse is followed by inter-chain association and nucleation of globule clusters that subsequently grow in size through a poorly controlled aggregation process, as evidenced from an increase in the polydispersity of particle size. Kinetic studies have shown that these small globules are formed rapidly after application of the stimulus and also disaggregate within a minute. The aggregate sizes are heterogeneous and depend on formulation parameters (e.g. heating rate) and PNIPAAm molecular weight (number-averaged, M_n). Experiments with mixed polymer systems in a microfluidic device with a temperature gradient have shown that the initial formation of particles is kinetically controlled, followed thermodynamic rearrangement [15]. With dynamic light scattering experiments, we have found that for lower M_n PNIPAAm and concentrations in the nanomolar to picomolar range, kinetically-controlled formation of small hydrophobic globules is followed by further aggregation over time to produce large, non-uniform particles [21].

16.2. SMART MESO-SCALE PARTICLE SYSTEMS

16.2.1. Introduction

The unique properties of nanoscale particles have been exploited to develop new biotechnology applications by coupling the particles to biomolecules [22–25]. Nanoparticles modified with oligonucleotides have been used to detect DNA targets, and protein-modified nanoparticles have been used in drug delivery [26, 27]. Polymers have often been used as building blocks for these particles. For example, block copolymers of hydrophilic poly(ethylene oxide) and hydrophobic poly(ethylethylene) have been used to self-assemble micelles, cylindrical rods and vesicles in aqueous phases [28]. The bottom-up strategy used in making these particles exploits the hydrophobic effect as a driving force for self-assembly [29].

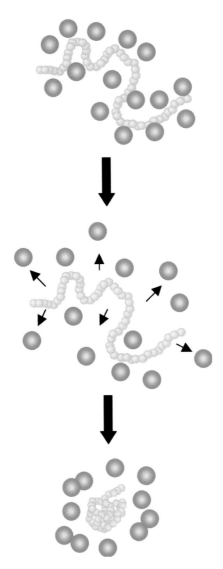

FIGURE 16.1. Temperature sensitive polymer, PNIPAAm, is fully solvated below the LCST of the polymer. When the solution is heated to temperatures above the LCST, PNIPAAm becomes hydrophobic, and the water around PNIPAAm is disordered. Finally, the polymer collapses and intrachain aggregation is seen within seconds.

We have developed a particle system in which PNIPAAm, in its hydrophobic state, directs assembly of polymer-protein conjugates into particles of controlled sizes. By using a smart polymer, the reversible hydrophilic to hydrophobic phase transition of smart polymers can be exploited to develop a reversible particle system in the nano- and mesoscale size range. When a solution of PNIPAAm is heated above its LCST, it transitions to a hydrophobic state that is driven by partial dehydration. The PNIPAAm-biomolecule conjugates in this state exhibit a nucleation and growth process into particles with narrow size polydispersity.

On reversal of the stimulus, the particles redissolve rapidly to the free conjugate species. The different properties of the free versus particulate species can then be exploited to reversibly capture and release target analytes, turn enzymes on and off, and reversibly adhere targets to device surfaces.

16.2.2. Preparation of PNIPAAm-Streptavidin Particle System

The synthesis of PNIPAAm by reversible addition fragmentation chain transfer (RAFT) polymerization provides precise control over the molecular weight of the polymer [30, 31]. The RAFT technique allows the synthesis of polymers with very narrow molecular weight distribution, which is important in determining the effect of molecular weight on particle size. Previous studies have shown that in a polydisperse mixture of PNIPAAm polymers, the higher molecular weight polymers phase separate first followed by the smaller polymers [17, 20]. Studies by Mao et al. have, however, shown that the polydispersity of the polymer solution does not affect the LCST [15]. The effect of molecular weight on particle size is not clear, although preliminary results indicate that for bioconjugate particle formation, the molecular weight has to be greater than a threshold level. Another advantage of RAFT is that it yields PNIPAAm with a reactive sulfhydryl containing end-group, which can then be modified using any of several commercially available sulfhydryl-reactive biotin molecules, such as the iodoacetyl biotin or maleimide biotin.

The initial characterization of the stimuli-responsive particles was conducted on molecular conjugates of streptavidin and PNIPAAm. The PNIPAAm molecule is first biotinylated at the end group using maleimide conjugation chemistry and then mixed with streptavidin to form a PNIPAAm-streptavidin complex. Because the four biotin binding sites of streptavidin are located in proximal pairs on opposite faces of the streptavidin tetramer, steric constraints limit the complexation of the biotinylated PNIPAAm chain to one site per face.

16.2.3. Mechanism of Aggregation

Below the LCST, the polymer and the protein are both hydrophilic and the conjugate remains in solution. Above the LCST, however, the bioconjugate undergoes an interesting transition into an amphiphile-like conjugate, with streptavidin as the hydrophilic block and the PNIPAAm as the hydrophobic block. This drives the aggregation of the conjugate into uniform sized particles as depicted in Figure 16.2. We have previously observed that PNIPAAm nucleates and grows over time into continuously larger aggregates in aqueous solution above the LCST [21]. It was also seen that in the presence of surfactant, the aggregates are more stable and uniform sized [13, 32]. Similar to the effect of surfactants, particles formed from the biohybrid conjugates are stabilized by the charged and hydrophilic streptavidin, and remain stable even at higher concentrations. The protein thus serves as a hydrophilic capping agent to limit the size of the bioconjugate particle.

16.2.4. Properties of PNIPAAm-Streptavidin Particle System

The most remarkable feature of this particle system is the rapid and reversible dissolution of the particles. A study of the kinetics of particle formation and growth revealed that stable particles formed within a minute of stimulus application. For PNIPAAm alone,

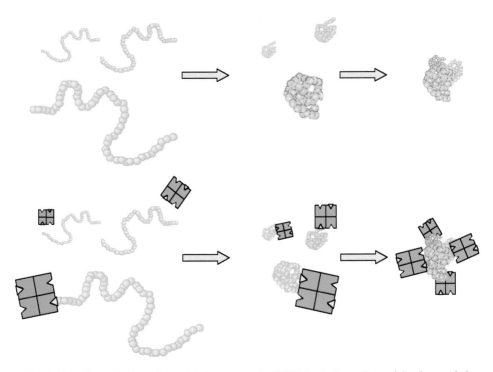

FIGURE 16.2. The mechanism of mesoglobule formation by PNIPAAm is shown. In panel A, when a solution of PNIPAAm is heated above the LCST, the polymer collapses. This is followed by interchain aggregation to form particles with wide distribution. PNIPAAm globules keep on aggregating resulting in unstable particles. In panel B, when PNIPAAm-streptavidin conjugate is heated, the polymer collapses and directs aggregation with other conjugates. However, after initial aggregate formation, the globules do not aggregate further due to the surfactant-like effect of streptavidin.

the aggregates formed in two stages: initial rapid aggregation into uniform sized particles within one minute, followed by continuous further aggregation of the particles. For the streptavidin-PNIPAAm molecules, there was no further aggregation after the initial rapid aggregation due to the presence of the hydrophilic streptavidin on the surface of the particles. Another striking feature is the control over the particle size by various 'engineering handles'. The particle size was seen to be strongly dependent on the concentration of the conjugate, molecular weight of the polymer used, the stimulus applied and the rate of application of the stimulus. By manipulating these features of the conjugate system, the particle size could be 'tuned' between 200 and 900 nm for the PNIPAAm-streptavidin system.

16.2.5. Protein Switching in Solution using Aggregation Switch

An interesting property of the protein-polymer particle system is the effect of particle formation on the activity of the protein. By engineering precisely the site of attachment of the polymer onto an enzyme surface, the activity of the enzyme could be turned on or off in the particle state. If the polymer was attached at a position away from the active site, the enzyme was active in both the free and particle state. However, if the polymer was attached

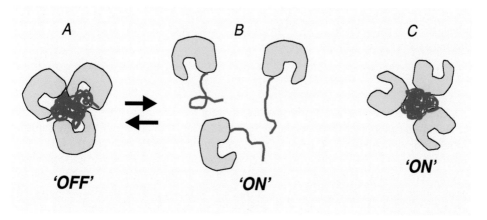

FIGURE 16.3. Enzyme endoglucanase is conjugated at various sites to temperature and light sensitive polymer, PDMAAm. When the PDMAAm is conjugated close to the active site of the enzyme, the aggregate formed on heating occludes the access of the substrate to the binding site. If the PDMAAm would be conjugated away from the active site, it would form a biofunctional particle on heating above the LCST or providing the appropriate light stimulus. Thus the 'aggregation switch' can be controlled site-specifically.

near the active site, the enzyme was active in the free state, but turned off in the particle state. This suggested that the orientation of the enzyme active site could be controlled in the particle such that substrate binding was allowed or blocked, depending on the orientation of the polymer relative to the active site (Figure 16.3). This application was demonstrated with both thermal and photo-responsive polymers that served as molecular switches to control the activity of the commercially important enzyme, endoglucanase 12A [2, 33, 34].

A unique photo- and temperature-responsive smart polymer, poly(N, N-dimethylacrylamide-co-N-4-phenylazophenylacrylamide) (PDMAAm) was synthesized using free-radical polymerization. Normally, the polymer poly(dimethylacrylamide) has an intrinsic LCST in aqueous solutions well above 100 °C. By copolymerizing the dimethylacrylamide (DMA) with the hydrophobic phenylazophenyl monomer, the LCST of the polymer was reduced to 37 °C. By illuminating the polymer with far UV light at approximately 360 nm, a cis to trans change in configuration at the azo groups is induced. This configuration change affects the LCST of the polymer, shifting it to 43 °C. The ratio of DMA to phenylazophenyl monomer can be manipulated to tune the LCST properties such that an optimal temperature for a given enzyme can be selected where photo-stimulation leads to reversible expansion and collapse of the polymer chain. Furthermore, the LCST values are also affected by the ionic strength of the solution, providing a system with several control levers.

As illustrated in Figure 16.3 (configuration A), a site on the endoglucanase enzyme surface was chosen to be near the binding pocket groove. A surface exposed asparagine residue (N55) was chosen and mutated using recombinant DNA techniques to a cysteine that contains a sulfhydryl group. Using vinyl sulfone functionalized DMAAm, the polymers were conjugated to the endoglucanase. The switching of the polymer was demonstrated using different stimuli, temperature and light, and in both, immobilized or free solution formats. When immobilized on the beads, a molecular switching phenomenon occurs, where each enzyme is modulated by the physical state of the single attached polymer. When the conjugates are free in solution, it is the formation of the particle that causes the enzyme to turn off, as previously described.

16.2.6. Potential uses of Smart Polymer Particles in Diagnostics and Therapy

The properties of this reversible particle system provide interesting opportunities for technology development in the bioanalytical field. The reversibility of the bioconjugate particle formation allows molecular recognition steps to be performed with the free conjugates, while then enabling separation and concentration steps based on particle properties, or vice versa. Many diagnostic assays are based on antibody or DNA markers that are present at very low concentration in samples such as blood and urine. This system could be used to concentrate and purify such targets to enhance analytical sensitivities in these tests. For example, smart polymer conjugates containing an affinity moiety can be used to bind the analyte of interest in the test solution. The binding is followed by the application of a temperature stimulus to trigger the formation of the mesoscale globules of fixed size. Based on the size, appropriate filters can be used to capture the analyte of interest. By using different polymers, both in composition and size, it is possible to capture two different analytes sequentially by inducing the formation of particles of different size. Additionally, filters have been developed, based on hydrophobic interactions, which bind to the smart polymers only at higher temperatures. These preconcentration techniques are particularly useful in the development of DNA or RNA based diagnostics where the concentration of the DNA and RNA in the blood is low.

Smart polymers can be used as efficient tools for affinity separation of biomolecules and has been demonstrated in several systems. For example, the streptavidin conjugates can be used to capture biotinylated antibodies and their targets in homogeneous solution, but can then be filtered as the mesoscale beads after temperature-triggered particle formation. After isolation, the bioconjugates can be released back as free molecular species with simple temperature reversal. Alternatively, the different diffusive properties of the particles might be exploited in microfluidic devices to separate particle-bound targets from freely diffusing contaminants, with the benefit again of being able to regenerate the bioconjugate species at controlled places in the channel device for capture or release.

16.3. SMART BEAD BASED MICROFLUIDIC CHROMATOGRAPHY

16.3.1. Introduction

Smart polymers have been used to engineer the properties of surfaces for various biomedical applications [35–40]. By changing the physical state of a smart polymer coated on the surface of a material, the surface can be switched from hydrophilic to hydrophobic using an environmental stimulus. In the hydrophobic state, such surfaces have been found to adhere to proteins and cells and have been used to pattern proteins and culture bacterial and human cells [41, 42]. The unique reversal property of smart polymers allows a transition back to hydrophilic state, releasing the patterned molecules or cells, a property that has been exploited to engineer human tissue [43].

Our group has recently described an approach to grafting smart polymers onto the surfaces of nanobeads to create a class of 'smart beads' with stimuli-responsive properties. These smart polymer coated beads transition from a hydrophilic surface to a hydrophobic surface in a stimuli responsive way. This transition can be observed as a change in the hydrodynamic radius of the beads and exhibit a propensity to aggregate with other hydrophobic beads. Beads coated with other modified smart polymers, such as a copolymer of PNIPAAm and poly(acrylic acid) show other responses such as change in the surface charge (zeta potential) [44]. Similarly, optical properties of particles have been changed by transitions of temperature sensitive polymers [45].

Our group has used PNIPAAm to engineer the surface of polystyrene latex nanobeads to create 'smart' beads with temperature-dependent aggregation properties [46]. At low temperatures, the PNIPAAm molecules on the surface of these beads are fully solvated and prevent sticking of the beads due to steric hindrance. However, on raising the temperature of these beads, the PNIPAAm on the surface becomes hydrophobic or 'sticky', causing aggregation into large clumps in solution. Some researchers have utilized this aggregation property to develop gold nanoparticle based biosensors, in which particle aggregation results in rearrangement of the electron clouds around the particles, causing a color shift. Kawaguchi's group has utilized the thermal flocculation property of latex beads co-functionalized with PNIPAAm and an enzyme, to reversibly control the activity of enzymes [44]. On a molecular level, the change in the physical state of the polymer has been used to control substrate access to an enzyme co-immobilized with PNIPAAm on a latex 'hairy' particle.

Besides the property of thermal aggregation in solution, the PNIPAAm-coated beads were found to adhere to the walls of a microfluidic device made from mylar and form a matrix. We have utilized this property of the beads to develop a microfluidics-based chromatography system and a diagnostic immunoassay, both described in the following sections.

16.3.2. Preparation of Smart Beads

The use of functionalized beads has gained rapid acceptance in a large number of biological assays, and beads with a wide variety of functional groups on their surface can now be purchased commercially. There are two broad strategies for modifying beads or particles with polymer. One strategy is to functionalize the bead with an initiator first, followed by in situ polymerization on the beads. This method has been used by Kawaguchi et al., to create beads with a copolymer mixture of PNIPAAm and poly(acrylic acid) (PAAc) [44]. The other strategy is to perform the polymerization separately and generate polymers with functional end groups, and then conjugating to the bead by forming covalent bonds with the surface groups on the beads. An important advantage of the latter method is that mixed modification of the beads by several polymers is easier.

Malmstadt et al., starting from amine-coated polystyrene latex beads, used PNIPAAm functionalized with an N-hydroxysuccinimide end group to modify the amine groups on the beads [46]. A similar chemistry was used to doubly modify the beads with a hydrophilic polymer, poly(ethylene glycol) (PEG), with a biotin cap, to create 'doubly smart' beads. One, the beads have a thermosensitive property of aggregating at higher temperatures, and two, they have a biological recognition moiety, biotin, on their surface. Careful choice of molecular weights of the polymers used and ratios of conjugation are required to optimize functionality of the dual beads.

16.3.3. Microfluidic Devices for Bioanalysis

The ability to manipulate nanoliter and picoliter volumes of fluid in microscale channels and chambers has enabled the development of integrated analysis systems on a microchip, devices more commonly referred to as lab-on-a-chip or microfluidic devices. These devices have great potential in biological assays, particularly due to their portability, mass producibility and small volume of reagents used. For example, a micro total analysis system developed by Burns et al. can perform the operations of polymerase chain reaction, chromatographic separation and DNA detection on a single microchip [47]. Lab-on-a-chip systems are ideal platforms for the development of integrated diagnostic devices, combining several operations of the test onto a single chip. A typical sandwich diagnostic test involves preconcentration of analyte, analyte binding to first antibody, followed by secondary antibody binding, catalysis by the attached enzyme followed by optical detection. Combining all these steps onto a single platform, where the only manual operation needed is introduction of sample into a chip, would allow for convenient and easy deployment of the test.

Microfluidic devices are also being investigated for use in drug delivery, detection of pathogens and immunoassays. One of the challenges that microfluidic systems face, however, is the issue of biological sample handling. Biological fluids typically used in bioassays such as saliva and blood are complex mixtures of various biomolecules. For lab-on-a-chip systems to be widely adopted, a robust system for preconcentration and sorting of biomolecules must be developed. For example, a microfluidic chip based multiplexed detection device for sexually transmitted (STD) diseases (replaces the currently available STD panel), would have to be able to separate the desired antigens or antibodies from the blood sample, and transfer them to different regions on the chip for analysis. If the original blood sample is used for detection, the background noise is high, leading to low sensitivity. Several strategies have been explored for preconcentrating the desired biomolecule. Most of them have been via capture of the analyte of interest with a biological recognition moiety immobilized in the microfluidic device. Microfabrication techniques usually employ acid etches and high temperatures, which are incompatible with biomolecules. Downstream conjugation of biomolecules usually has low efficiency and is irreversible, and stability problems limit the shelf-life of the device. Additionally, each application will require specific conjugation procedures, limiting applicability and increasing costs.

Using the smart beads described above, we have developed a system for the reversible and facile immobilization of biomolecules for separation and detection in microfluidic devices [46, 48]. The biomolecules immobilized in this fashion can be used as a reversible affinity chromatography matrix, for capture and release of the biomolecule of interest, or used in diagnostic immunoassays, where the analyte of interest is captured and detected. Both embodiments are described in the following sections.

16.3.4. Microfluidic Affinity Chromatography Using Smart Beads

While researching the behavior of PNIPAAm-coated beads in microfluidic devices made of mylar, it was discovered that the beads adhered to the mylar above the LCST

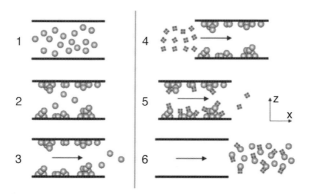

FIGURE 16.4. Schematic of the experimental protocol for streptavidin affinity chromatography (Borrowed from Ref. *46*). The channel is initially filled at room temperature with a suspension of biotinylated, PNIPAAm-coated beads (1). The temperature in the channel is then raised to 37 °C, and the beads aggregate and adhere to the channel walls (2). Buffer is then pumped through the channel (the presence of flow is indicated in this schematic by an arrow), washing out any unbound beads (3). A fluorescently labeled streptavidin sample is then introduced into the flow stream (4). Streptavidin binds the beads, and any unbound streptavidin is washed out of the channel (5). Finally, the temperature is reduced to room temperature, leading to the breakup of the bead aggregates. Beads, bound to labeled streptavidin, elute from the channel (6).

[46]. In fact, temperature could in some ways be considered a 'flow switch' of the beads in the device. Below the LCST, the beads easily flow through the narrow channels, but immediately upon entering a region of high temperature, they simultaneously aggregate with other beads and also stick to the walls of the channel in that localized region, forming a mesh. Interestingly, the beads remain enmeshed stably under flow conditions, but start flowing again within seconds upon removal of the temperature stimulus. A natural application for this mesh system is chromatography, where the mesh can be functionalized with moieties that can capture desired molecules from a flowstream. This versatile chromatography 'net' can be cast at any desired region of the microchip, and can be used for preconcentration, analysis by immunoassays or competition assays [46, 48].

Microfluidic devices are appealing for chromatographic separations, due to their high surface area to volume ratios, especially for purification of biological species available at low concentrations. A schematic of smart beads as a chromatography matrix is shown in Figure 16.4. Beads modified with both PNIPAAm and PEG biotin have been used to create the reversible chromatographic matrix. Stacked Mylar, or poly(ethylene terephthalate), sheets were used to fabricate a microfluidic device with a narrow channel, only 300 μm deep and 1000 microns wide. While the PNIPAAm-coated beads are heated and form the mesh in the channel, the PEG-biotin is soluble and projects into the flowstream. To demonstrate affinity chromatography, streptavidin is captured from the flow stream employing its high affinity binding to biotin. A primary advantage of this system is its versatility, because the chromatography matrix can be generated at any location in a complex device, and can be reversibly dissolved after capture and separation. The smart beads can be modified with several affinity capture agents, making it an attractive tool for multiplexing. Multiplexing can also be achieved by mixing different kinds of beads. Another appealing feature of the 'mobile' matrix platform is the ease of adaptability to various separation or detection

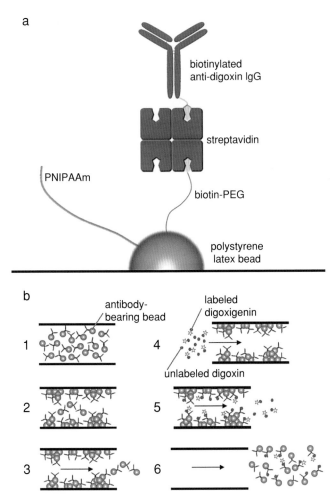

FIGURE 16.5. Smart bead immunoassay system (Borrowed from Ref. 48). Panel a: The assembled smart bead construct. A 100 nm diameter latex nanobead bead is surface-conjugated with biotin-PEG and PNIPAAm. Streptavidin is bound to the exposed biotin, providing a binding site for the biotinylated anti-digoxin IgG. Panel b: A schematic of the experimental protocol. Suspended smart beads are loaded into the PET microfluidic channel (1). The temperature in the channel is then increased from room temperature to 37 °C, resulting in aggregation and adhesion of the beads to the channel wall (2). Flow is initiated (the presence and direction of flow is indicated by an arrow in this diagram), washing unadsorbed beads out of the channel (3). A mixture of fluorescently labeled digoxigenin (at a fixed concentration) and digoxin (at varying concentrations) is flowed into the channel (4); components of this mixture that fail to bind the immobilized antibodies are washed through (5). Since digoxin and labeled digoxigenin compete for antibody binding, the higher the concentration of digoxin, the greater the amount of labeled digoxigenin that will flow through at this step. Finally, the temperature in the channel is reduced, and the aggregation/absorption process reversed as antigen-bound beads leave the channel with the flow stream (6).

needs. Whereas the same mylar device with the heating module can be used for diverse applications, the functionalized beads can be custom-built for specific end-uses.

16.3.5. Microfluidic Immunoassay Using Smart Beads

One of the 'grand' challenges in healthcare, especially in developing countries, is the development of point-of-care diagnosis of multiple pathogens and disease conditions [49]. Microfluidic devices or microchips are appealing platforms for overcoming this barrier, given the potential low cost of chip manufacturing (priority for developing countries) and easy portability. A primary obstacle in applying microchips for biological uses is the difficulty in incorporating biologicals into these devices. The smart polymer system provides new opportunities for utilizing biomolecules. For example, the smart beads can decouple the bio-immobilization step from the microfabrication steps, as the bead system can be utilized to introduce the desired biomolecules.

The use of this system in an immunoassay format has been demonstrated using a model digoxin antibody-antigen system [48]. The PNIPAAm and PEG-biotin coated beads were prepared as described above. These beads were then bound with a layer of streptavidin, and biotinylated anti-digoxin antibody was subsequently immobilized via the biotin linkage. This digoxin capture system was immobilized in a locally heated region of a microfluidic channel, and a competition immunoassay was performed with digoxin and a fluorescent analogue (Figure 16.5). The small molecules were found to bind to the immobilized antibody, and by measuring the amount of the unbound fluorescent analogue, the concentration of digoxin was determined. Optimization of the optical detection set-up can potentially increase the sensitivity to the nanomolar range. While multiplexing is an attractive opportunity, even in single assay formats, the ability to replenish the capture agent for each assay in an effortless way makes it a powerful technique.

16.3.6. Smart Polymer Based Microtechnology—Future Outlook

The approaches outlined in this section demonstrate the potential of smart polymers to serve as antennae and actuators that control biomolecule and bioanalytical bead functionality in diagnostic devices. Molecular switches that control recognition and enzyme activity can be used to turn proteins and enzymes on and off at defined zones and positions in the microfluidic streams. The reversible capture and release of targets through utilization of smart bioconjugates and beads that also can be controlled spatially and temporally provides important new capabilities for sample preconditioning and concentrating. In a general sense, this technology allows signals to be sent to specific biomolecules and beads in a multi-plexed system through the use of polymers with different stimuli-responsive properties. Approaches of this sort should bring the biomolecular componentry side of diagnostics to a new level of sophistication that matches that of the microfabricated devices.

ACKNOWLEDGEMENTS

The authors gratefully acknowledge NIH grant No. EB00252 and the NSF-funded Center for Nanotechnology, University of Washington (UIF fellowship to S.K.).

REFERENCES

[1] M. Heskins and J.E. Guillet. Solution properties of poly(N-isopropylacrylamide). *J. Macromol. Sci.-Chem.*, 1441–1455, 1968.

[2] T. Shimoboji, E. Larenas, T. Fowler, S. Kulkarni, A.S. Hoffman, and P.S. Stayton. Photoresponsive polymer-enzyme switches. *Proc. Natl. Acad. Sci. U.S.A.*, 99:16592–16596, 2002.

[3] A.S. Hoffman, P.S. Stayton, O. Press, N. Murthy, C.A. Lackey, C. Cheung, F. Black, J. Campbell, N. Fausto, T.R. Kyriakides, and P. Bornstein. Design of "smart" polymers that can direct intracellular drug delivery. *Poly. Adv. Technol.*, 13:992–999, 2002.

[4] M. Harimoto, M. Yamato, A. Kikuchi, and T. Okano. Cell sheet engineering: intelligent polymer patterned surfaces for tissue engineered liver. *Macromol. Symp.*, 195:231–235, 2003.

[5] I.Y. Galaev and B. Mattiasson. Smart polymers and what they could do in biotechnology and medicine. *Trends Biotechnol.*, 17:335–340, 1999.

[6] A.S. Hoffman, P.S. Stayton, V. Bulmus, G.H. Chen, J.P. Chen, C. Cheung, A. Chilkoti, Z.L. Ding, L.C. Dong, R. Fong, C.A. Lackey, C.J. Long, M. Miura, J.E. Morris, N. Murthy, Y. Nabeshima,T.G. Park, O.W. Press, T. Shimoboji, S. Shoemaker, H.J. Yang, N. Monji, R.C. Nowinski, C.A. Cole, J.H. Priest, J.M. Harris, K. Nakamae, T. Nishino, and T. Miyata. Really smart bioconjugates of smart polymers and receptor proteins. *J. Biomed. Mater. Res.*, 52:577–586, 2000.

[7] K. Auditorehargreaves, R.L. Houghton, N. Monji, J.H. Priest, A.S. Hoffman, and R.C. Nowinski. Phase-separation immunoassays. *Clin. Chem.*, 33:1509–1516, 1987.

[8] R.B. Fong, Z.L. Ding, C.J. Long, A.S. Hoffman, and P.S. Stayton. Thermoprecipitation of streptavidin via oligonucleotide-mediated self-assembly with poly(N-isopropylacrylamide). *Biocon. Chem.*, 10:720–725, 1999.

[9] C.Y. Cheung, N. Murthy, P.S. Stayton, and A.S. Hoffman. A pH-sensitive polymer that enhances cationic lipid-mediated gene transfer. *Biocon. Chem.*, 12:906–910, 2001.

[10] A.S. Hoffman, P.S. Stayton, Z. Ding, T. Shimoboji, V. Bulmus, and R. Fong. Site-specific conjugates of stimuli-sensitive polymers and genetically-engineered mutant proteins. *Ab. Papers Am. Chem. Soc.*, 222:U418–U418, 2001.

[11] C. Boutris, E.G. Chatzi, and C. Kiparissides. Characterization of the LCST behaviour of aqueous poly(Nisopropylacrylamide) solutions by thermal and cloud point techniques. *Polymer*, 38:2567–2570, 1997.

[12] J. Ricka, M. Meewes, R. Nyffenegger, and T. Binkert. Intermolecular and intramolecular solubilization—collapse and expansion of a polymer-chain in surfactant solutions. *Phys. Rev. Lett.*, 65:657–660, 1990.

[13] H.G. Schild and D.A. Tirrell. Microheterogeneous solutions of amphiphilic copolymers of Nisopropylacrylamide—an investigation via fluorescence methods. *Langmuir*, 7:1319–1324, 1991.

[14] H. Ringsdorf, J. Venzmer, and F.M. Winnik. Fluorescence studies of hydrophobically modified poly(Nisopropylacrylamides). *Macromolecules*, 24:1678–1686, 1991.

[15] H.B. Mao, C.M. Li, Y.J. Zhang, D.E. Bergbreiter, and P.S. Cremer. Measuring LCSTs by novel temperature gradient methods: evidence for intermolecular interactions in mixed polymer solutions. *J. Am. Chem. Soc.*, 125:2850–2851, 2003.

[16] H. Feil, Y.H. Bae, F.J. Jan, and S.W. Kim. Effect of comonomer hydrophilicity and ionization on the lower critical solution temperature of N-isopropylacrylamide copolymers. *Macromolecules*, 26:2496–2500, 1993.

[17] H.G. Schild and D.A. Tirrell. Microcalorimetric detection of lower critical solution temperatures in aqueous polymer-solutions. *J. Phys. Chem.*, 94:4352–4356, 1990.

[18] S. Fujishige, K. Kubota, and I. Ando. Phase-transition of aqueous-solutions of poly(NIsopropylacrylamide) and poly(N-Isopropylmethacrylamide). *J. Phys. Chem.*, 93:3311–3313, 1989.

[19] H.G. Schild. Poly (N-Isopropylacrylamide)—experiment, theory and application. *Prog. Poly. Sci.*, 17:163–249, 1992.

[20] C. Wu and S.Q. Zhou. Laser-light scattering study of the phase-transition of poly(NIsopropylacrylamide) in water .1. single-chain. *Macromolecules*, 28:8381–8387, 1995.

[21] S. Kulkarni, C. Schilli, A.H. Muller, A.S. Hoffman, and P.S. Stayton. Reversible meso-scale smart polymer-protein particles of controlled sizes. *Bioconj. Chem.*, 15(4):747–453, 2004.

[22] Y.W.C. Cao, R.C. Jin, and C.A. Mirkin. Nanoparticles with Raman spectroscopic fingerprints for DNA and RNA detection. *Science*, 297:1536–1540, 2002.

[23] S.R. Sershen, S.L. Westcott, N.J. Halas, and J.L. West. Temperature-sensitive polymer-nanoshellcomposites for photothermally modulated drug delivery. *J. Biomed. Mater. Res.*, 51:293–298, 2000.

[24] C.M. Niemeyer. Nanoparticles, proteins, and nucleic acids: biotechnology meets materials science. *Angewandte Chemie-International Ed.*, 40:4128–4158, 2001.

[25] S.E. Stiriba, H. Frey, and R. Haag. Dendritic polymers in biomedical applications: from potential to clinical use in diagnostics and therapy. *Angewandte Chemie-International Ed.*, 41:1329–1334, 2002.

[26] W.C.W. Chan. and S.M. Nie. Quantum dot bioconjugates for ultrasensitive nonisotopic detection. *Science*, 281:2016–2018, 1998.

[27] C.A. Mirkin, R.L. Letsinger, R.C. Mucic, and J.J. Storhoff. A DNA-based method for rationally assembling nanoparticles into macroscopic materials. *Nature*, 382:607–609, 1996.

[28] B.M. Discher, Y.Y. Won, D.S. Ege, J.C.M. Lee, F.S. Bates, D.E. Discher, and D.A. Hammer. Polymersomes: tough vesicles made from diblock copolymers. *Science*, 284:1143–1146, 1999.

[29] G.M. Whitesides, J.P. Mathias, and C.T. Seto. Molecular self-assembly and nanochemistry—a chemical strategy for the synthesis of nanostructures. *Science*, 254:1312–1319, 1991.

[30] J. Chiefari, Y.K. Chong, F. Ercole, J. Krstina, J. Jeffery, T.P.T. Le, R.T.A. Mayadunne, G.F. Meijs, C.L. Moad, G. Moad, E. Rizzardo, and S.H. Thang. Living free-radical polymerization by reversible addition-fragmentation chain transfer: the RAFT process. *Macromolecules*, 31:5559–5562, 1998.

[31] C. Schilli, M.G. Lanzendorfer, and A.H.E. Muller. Benzyl and cumyl dithiocarbamates as chain transfer agent in the RAFT polymerization of N-isopropylacrylamide. In situ FT-NIR and MALDI-TOF MS investigation. *Macromolecules*, 35:6819–6827, 2002.

[32] M. Meewes, J. Ricka, M. Desilva, R. Nyffenegger, and T. Binkert. Coil globule transition of poly(NIsopropylacrylamide) —a study of surfactant effects by light-scattering. *Macromolecules*, 24:5811–5816, 1991.

[33] T. Shimoboji, Z.L. Ding, P.S. Stayton, and A.S. Hoffman. Photoswitching of ligand association with a photoresponsive polymer-protein conjugate. *Bioconj. Chem.*, 13:915–919, 2002.

[34] T. Shimoboji, E. Larenas, T. Fowler, A.S. Hoffman, and P.S. Stayton. Temperature-induced switching of enzyme activity with smart polymer-enzyme conjugates. *Bioconj. Chem.*, 14:517–525, 2003.

[35] M. Ebara, M. Yamato, T. Aoyagi, A. Kikuchi, K. Sakai, and T. Okano. Temperature-responsive cell culture surfaces enable "on-off" affinity control between cell integrins and RGDS ligands. *Biomacromolecules*, 5:505–510, 2004.

[36] M. Yamato, C. Konno, M. Utsumi, A. Kikuchi, and T. Okano. Thermally responsive polymer-grafted surfaces facilitate patterned cell seeding and co-culture. *Biomaterials*, 23:561–567, 2002.

[37] N. Nath and A. Chilkoti. Creating Smart surfaces using stimuli responsive polymers. *Adv. Mater.*, 14:1243-+, 2002.

[38] J. Hyun and A. Chilkoti. Micropatterning biological molecules on a polymer surface using elastomeric microwells. *J. Am. Chem. Soc.*, 123:6943–6944, 2001.

[39] H. Ringsdorf, J. Venzmer, and F.M. Winnik. Interaction of hydrophobically-modified poly-N-Isopropylacrylamides with model membranes—or playing a molecular accordion. *Angewandte Chemie-International Edition in English*, 30:315–318, 1991.

[40] Y.V. Pan, R.A. Wesley, R. Luginbuhl, D.D. Denton, and B.D. Ratner. Plasma polymerized nisopropylacrylamide:synthesis and characterization of a smart thermally responsive coating. *Biomacromolecules*, 2:32–36, 2001.

[41] Nath and A. Chilkoti. Fabrication of a reversible protein array directly from cell lysate using a stimuliresponsive polypeptide. *Anal. Chem.*, 75:709–715, 2003.

[42] B.D. Ratner, X.H. Cheng, Y.B. Wang, Y. Hanein, and K. Bohringer. Temperature-responsive polymeric surface modifications by plasma polymerization: cell and protein interactions *Ab. Papers Am. Chem. Soc.*, 225:U582–U582, 2003.

[43] T. Shimizu, M. Yamato, A. Kikuchi, and T. Okano. Cell sheet engineering for myocardial tissue reconstruction.*Biomaterials*, 24:2309–2316, 2003.

[44] S. Takata, M. Shibayama, R. Sasabe, and H. Kawaguchi. Preparation and structure characterization of hairy nanoparticles consisting of hydrophobic core and thermosensitive hairs. *Polymer*, 44:495–501, 2003.

[45] N. Nath and A. Chilkoti. A colorimetric gold nanoparticle sensor to interrogate biomolecular interactions in real time on a surface. *Anal. Chem.*, 74:504–509, 2002.

[46] N. Malmstadt, P. Yager, A.S. Hoffman, and P.S. Stayton. A smart microfluidic affinity chromatography matrix composed of poly(N-isopropylacrylamide)-coated beads. *Anal. Chem.*, 75:2943–2949, 2003.

[47] M.A. Burns, B.N. Johnson, S.N. Brahmasandra, K. Handique, J.R. Webster, M. Krishnan, T.S. Sammarco, P.M. Man, D. Jones, D. Heldsinger, C.H. Mastrangelo, and D.T. Burke. An integrated nanoliter DNA analysis device. *Science*, 282:484–487, 1998.

[48] N. Malmstadt, A.S. Hoffman, and P.S. Stayton. Smart mobile affinity matrix for microfluidic immunoassays. *Lab on a chip*, (In Press), 2004.

[49] H. Varmus, R. Klausner, E. Zerhouni, T. Acharya, A.S. Daar, and P.A. Singer. Grand challenges in global health. *Science*, 302:398–399, 2003.

17

Supported Lipid Bilayers as Mimics for Cell Surfaces and as Tools in Biotechnology

Jay T. Groves

Department of Chemistry, University of California Berkeley, Berkeley CA 94720

17.1. INTRODUCTION

The phospholipid bilayer is a universal molecular architecture that is shared by all cell membranes. The canonical vision of this material is a fluid mosaic [1] of proteins, lipids, and cholesterol that exhibits both a high degree of lateral fluidity and heterogeneity [2, 3]. There is widespread evidence for miscibility phase separation in membranes [4–6]. The membrane thus consists of a nanoemulsion of two-dimensional phase domains, often referred to as rafts, which are dynamic entities populated with distinctive and changing sets of proteins. The localization and cooperative arrangement of membrane components, especially in the 1–1000 nm size range, is broadly implicated in many facets of cell membrane function. There is tremendous interest, both from academy and industry, in membrane lateral structure and the way it regulates the activity of constituent membrane proteins.

Despite their notorious complexity, membranes exhibit a fortuitous and general tendency to spontaneously assemble on surfaces such as silica and various polymers [7–9]. The resulting supported membrane is typically separated from the underlying substrate by a thin (~1 nm) film of water, which preserves a high degree of lateral fluidity within the membrane bilayer along with much of its natural physical structure. Remarkably, significant biological functionality can be preserved in supported membranes as well. This is perhaps most notably demonstrated by the formation of elaborately patterned immunological synapses between living T lymphocytes and supported membranes displaying the appropriate cognate receptors from the antigen presenting cell membrane [10, 11]. Such

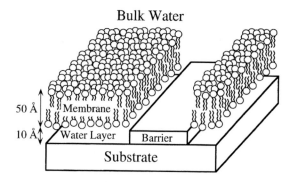

FIGURE 17.1. Schematic of a supported membrane on a patterned substrate. Continuous bilayer membrane coats the substrate while patterns of barrier materials impose boundaries on the fluid membrane. A thin layer of water between the membrane and substrate helps to preserve lateral fluidity of membrane components. Adapted from [74].

hybrid live cell—supported membrane interfaces offer great scientific and technological potential as a means of forming functional interfaces between living and nonliving systems.

In recent years, there has been rapid development of techniques and methodologies to physically pattern and manipulate supported membranes. These generally combine membrane self-assembly processes with various conventional and novel forms of hard and soft lithographic techniques (Figure 17.1) [12, 13]. The resulting barrage of patterned supported membrane structures and devices, which have recently emerged, provide a wealth of opportunities for the integration of biological functionality into micro- and nano-devices.

In the following, an overview of the current supported membrane technology is sketched. This begins with a brief examination of the salient physical characteristics of membranes, such as fluidity and phase separation properties. The degree to which these various properties can be preserved in a variety of supported membrane configurations is discussed. Next, the practical aspects of supported membrane fabrication and patterning are reviewed. Capabilities and limitations are discussed in parallel, in an effort to provide a usable reference for the design and implementation of supported membrane systems and devices. Lastly, several representative applications are mentioned. These include specific examples illustrating the biological functionality of supported membranes as well some more bioanalytical embodyments. While these examples are by no means intended to constitute a comprehensive survey of uses of supported membrane systems, they are intended to provide a sense of the current state of the field. Ultimately, the implementation of supported membrane technology into devices is in the early stages of development, and it is hoped that this review provides background information for many new applications of this emerging technology.

17.2. PHYSICAL CHARACTERISTICS

A definitive feature of supported membranes is the long-range lateral mobility of membrane lipids. The natural fluidity of free lipid bilayers is preserved in the supported membrane configuration. This distinguishes supported membranes from the vast repertoire of immobile

FIGURE 17.2. Fluorescence recovery after photobleaching (FRAP) experiment on a supported membrane. First panel: Initially uniform fluorescence from the yet unbleached membrane. Second panel: Bleach pattern shortly after exposure. Subsequent panels: Time sequence of images illustrating diffusive mixing of the bleach pattern with nearby unbleached regions of membrane. Adapted from [80].

surface coatings and thin films, such as silane monolayers, polymer layers, adsorbed proteins etc. The consequences of lateral mobility are multifold. At the most basic level, free movement of lipids enables the membrane to react to the presence of proteins, charges, and physical forces in a dynamic and responsive manner. This characteristic ability of fluid membranes to reorganize upon interaction with external perturbations is proving to be paradigmatic in the functionality of live cell membranes [14–17]. Lateral rearrangement of membrane components, in turn, enables higher levels of behavior, ranging from basic physical chemical phenomena of membranes, such as miscibility phase separation (raft formation) [18–20], to bulk transport, such as membrane microelectrophoresis [21–26]. Preservation of lateral fluidity in supported membranes tremendously enhances their biological functionality as well; this important aspect of the technology is discussed in greater detail below.

The fluidity of supported membranes can be quantitatively characterized by a number of optical techniques. Fluorescence recovery after photobleaching (FRAP) is one of the most common ways of measuring molecular diffusion coefficients in membranes [27, 28]. FRAP measurements rely on doping a small quantity of fluorescent probe molecules, usually covalently coupled to lipids, into the membrane. A brief burst of intense excitation light is projected onto the membrane, photobleaching probe molecules within the exposure zone. This bleach pulse must form a spatial pattern on the sample, which then provides the basis to monitor diffusive mixing over time. An example of a photobleached pattern in a supported membrane at various stages of diffusive recovery is illustrated in Figure 17.2. The first panel of the sequence depicts uniform fluorescence from probe molecules prior to the bleach. The second panel depicts the bleach pattern shortly after exposure. In this example, excitation light was projected through a photomask to produce the pattern. For a basic FRAP measurement, a single bleach spot will suffice and the field stop aperture on conventional fluorescence microscopes is frequently used for this purpose [29]. Subsequent panels in Figure 17.2 illustrate the transformation of the bleach pattern by diffusive mixing. Quantitative analysis of fluorescence intensity in a diffusing pattern can yield accurate measurements of the diffusion coefficient for the fluorescently labeled species. Measurements of lipid diffusion coefficients in supported membranes typically range from 1–10 $\mu m^2 s^{-1}$ with similar rates observed for membrane-linked proteins. In recent years, fluorescence correlation spectroscopy (FCS) and single particle or molecule tracking have been gaining popularity as means of assaying molecular mobility in membranes [30–39]; these techniques can be readily applied to supported membranes as well.

Looking beyond the mobility of individual lipids, phase separation and the collective mobility of phase-separated domains is also an important structural aspect of membranes. Innumerable studies of cell membrane rafts and raft-like phase separated domains in model membranes have emerged recently [3–6]. Miscibility phase transitions can also be seen in supported bilayers, however influences of the supporting substrate can be significant [19, 40, 41]. Substrate influences generally immobilize transmembrane proteins in supported membranes as well, although individual protein functionality is not necessarily destroyed [29].

In efforts to minimize the influence of the supporting substrate on membrane structure, a variety of polymeric substrate materials have been investigated [9]. These have included polyacrylamide [42], polyethylenimine (PEI) [43, 44], dextran [45], trimethylsilylcellulose (TMSC) [46–48], chitosan [49], and hyaluronic acid [50]. Additionally, several tethering strategies involving silane-polyethyleneglycol-lipid [51], lipopolymers consisting of poly(ethyloxazoline-*co*-ethyleneimine) (PEOX-PEI) with alkyl-chain side groups [52, 53], and streptavidin protein coupling between biotinylated lipid and biotin-derivatized substrates [54], have been developed to stabilize the membrane polymer interface. Tethering of membranes directly to a solid substrate, without intervening polymer, has also been employed in efforts to control the aqueous layer between the membrane and substrate [55–57]. This collection of polymer-supported membrane systems offers a range of differing physical characteristics. Notably, the silane-polyethyleneglycol-lipid tethered polymer-supported membrane can allow significant lateral diffusion of integral membrane proteins (cytochrome b_5 and t-SNARE) [51, 58]. However, no one system has yet emerged as clearly superior and new strategies continue to be developed.

One such alternative involves using a conventional supported membrane as the foundation on top of which a second membrane is deposited (Figure 17.3). The second membrane can be deposited by rupture of a giant unilamellar vesicle (GUV) onto a preformed supported membrane [59, 60] or by successive transfer of four monolayers from the air-water interface [61, 62]. These supported membrane junctions have utility in the study of membrane-membrane interactions [63], such as occur at intercellular interfaces, but for the moment we focus on the unique physical characteristics of the upper membrane in the junction for purposes of comparison with membranes supported on solid substrates.

Resolution of membrane structure on surfaces can be achieved using a combination of fluorescence techniques, which provide real-time imaging of membrane topographical patterns. Intermembrane fluorescence resonance energy transfer (FRET) occurs between membranes, which have been doped with complementary fluorescent probes, when the intermembrane spacing is comparable to the Förster distance for the probe pair ($\sim$5 nm). Quantitative analysis of FRET efficiency provides measurement of intermembrane spacing with sub-nanometer precision in closely spaced membrane junctions [59, 60]. Measurements of intermembrane spacings beyond the range of FRET can be achieved using fluorescence interference contrast microscopy (FLIC) [19, 59, 64, 65]. This technique exploits the spatial intensity variation within an optical standing wave to modulate the fluorescence intensity of probes as a function of their position along the optical axis, which is perpendicular to the interface in this configuration. FLIC can resolve topographical structures extending hundreds of nanometers from the primary plane with nanometer resolution. Reflection interference contrast microscopy (RICM) can provide similar topographical information as FLIC, and has been successfully applied to supported membrane systems [66–68]. RICM requires transparent substrates whereas FLIC requires reflective substrates, which may or

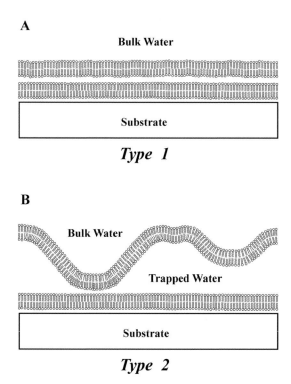

FIGURE 17.3. Schematic illustrating two types of supported membrane junctions. Adapted from [19].

may not be transparent. For a review of optical techniques for imaging membrane surface topography, see reference [69].

The most notable feature of the upper membrane in a supported membrane junction is the existence of two distinctly different states of adhesion to the lower bilayer membrane. The first, referred to as *Type 1*, is characterized by uniform intermembrane FRET, which indicates intermembrane separation distances within a few nanometers. In the second state (*Type 2*), large intermembrane spacings (~50 nm) are maintained by a balance between Helfrich (entropic) repulsion [70] and occasional adhesion sites that pin the two membranes together. No intermembrane FRET is visible in *Type 2* junctions, however FLIC reveals large-scale thermal undulations. FLIC images of the two junction types are illustrated in Figure 17.4. The existence of multiple states in membrane junctions reveals some of the range of possibilities for the association of membranes with surfaces. At present, *Type 2* junctions have only been formed by rupture of GUVs onto supported membranes. Presumably the extremely weak interaction between the two membranes in this configuration precludes monolayer transfer techniques. Further development of secondary membrane deposition methods is required before these systems can be implemented with the full range of supported membrane patterning techniques described below.

The physical properties of membranes in the different types of junction have been compared using a phase-separating mixture of phosphatidylcholine (PC), cholesterol, and sphingolipid. Below the miscibility transition temperature, the mixture separates into coexisting liquid phases, which can be observed directly by fluorescence microscopy. Miscibility

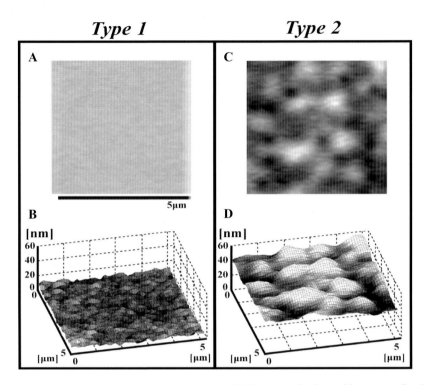

FIGURE 17.4. Fluorescence interference contrast images (FLIC), A and C, along with corresponding height profiles, B and D, for *Type 1* and *Type 2* membrane junctions systems. Adapted from [19].

phase transition temperatures are unaffected by the state of adhesion. This indicates that close apposition to another membrane surface in *Type 1* junctions does not substantially affect mixing-demixing thermodynamics at the molecular level. However, collective motion of phase separated domains in the two junction types differ substantially. Domains in the upper membrane of *Type 2* junctions exhibit rapid Brownian motion while similar domains in *Type 1* junctions and primary supported membranes remain nearly fixed in place. In contrast, lateral diffusivities of individual molecules are of similar magnitude in each of these three configurations. These results reveal the influence of the separation distance between a membrane and a supporting substrate on the size-scaling characteristics of diffusion coefficient. Correspondingly, non-equilibrium size distributions of phase-separated domains can be stabilized in supported membranes and *Type 1* junctions relative to free membranes, despite rapid diffusion of individual molecules. It is important to consider these facts in the design and study of supported membranes containing phase separating mixtures of lipids.

17.3. FABRICATION METHODOLOGIES

Supported membranes can be most easily formed by spontaneous adsorption and fusion of small unilamellar vesicles (SUVs) with an appropriate substrate [7, 11 , 12, 29]. The vesicle fusion process is quite general and accommodates a wide range of lipid and protein compositions. Clean silica surfaces have proven to be one of the best substrates although a variety of polymeric substrates can also be used [9], as discussed above. Other methods of membrane deposition include monolayer transfer from an air-water interface [71] and membrane spreading [72], both of which can also yield high quality supported membranes. Alternatively, detergent-solubalized membrane proteins, in one case G-protein coupled receptors (GPCRs), and lipid have been assembled into supported membranes from solution on carboxylated dextran surfaces with modified alkyl groups [73]. For membranes assembled on silica by the vesicle fusion process, a variety of forces including electrostatic, van der Waals, and hydration are known to tightly trap the freely supported membrane in a plane, separated from the solid surface by a nanometer layer of water [8]. This water layer prevents the substrate from interfering with the responsive membrane bilayer structure, thus preserving physical attributes of the membrane such as lateral fluidity. The membrane patterning techniques and applications described below have been implemented with silica-supported membranes, unless otherwise noted.

The fluid nature of supported membranes presents intrinsic difficulties with respect to the formation of membrane patterns on surfaces. A number of solutions have now been developed that enable membrane patterning into a wide range of configurations [12, 13]. One general theme revolves around the use of solid-state patterns on the substrate to impose structure onto the supported membrane [74]. This strategy is quite simple: materials that do not readily support continuous membranes can function as barriers to lateral fluidity when patterned onto a membrane compatible substrate. A wide range of materials including metals (Au, Al, Cr, Ti, etc.), metal oxides (Al_2O_3, TiO_2), and some polymers (including proteins patterned by microcontact printing [75]) have proven to function as effective barriers. Lipid vesicles readily adsorb onto a wide range of surfaces, including many barrier-forming materials. Formation of supported membranes is a two-step process in which vesicles first adsorb on the surface and subsequently fuse together to form a single, continuous supported membrane [76–78]. The distinctive characteristic that renders a material a good barrier to lateral diffusion in supported membranes is the tendency to inhibit the fusion step of supported membrane formation. Thus, although lipids are frequently present on the surface of barrier materials in the form of adsorbed, but unfused vesicles, long-range lateral diffusion is arrested.

Grids of barriers can partition a membrane into an array of isolated fluid corrals. In this configuration, membrane within each corral is fully connected and fluid. However, the barriers prevent mixing between separate corrals. This is the basis of the membrane microarray [74, 79]. Several strategies for filling each corral of the array with a different membrane composition have been explored. These include photolithographic modification [80], microcontact printing [81, 82], and direct micro-deposition [83, 84]. An array of membrane compositions within a grid of diffusion barriers is illustrated in Figure 17.5. In this example, a light pattern was projected onto the membrane in registry with the corrals. Each corral thus receives an individual dose, specified by the exposure area, and is photochemically altered accordingly. Diffusive mixing within each corral leads to a uniformly blended composition; thus a continuous composition scale can be achieved with a binary exposure.

Membranes can be directly deposited by microcontact printing [81, 82]. In this strategy, a polymeric stamp such as poly-(dimethylsiloxane) (PDMS) is used to deposit sections of

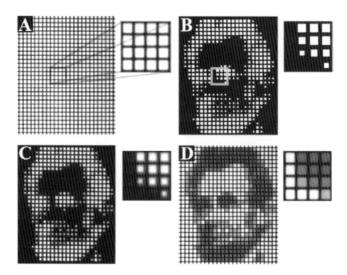

FIGURE 17.5. Array of $10 \times 10\mu m$ fluid membrane corrals of differing compositions. A. Original membrane array prior to photolithographic patterning. B. Photomask. C. Projected illumination image on membrane array. D. Resulting array of photochemically altered membrane corrals. Adapted from [80].

membrane with geometry defined by the topography of the stamp. The stamp is "inked" with membrane by vesicle fusion to the freshly oxidized PDMS. A supported membrane forms on the PDMS. This membrane can then be transferred as a unit to a silica substrate by bringing the stamp into contact with the substrate. Membranes are thus literally printed. An important point, which is well illustrated by this method, is that material barriers on the substrate are not required to maintain separation between patches of membrane. If deposited with sufficient separation, separated patches of membranes tend to remain nearl fixed in place and will not mix. In this case, the barrier consists of an uncoated region of the original substrate between patches of supported membrane. Microcontact printing is particularly convenient for applications in which a small number of distinct membrane compositions are needed; for example, to characterize the way cells discriminate between two membrane types. This technique is easily implemented by hand. However, scale-up to automated printing would be challenging due to the sensitivity of the membrane transfer process to applied force. The total force needed for good transfer is pattern dependent.

The ability to print supported membrane patterns that partially cover a substrate allows for subsequent deposition of a second membrane composition into the uncoated regions of the substrate. By combining this strategy with prepatterned grids of diffusion barriers, composition arrays consisting of numerous blending ratios of the two original membrane compositions can be created [81, 82]. The starting substrate is prepatterned with a grid of diffusion barriers. The first membrane composition is then printed onto the substrate in registry with the grid. Different amounts (area) of this first membrane are contained within the various corrals. The second membrane is then deposited by blanket vesicle fusion over the whole substrate, filling in the exposed areas. Diffusive mixing within individual corrals produces uniformly blended compositions of the two membranes in ratios defined by the fractional areas of each within the corral.

Intermembrane adhesion and molecular-exchange provides a mechanism of altering the composition of fully preformed supported membranes. The adsorption and fusion of SUVs (25–250 nm diameter) is generally self-limiting at a single bilayer. However, when there is sufficient adhesive interaction between two membranes, fusion followed by diffusive mixing of membrane components can occur. This process is widespread in biological systems where selectivity and regulation is mediated by proteins such as the SNAREs [85]. Selective intermembrane exchange with supported membranes has been achieved using electrostatic complementarity between the delivery membrane and the target membrane [86]. The method has been implemented with SUVs and membrane-coated silica beads serving as the delivery agents. An attractive aspect of using membrane-coated beads is that they can be recaptured to allow retrieval of molecular components from supported membranes. These methodologies can also be implemented within microfluidic device environments.

Membrane patterns similar to those achievable by microcontact printing, though with somewhat higher spatial resolution, have been produced using a polymer lift-off technique [87]. In this process, a thin film of parylene is first patterned on the substrate, photolithographically. A membrane is then deposited, uniformly, over the substrate. The parylene thin film can then be peeled off, leaving membrane patterns stenciled onto the bare substrate. This variation of membrane patterning is expected to leave particularly clean regions between the membranes, and its application in live cell experiments suggests this is the case [87]. A drawback of this technique in its present form is that only one membrane composition may be patterned.

A very general method, suitable for the formation of large scale patterns and arrays of membranes, is by direct deposit. Though tedious when done manually, it is possible to fill the corrals of a membrane array by directly depositing droplets of vesicle suspension [83, 84]. Commercial robotic arraying systems, originally developed for DNA array fabrication, can be adapted to print membrane arrays.

An elegant example of membrane patterning, which combines microfluidic flow channels with grids of diffusion barriers, was introduced by Kam and Boxer [88, 89]. Two or more vesicle suspensions are allowed to flow down a channel under partial mixing conditions. Vesicles fuse with the substrate and form a supported membrane, the composition of which represents the particular blend of vesicle compositions that was present above that region of the surface. Grids of diffusion barriers on the substrate effectively bin the captured membrane, resulting in a permanent composition array. The fact that this method produces membrane arrays inside of microfluidic systems is a significant advantage in light of recent advances in microfluidics [90, 91].

17.4. APPLICATIONS

17.4.1. Membrane Arrays

The ability to create arrays of discrete fluid membrane corrals on surfaces rapidly led to application of membrane array technology to drug discovery [13, 92]. In this context, the technology offers two levels of advantage. At one level there is the intrinsic scaling benefit common to all array technologies. DNA and protein arrays, for example, have found a niche based largely on scaling benefit [93–96]. New scientific capabilities, such as comprehensive gene expression or proteome profiling, are enabled by the sheer quantity of information that can be collected with the array. Supported membrane systems, in arrays or as individual membranes, additionally offer greatly enhanced levels of biological functionality. This latter point is of critical importance in drug discovery, and has fueled the development of supported membranes for a number of applications.

The G protein-coupled receptors (GPCRs) are a class of membrane proteins well known to be a rich source of drug targets [97]. Correspondingly, they have been among the most desired targets for integration in supported membrane systems and a number of strategies have been developed [57, 73, 98]. Although detailed structural data on the state of the GPCR membranes in these examples is not available, the reported ligand binding data illustrate that some of the surface adsorbed protein remains functional nonetheless. Incorporation of large transmembrane proteins into uniform and fully fluid supported membranes has been achieved previously [29]. Thus it is likely that the GPCR technology will continue to advance.

An array based immunoassay for monitoring the binding of protein to fluid membranes has been introduced in a microfluidic format [90, 99]. In one configuration, each microchannel of the array is coated with the same membrane. Different protein concentrations can then be run down each channel while binding is measured by fluorescence. Protein binding was monitored with a total internal reflection fluorescence (TIRF) system. Conventional fluorescence imaging systems could be used as well by including a rinse step. Alternatively, this configuration is also compatible with surface plasmon resonance (SPR) detection methods, which offer the added advantage of detection without the need to fluorescently label the protein of interest. By imaging an array of channels simultaneously, an entire binding curve can be read out in a single measurement.

17.4.2. Membrane-Coated Beads

Membranes supported on colloidal silica or polymer particles have proven to be an effective format for bioanalytical applications. Membranes can be assembled on the surface of silica beads by the vesicle fusion procedure and are essentially equivalent to planar supported membranes [100, 101]. The bead format is readily compatible with high-throughput screening and is a commercial product [102, 103]. The colloidal behavior of a population of membrane coated beads offers intriguing methods of detection and readout [104]. The behavior of a colloidal system is driven by the pair interaction potential between particles. In the case of membrane-derivatized silica beads, the pair potential is dominated by membrane-membrane interactions. Hence, the collective phase behavior of the system is responsive to details of the interaction between membranes. The strength of this interaction can be tuned by adjusting the membrane composition. Positioning the system near a phase transition sensitizes it to small perturbations of the membrane surface. Thus the collective phase behavior serves as a cooperative amplifier that produces a readily detectable response from a small number of molecular events on the membrane surface. For example, protein binding to membrane-associated ligand at densities as low as 10^{-4} monolayer (corresponding to $\sim$10 molecules per interface) has been observed to trigger a phase transition.

A typical colloidal phase transition triggered by protein binding to membrane surface ligand is depicted in Figure 17.6. To perform the assay, membrane-derivatized beads are dispersed, underwater, where they settle gravitationally onto the underlying substrate and

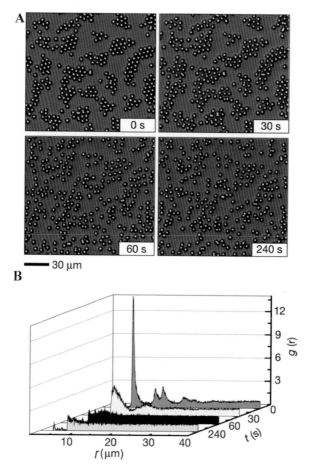

FIGURE 17.6. Protein binding-triggered colloidal phase transition of membrane-coated beads. A. Time sequence of images depicting the transition from a condensed to a dispersed colloidal phase, triggered by addition of. Transitions were triggered only when the appropriate ligand was also incorporated into the membrane. B. Corresponding plots of $g(r)$ for the time sequence in A. Adapted from [104].

form a two-dimensional colloid. The beads exhibit free lateral diffusion and the system behaves as an ergodic fluid. For highly sensitive assays, the membrane composition is tuned so that the colloid weakly condenses, as seen in the first panel of 6A. The condensed distribution is dynamic, with individual beads continuously evaporating and recondensing into clusters. However, the overall structure is invariant. Protein binding to membrane surfaces generally triggers a condensed to dispersed phase transition. Figure 17.6 depicts a time sequence of a phase transition triggered by addition of protein at $t = 0$ s. These experiments were performed with ~300 µl solution in ~5 mm round wells of a 96-well plate. Within 30 s of adding a drop of protein solution to the top of the well, uniform disruption of the condensed phase was discernable. Within 60 s, the colloid attained a dispersed distribution. Individual bead mobility is unaffected by protein binding. Exposure to a particular protein of interest triggered a phase transition only when the appropriate cognate ligand was incorporated into the colloid membrane.

Quantitative analysis of the colloidal phases is performed by extracting the pair distribution function, $g(r)$. Bead positions are measured from wide-field (~ 1 mm^2) images by an object locating algorithm to a precision of ~ 15 nm. Experiments are typically carried out at bead area fractions of $\phi \approx 0.2$, corresponding to $\sim 10^4$ individual beads per image; $g(r)$ is then generated from $> 10^8$ measured pair distances. Higher area fractions allow determination of $g(r)$ with higher precision, however the system is more likely to become trapped in non-equilibrium configurations. Plots of $g(r)$ for different time points during a phase transition triggered by protein binding are illustrated in Figure 17.6B. The condensed phase $g(r)$ is characterized by a large peak at the nearest neighbor separation distance of one bead diameter (r_0) and secondary peaks occurring at $r = \sqrt{3}r_0$ and $2r_0$, corresponding to next nearest neighbors in the hexagonal crystallites. Independent measurements of $g(r)$ were highly consistent. Standard deviations in the magnitude of the r_0 peak determined from separate colloidal preparations were generally less than 5%. Dispersed phases, consisting of random distributions and correspondingly flat $g(r)$ functions, are visibly distinguishable from condensed phases. Quantitative determination of $g(r)$ additionally distinguishes a range of intermediate distributions. These can be transient, such as the dispersing crystallites in Figure 17.6, $t = 30$ s. Intermediate degrees of order can also observed in near equilibrium distributions, corresponding with differing amounts of protein binding on the membrane surface. The colloid detection strategy can thus be used to extract binding affinity constants for membrane associated molecules, without the use of labels or sophisticated instrumentation. A low-magnification imaging system and some image analysis is all that is required.

17.4.3. Electrical Manipulation

Ion channels are a very important class of membrane proteins, and measurements of electrical conductance through these channels has been an major area of study for many years. Conventional techniques such as patch clamping [105], in which a micropipette is carefully brought into contact with a membrane, and the black lipid membrane (BLM) method [106], which consists of a reconstituted free standing bilayer membrane spanning an aperture between two aqueous chambers, have yielded a wealth of information. However, these are challenging techniques, which require a high degree of craftsmanship to perform successfully. Consequently, there has been much interest in developing supported membrane systems on conductive substrates to produce chip-based versions of the patch clamp or BLM techniques.

For electrical measurements, it is critical that the membrane be highly insulating. Indeed the success of patch clamp and BLM techniques is based largely on the extremely high electrical resistance achievable in these configurations (> 10MΩ cm^2). Whereas freestanding bilayers are self-healing and virtually defect free, the proximity of a solid substrate can stabilize defects in supported membranes. This is particularly true for highly hydrated polymeric substrates. Nonetheless, working systems have been achieved with indium-tin-oxide (ITO), doped silicon, and gold (when a tethering strategy is also employed) serving as conductive substrates for supported membranes [55, 107, 108]. Typical values of the electrical resistance in these supported membrane systems range from 10^{-3}–1 MΩ cm^2.

Electrical measurements are generally made by impedance spectroscopy in the 10^{-1}–10^5 Hz range with ~50 mV applied voltage using conventional impedance analyzers.

The fluid nature of the supported membrane permits dynamic rearrangement of membrane components over macroscopic length scales. Electrophoretic forces, resulting from application of electric fields tangent to the membrane, can induce bulk transport of charged and uncharged membrane components [21]. When this supported membrane microelectrophoresis is combined with patterns of diffusion barriers, several novel types of membrane manipulation are enabled [12, 22, 74].

The electrophoretic drift velocities of molecules in a membrane can be used as a basis for molecular separation. Subtle differences in drift velocity can be amplified by driving the membrane through a geometrical Brownian ratchet, consisting of a two-dimensional labyrinth of asymmetrical diffusion barriers [26]. Percolation of molecules through the array of barriers results in a net drift perpendicular to the field, which depends on the ratio of the parallel electrophoretic drift velocity to the diffusion coefficient. Attractive features of this type of separation are that it is continuous process and that it can be performed in a membrane environment.

Electrophoretic rearrangement membranes entirely confined within a boundary of diffusion barriers can lead to equilibrium concentration profiles [12, 22, 24, 25, 109]. Electric field strengths in the 10–100 V/cm range typically produce concentration profiles on the 100 μm size scale. These can be readily imaged by fluorescence microscopy when membranes are doped with fluorescent probes (Figure 17.7). The shape of the concentration profile results from a balance between the applied electrophoretic forces (including both electrical and electroosmotic) and entropy. Concentration profile imaging has proven to be a highly informative analytical probe for weak interactions among molecules in membranes. It also provides a rapid assay for membrane lateral fluidity. The ability to locally concentrate membrane components by electrophoretic rearrangement can be of utility as well [23].

17.4.4. Live Cell Interactions

One of the most enabling applications of supported membrane technology is its usage to create phantom cell surfaces for interactions with living cells. Some of the earliest implementations of supported membranes were for purposes of live cell studies, and this continues to be a promising growth area [7, 110]. Supported membrane systems are uniquely poised to recapture the rich biological functionality that emerges when cell surface proteins are permitted to move freely about. A hallmark example of this functionality is illustrated by the formation of a T cell immunological synapse between a living T cell and a supported membrane, which displays the relevant proteins (Figure 17.8) [10]. In addition to achieving antigen specificity, proteins in both the T cell and the supported membrane undergo large-scale spatial rearrangements in their respective membranes to form highly organized patterns. Supported membrane technology has been instrumental in the study of these immune recognition processes and continues to offer new insights [11]. The ability to fabricate patterns of diffusion barriers with sub-cellular (sub-micron) spatial dimensions provides a method of imposing constraints on protein motion and transport in the supported membrane. Through engagements with cell surface proteins on the living cells, these freedom-of-motion constraints can be transmitted to living signal transduction networks. Patterned supported membrane systems thus offer a method of imposing spatial constraints on biochemical

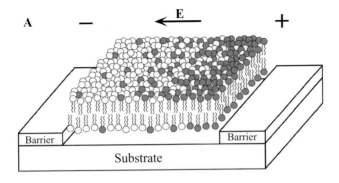

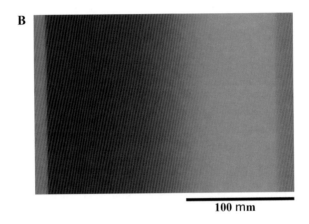

100 mm

FIGURE 17.7. A. Schematic of a electric field—induced concentration gradient in a confined patch of supported membrane. B. Image of a field-induced concentration gradient of the negatively-charged fluorescent probe (Texas Red DPPE) in an otherwise neutral supported membrane. Adapted from [12].

signaling networks in living cells. Application of these concepts can be expected to yield new information on the function of live cell signaling systems, and may further offer new ways of transmitting specific signals to cells from non-living materials.

At a more industrial level, live cell assays are used extensively within the pharmaceutical industry. Hybrid live cell-supported membrane assays, such as the T cell immune synapse formation mentioned above, offer many unique capabilities, which may be of significant utility in drug discovery applications. As one example, consider the ability to precisely control the specific protein composition of each membrane. In addition, hybrid live cell-supported membrane assays can be executed in array format.

An interesting aspect of performing live cell experiments on membrane arrays is that different results can be observed compared with cells assayed on substrates consisting of only a single composition [13, 84]. For example, phosphatidylserine (PS) is known to promote the pathological adhesion of erythrocytes (abnormally expressing PS in the outer leaflet of their membrane) to endothelial cells in diseases such as malaria and diabetes. Membrane arrays displaying different mixtures of lipids have been used as cell culture

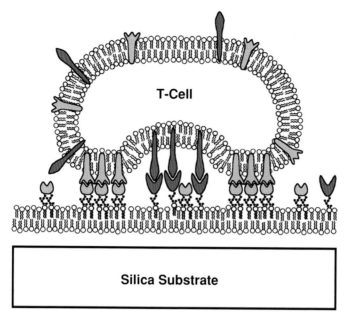

FIGURE 17.8. Schematic of a hybrid live cell—supported membrane synapse, modeled after actual synapses formed between living T cells and supported membranes displaying majorhistocompatibility protein (MHC) and intercellular adhesion molecule (ICAM). Adapted from [11].

substrates to characterize PS-mediated cell adhesion. The cells clearly distinguish between lipid composition, selectively adhering to and growing on PS containing membrane corrals. Less adhesion to PS-free membrane was observed when the cells had the option of growing on PS-containing membrane. Although the reasons for this are not clear, it suggests that side-by-side cell discrimination experiments are likely to provide more insightful results than can be obtained form the same experiments performed in isolated wells of a cell culture plate. A key feature of the assay is that the spacing of the different membrane corrals is sufficiently small that cells can sample multiple membranes before settling down; 200-micron square corrals were used in the experiments mentioned above.

17.5. CONCLUSION

Supported membrane technology has received increasing interest in recent years. This interest is fueled by prospects of creating cell mimetic surfaces for the study of biological membranes and membrane processes as well as the construction of biologically inspired devices. Substantial progress has been made in areas of membrane patterning, electrical integration, support modification, and biological functionalization. Further advancement can certainly be expected. Supported membrane systems combine elements of surface chemistry, soft condensed matter materials science, and biology. This is a complex intersection of scientific disciplines, which at the same time presents promising and intriguing prospects for creating new interfaces between living and non-living systems.

REFERENCES

[1] S.J. Singer and G.L. Nicolson. The fluid mosaic model of cell membranes. *Science*, 175:720–731, 1972.

[2] K. Jacobson, E.D. Sheets, and R. Simson. Revisiting the fluid mosaic model of membranes. *Science*, 268:1441–1442, 1995.

[3] R.G.W. Anderson and K. Jacobson. A role for lipid shells in targeting proteins to caveolae, rafts, and other lipid domains. *Science*, 296:1821–1825, 2002.

[4] K. Simmons and E. Ikonen. Functional rafts in cell membranes. *Nature*, 387:569–572, 1997.

[5] H.M. McConnell and M. Vrljic. Liquid-liquid immiscibility in membranes. *Annu. Rev. Biophys. Biomol. Struct.*, 32:469–492, 2003.

[6] S.L. Veatch and S.L. Keller. Organization in lipid membranes containing cholesterol. *Phys. Rev. Lett.*, 89(26):268101, 2002.

[7] A.A. Brian and H.M. McConnell. Allogenic stimulation of cytoxic T cells by supported planar membranes. *Proc. Natl. Acad. Sci. U.S.A.*, 81:6159–6163, 1984.

[8] E. Sackmann Supported membranes: Scientific and practical applications. *Science*, 271:43–48, 1996.

[9] E. Sackmann and M. Tanaka. Supported membranes on soft polymer cusions: fabrication, characterization, and applications. *TIBTECH*, 18:58–64, 2000.

[10] A. Grakoui et al. The immunological synapse: A molecular machine controlling T cell activation. *Science*, 285:221–227, 1999.

[11] J.T. Groves and M.L. Dustin. Supported planar bilayers in studies of immune cell adhesion and communication. *J. Immunol. Methods*, 278:19–32, 2003.

[12] J.T. Groves and S.G. Boxer. Micropattern formation in supported lipid membranes. *Acc. Chem. Res.*, 35:149–157, 2002.

[13] J.T. Groves. Membrane array technology for drug discovery. *Curr. Op. Drug Disc. Dev.*, 5(4):606–612, 2002.

[14] Bray, M.D. Levin, and C.J. Morton-Firth. Receptor clustering as a cellular mechanism to control sensitivity. *Nature*, 393(7):85–88, 1998.

[15] G. Maheshwari et al. Cell adhesion and motility depend on nanoscale RGD clustering. *J. Cell Sci.*, 113:1677–1686, 2000.

[16] J.E. Gestwicki and L.L. Kiessling. Inter-receptor communication through arrays of bacterial chemoreceptors. *Nature*, 415(3):81–84, 2002.

[17] P.A. vanderMerwe and S.J. Davis. The immunological synapse—a multitasking system. *Science*, 295:1479–1480, 2002.

[18] C. Dietrich et al. Lipid rafts reconstituted into model membranes. *Biophys. J.*, 80:1417–1428, 2001.

[19] Y. Kaizuka and J.T. Groves. Structure and dynamics of supported intermembrane junctions. *Biophys. J.*, 86:905–912, 2004.

[20] C. Dietrich et al. Partitioning of Thy-1, GM1, and cross-linked phsopholipid analogs into lipid rafts reconstituted in supported model membrane monolayers. *Proc. Natl. Acad. Sci. U.S.A.*, 98(19):10642–10647, 2001.

[21] M. Stelzle, R. Miehlich, and E. Sackmann. Two-dimensional microelectrophoresis in supported lipid bilayers. *Biophys. J.*, 63:1346–1354, 1992.

[22] J.T. Groves and S.G. Boxer. Electric field-induced concentration gradients in planar supported bilayers. *Biophys. J.*, 69:1972–1975, 1995.

[23] J.T. Groves, C. Wülfing, and S.G. Boxer. Electrical manipulation of glycan-phosphatidyl inositol-tethered proteins in planar supported bilayers. *Biophys. J.*, 71:2716–2723, 1996.

[24] J.T. Groves, S.G. Boxer, and H.M. McConnell. Electric field-induced reorganization of two-component supported bilayer membranes. *Proc. Natl. Acad. Sci. U.S.A.*, 94:13390–13395, 1997.

[25] J.T. Groves, S.G. Boxer, and H.M. McConnell. Electric field-induced critical demixing in lipid bilayer membranes. *Proc. Natl. Acad. Sci. U.S.A.*, 95:935–938, 1998.

[26] A. van Oudenaarden and S.G. Boxer. Brownian ratchets: molecular separations in lipid bilayers supported on patterned arrays. *Science*, 285:1046–1048, 1999.

[27] D. Axelrod et al. Mobility measurement by analysis of fluorescence photobleaching recovery kinetics. *Biophys. J.*, 16:1055–1069, 1976.

[28] L.K. Tamm and E. Kalb. Microspectrofluoremetry of supported planar membranes. In S.G. Schulman (ed.), *Molecular Luminescence Spectroscopy*. John Wiley & Sons Inc., pp. 253–305, 1993.

[29] J. Salafsky, J.T. Groves, and S.G. Boxer. Architecture and function of membrane proteins in planar supported bilayers: a study with photosynthetic reaction centers. *Biochemistry*, 35:14773–14781, 1996.

[30] D. Magde, E. Elson, and W.W. Webb. Thermodynamic fluctuations in a reacting system-measurement by fluorescence correlation spectroscopy. *Phys. Rev. Lett.*, 29(11):705–708, 1972.

[31] K. Bacia and P. Schwille. A dynamic view of cellular processes by in vivo fluorescence auto and crosscorrelation spectroscopy. *Methods*, 29:74–85, 2003.

[32] E. Haustein and P. Schwille. Ultrasensitive investigations of biological systems by fluorescence correlation spectroscopy. *Methods*, 29:153–166, 2003.

[33] N. Kahya et al. Probing lipid mobility of raft-exhibiting model membranes by fluorescence correlation spectroscopy. *J. Biol. Chem.*, 278:28109–28115, 2003.

[34] J. Korlach et al. Characterization of lipid bilayer phases by confocal microscopy and fluorescence correlation spectroscopy. *Proc. Natl. Acad. Sci. U.S.A.*, 96:8461–8466, 1999.

[35] P. Schwille, J. Korlach, and W.W. Webb. Fluorescence correlation spectroscopy with single-molecule sensitivity on cell and model membranes. *Cytometry*, 36:176–182, 1999.

[36] E.D. Sheets, R. Simson, and K. Jacobson. New insights into membrane dynamics from analysis of cell surface interactions by physical methods. *Curr. Op. Cell Biol.*, 7:707–714, 1995.

[37] M. Vrljic et al. Translational diffusion of individual class II MHC membrane proteins in cells. *Biophys. J.*, 83:2681–2692, 2002.

[38] C. Dietrich et al. Relationship of lipid rants to transient confinement zones detected by single particle tracking. *Biophys. J.*, 82:274–284, 2002.

[39] M.J. Saxton and K. Jacobson. Single particle tracking: applications to membrane dynamics. *Annu. Rev. Biophys. Biomol. Struct.*, 26:373–399, 1997.

[40] B.L. Stottrup, S.L. Veatch, and S.L. Keller. Nonequilibrium behavior in supported lipid membranes containing cholesterol. *Biophys. J.*, 86:2942–2950, 2004.

[41] J.M. Crane and L.K. Tamm. Role of cholesterol in the formation and nature of lipid rafts in planar and spherical model membranes. *Biophys. J.*, 86:2965–2979, 2004.

[42] M. Kühner, R. Tampé, and E. Sackmann. Lipid mono and bilayer supported on polymer films: composite polymer films on solid substrates. *Biophys. J.*, 67:217–226, 1994.

[43] J.Y. Wong et al. Polymer-cushioned bilayers. I. A structural study of various preparation methods using neutron reflectometry. *Biophys. J.*, 77:1445–1457, 1999.

[44] J.Y. Wong et al. Polymer-cusioned bilayers. II. An investigation of interaction forces and fusion using the surface forces apparatus. *Biophys. J.*, 77:1458–1468, 1999.

[45] G. Elander, M. Kühner, and E. sackmann. Functionalization of Si/SiO2 and glass surfaces with ultrathin dextran films and deposition of lipid bilayers. *Biosens. Bioelectro.*, 11(6/7):565–577, 1996.

[46] H. Sigl et al. Assembly of polymer/lipid composite films on solids based on hairy rod LB-films. *Eur. Biophys. J.*, 25:249–259, 1997.

[47] H. Hillebrandt et al. High electric resistance polymer/lipid composite films on indium-tin-oxide electrodes. *Langmuir*, 15(24):8451–8459, 1999.

[48] F. Rehfeldt and M. Tanaka. Hydration forces in ultrthin films of cellulose. *Langmuir*, 19:1467–1473, 2003.

[49] T. Baumgart and A. Offenhäusser. Polysaccharide-supported planar bilayer lipid model membranes. *Langmuir*, 19(5):1730–1737, 2003.

[50] K. Sengupta et al. Mimicking tissue surfaces by supported membrane coupled ultrathin layer of hyaluronic acid. *Langmuir*, 19(5):1775–1781, 2003.

[51] M.L. Wagner and L.K. Tamm. Tethered polymer-supported planar lipid bilayers for reconstitution of integral membrane proteins: silane-polyethyleneglycol-lipid as a cushion and covalent linker. *Biophys. J.*, 79:1400–1414, 2000.

[52] W.W. Shen et al. Polymer-supported lipid bilayers on benzophenone-modified substrates. *Biomacromolecules*, 2:70–79, 2001.

[53] C. Naumann et al. The polymer-supported phospholipid bilayer: tethering as a new approach to substrate-membrane stabilization. *Biomacromolecules*, 3:27–35, 2002.

[54] A. Berquand et al. Two-step formation of streptavidin-supported lipid bilayers by PEG-triggered vesicle fusion. Fluorescence and atomic force microscopy characterization. *Langmuir*, 19(5):1700–1707, 2003.

[55] B.A. Cornell et al. A biosensor that uses ion-channel switches. *Nature*, 387:580–583, 1997.

[56] B. Raguse et al. Tethered lipid bilayer membranes: formation and ionic reservoir characterization. *Langmuir*, 14:648–659, 1998.

[57] C. Bieri et al. Micropatterned immobilization of a G protein-coupled receptor and direct detection of G protein activation. *Nature Biotech.*, 17:1105–1108, 1999.

[58] V. Kiessling and L.K. Tamm. Measuring distances in supported bilayers by fluorescence interferencecontrast miscroscopy: polymer supports and SNARE proteins. *Biophys. J.*, 84:408–418, 2003.

[59] A.P. Wong and J.T. Groves. Topographical imaging of an inter-membrane junction by combined fluorescene interference and energy transfer microscopies. *J. Am. Chem. Soc.*, 123:12414–12415, 2001.

[60] A.P. Wong and J.T. Groves. Molecular topography imaging by intermembrane fluorescence resonance energy transfer. *Proc. Natl. Acad. Sci. U.S.A.*, 99(22):14147.

[61] T. Charitat et al. Adsorbed and free lipid bilayers at the solid-liquid interface. *Eur. Phys. J. B*, 8:583–593, 1999.

[62] G. Fragneto et al. A fluid floating bilayer. *Europhys. Lett.*, 53:100–106, 2001.

[63] R. Parthasarathy et al. Nonequilibrium adhesion patterns at lipid bilayer junctions. *J. Phys. Chem. B*, 108:649–657, 2004.

[64] A. Lambacher and P. Fromherz. Fluorescence interference-contrast microscopy on oxidized silicon using a monomolecular dye layer. *Appl. Phys. A*, 63:207–216, 1996.

[65] A. Lambacher and P. Fromherz. Luminescence of dye molecules on oxidized silicon and fluorescence ineterference contrast microscopy of biomembranes. *J. Opt. Soc. Am. B*, 19(6):1435–1453, 2002.

[66] J. Rädler and E. Sackmann. Imaging optical thicknesses and separation distances of phospholipid vesicles at solid surfaces. *J. Phys. II France*, 3:727–748, 1993.

[67] J.O. Rädler et al. Fluctuation analysis of tension-controlled undulation forces between giant vesicles and solid substrates. *Phys. Rev. E*, 51(5):4526–4536, 1995.

[68] R. Hirn et al. Collective membrane motions of high and low amplitude, studied by dynamic light scattering and micro-interferomtery. *Faraday Discuss.*, 111:17–30, 1998.

[69] R. Parthasarathy and J.T. Groves. Optical techniques for imaging membrane topography. *Cell Biochem. Biophys.*, 2004. (**in press.**)

[70] R. Lipowsky and E. Sackmann (eds.). Structure and dynamics of membranes. In *Handbook of Biological Physics*. Vol. 1, Elsevier Science Ltd., New York, 1995.

[71] L.K. Tamm and H.M. McConnell. Supported phospholipid bilayers. *Biophys. J.*, 47:105–113, 1985.

[72] J. Rädler, H. Strey, and E. Sackmann. Phenomenology and kinetics of lipid bilayer spreading on hydrophilic surfaces. *Langmuir*, 11:4539–4548, 1995.

[73] O.P. Karlsson and S. Löfås. Flow-mediated on-surface reconstitution of G-protein coupled receptors for applications in surface plasmon resonance biosensors. *Anal. Biochem.*, 300:132–138, 2002.

[74] J.T. Groves, N. Ulman, and S.G. Boxer. Micropatterning fluid lipid bilayers on solid supports. *Science*, 275:651–653, 1997.

[75] L.A. Kung et al. Patterning hybrid surfaces of proteins and supported lipid bilayers. *Langmuir*, 16:6773–6776, 2000.

[76] C.A. Keller et al. Formation of supported membranes from vesicles. *Phys. Rev. Lett.*, 84(23):5443–5446, 2000.

[77] C.A. Keller and B. Kasemo. Surface specific kinetics of lipid vesicle adsorprion measured with a quartz crystal microbalance. *Biophys. J.*, 75:1397–1402, 1998.

[78] E. Reimhult, F. Hääk, and B. Kasemo. Intact vesicle adsorption and supported biomembrane formation from vesicles in slution: Influence of surface chemistry, vesicle size, temperature, and osmotic pressure. *Langmuir*, 19(5):1681–1691, 2003.

[79] S.G. Boxer, N. Ulman, and J.T. Groves. *Arrays of independently-addressable supported fluid bilayer membranes.* The Board of Trustees of the Leland Stanford Junior University, United States, 2001.

[80] L. Kung et al. Printing via photolithography on micropartitioned fluid lipid membranes. *Adv. Mater.*, 12(10):731–734, 2000.

[81] J.S. Hovis and S.G. Boxer. Patterned barriers to lateral diffusion in supported lipid bilayer membranes by blotting and stamping. *Langmuir*, 16:894–897, 2000.

[82] J.S. Hovis and S.G. Boxer. Patterning and composition arrays of supported lipid bilayers by microcontact printing. *Langmuir*, 17:3400–3405, 2001.

[83] P.S. Cremer and T. Yang. Creating spatially addressed arrays of planar supported fluid phospholipid membranes. *J. Am. Chem. Soc.*, 121:8130–8131, 1999.

[84] J.T. Groves, L.K. Mahal, and C.R. Bertozzi. Control of cell adhesion and growth with micropatterned supported lipid membranes. *Langmuir*, 17(17):5129–5133, 2001.

[85] J.S. Bonifacino and B.S. Glick. The mechanisms of vesicle budding and fusion. *Cell*, 116:153–166, 2004.

[86] A.R. Sapuri, M.M. Baksh, and J.T. Groves. Electrostatically-targeted intermembrane lipid exchange with micropatterned supported membranes. *Langmuir*, 19(5):1606–1610, 2003.

[87] R.N. Orth et al. Mast cell activation on patterned lipid bilayers of subcellular dimensions. *Langmuir*, 19:1599–1605, 2003.

[88] L. Kam and S.G. Boxer. Formation of supported lipid bilayer composition arrays by controlled mixing and surface capture. *J. Am. Chem. Soc.*, 122:12901–12902, 2000.

[89] L. Kam and S.G. Boxer. Spatially selective manipulation of supported lipid bilayers by laminar flow: steps toward biomembrane microfluidics. *Langmuir*, 19:1624–1631, 2003.

[90] T. Yang et al. Fabrication of phospholipid bilayer-coated microchannels for on-chip immunoassays. *Anal. Chem.*, 73:165–169, 2001.

[91] T. Thorsen, S.J. Maerkl, and S.R. Quake. Microfluidic large-scale integration. *Science*, 298:580–584, 2002.

[92] Y. Fang, J. Lahiri, and L. Picard. G protein-coupled receptor microarrays for drug discovery. *DDT*, 8(16):755–761, 2003.

[93] S.P.A. Fodor et al. Light-directed, spatially addressable parallel chemical synthesis. *Science*, 251:767–773, 1991.

[94] G. MacBeath and S.L. Schreiber. Printing proteins as microarrays for high-throughput function determination. *Science*, 289:1760–1763, 2000.

[95] P. Mitchell. A perspective on protein arrays. *Nature Biotech.*, 20:225–229, 2002.

[96] P. Wagner and R. Kim. Protein biochips: an emerging tool for proteomics research. *Curr. Drug Disc.*, May 23–28, 2002.

[97] T. Haga and G. Berstein (eds.). *G-protein Coupled Receptors*. CRC Press, Boca Raton, 1999.

[98] Y. Fang, A.G. Frutos, and J. Lahiri. Membrane Protein Microarrays. *J. Am. Chem. Soc.*, 124(11):2394–2395, 2002.

[99] T. Yang et al. Investigations of bivalent antibody binding on fluid-supported phospholipid membranes: the effect of hapten density. *J. Am. Chem. Soc.*, 125:4779–4784, 2003.

[100] T.M. Bayerl and M. Bloom. Physical properties of single phospholipid bilayers adsorbed to micro glass beads. *Biophys. J.*, 58:357–362, 1990.

[101] T. Buranda et al. Biomimetic molecular assemblies on glass and mesoporous silica microbeads for biotechnology. *Langmuir*, 19:1654–1663, 2003.

[102] A. Loidl-Stahlhofen et al. Solid-supported biomolecules on modified silica surfaces—A tool for fast physicochemical characterization and high-throughput screening. *Adv. Mater.*, 13(23):1829–1834, 2001.

[103] T.M. Bayerl. A glass bead game. *Nature*, 427:105, 2004.

[104] M.M. Baksh, M. Jaros, and J.T. Groves. Detection of molecular interactions at membrane surfaces through colloid phase transitions. *Nature*, 427:139–141, 2004.

[105] B. Sackmann and E. Neher. Plenum Press, New York, 1985.

[106] P. Mustonen et al. Binding of cytochrome c to liposomes as revealed by the quenching of fluorescence from pyrene-labeled phospholipids. *Biochemistry*, 26:2991–2997, 1987.

[107] Gritsch et al. Impedence spectroscopy of porin and gramicidin pores reconstituted into supported lipid bilayers on indium-tim-oxide electrodes. *Langmuir*, 14:3118–3125, 1998.

[108] G. Wiegand et al. Electrical properties of supported lipid bilayer membranes. *J. Phys. Chem. B*, 106:4245–4254, 2002.

[109] J.T. Groves, S.G. Boxer, and H.M. McConnell. Electric field effects in multicomponent fluid lipid membranes. *J. Phys. Chem. B*, 104(1):119–124, 2000.

[110] T.H. Watts and H.M. McConnell. Antigen presentation by supported planar membranes conntaining purified major histocompatibility complex proteins. In B. Pernis and H. Vogel (eds.), *Processing and Presentation of Antigens*. Academic Press. pp. 143–155, 1988.

18

Engineering Cell Adhesion

Kiran Bhadriraju, Wendy F. Liu, Darren S. Gray, and
Christopher S. Chen

*Johns Hopkins University School of Medicine, Department of Biomedical Engineering,
Traylor 718, Baltimore, MD-21205*
*University of Pennsylvania, 125 S. 31st St. Translational Research Labs, Suite 1400,
Philadelphia, PA-19104*

18.1. INTRODUCTION

Cells exist within a complex and ever-changing environment, which includes soluble molecules such as growth factors, an extracellular matrix that includes adhesive proteins and carbohydrates, and other neighboring cells. They actively sense and respond to changes in this environment, existing in a state of physiological equilibrium with it. Thus, it has been said, "...the unit of function in higher organisms is larger than the cell itself" [10]. The information content in the adhesive environment is encoded both in its composition and its organization on the nanometer to micrometer length scales. When taken out of this physiological context and cultured in plastic tissue culture dishes, cells lose the cues that maintain their in vivo identity or phenotype, and dedifferentiate. For example, hepatocytes—the principal cell type in the liver—perform several critical liver-specific functions such as production of bile, metabolism of urea, and the synthesis of important serum proteins such as albumin, fibrinogen, and transferrin [27]. When cultured in vitro and isolated from the liver microenvironment, they rapidly downregulate liver-specific phenotype [54]. Similarly, chondrocytes, which are required for the secretion and maintenance of cartilage, lose their differentiated function when cultured in vitro downregulating the synthesis and secretion of cartilage-specific collagens and proteoglycans [88]. Thus, tissue-specific cell function appears to be closely related to the microstructural organization of the tissue itself [11].

The realization that tissue structure and organization is critical to its function has led to the development of new tools to control, or engineer, such structures into cell culture systems [9]. Microfabricated tools adapted from the semiconductor industry appear to have the appropriate degree of spatial and chemical control in the fabrication process to recreate many of the key features cells use to determine their functional state.

The ability to create bioartificial tissues that recapitulate in vivo function using micro-engineering approaches would have many applications ranging from those with immediate clinical relevance to those for fundamental research [41]. For example, in drug develop-ment many compounds appear to have activity in cell culture experiments that later fail in animal or human studies. Drug candidates that fail in these later stages are largely re-sponsible for the ballooning costs of drug development. A recent estimate suggests that 1 in 13 compounds entering animal testing made it to the market in late 2000–2002, and the cost of successfully bringing such a drug to market is estimated at $1.7 billion [39]. The use of artificial tissues is likely to provide more faithful test models at the initial stages of testing that would eliminate false-positives candidates earlier rather than in the later, more expensive stages of drug testing [6]. Another application of microengineered tissues aims to address the problem of organ shortage through tissue engineering. According to www.optn.org (The Organ Procurement and Transplantation Network) and www.unos.org (United Network for Organ Sharing), in the year 2003 in USA, 23,357 organs were trans-planted, 83,195 candidates were on the waiting list, and 5559 patients died while waiting. As an alternative to transplantation, cells of specific tissues are expanded in culture and then transplanted into the body, incorporated into an extra corporeal support devices [1], or even used to recreate entire organs in the laboratory [33]. While some tissues such as skin or cartilage have relatively simple architectures, others contain many cells types and complex tissue architectures that play a critical role in tissue function. Recreation of this complexity will require micro- and nanofabrication technologies [9]. In addition to the primary market drivers of tissue engineering and drug discovery, many additional specialized applications are likely to emerge, such as the development of cell-based biosensors for bio-defense and environmental analysis where arrays of living cells act as detectors for toxins in the environ-ment [64]. Lastly, the process of learning how to construct such microengineered devices inevitably results in the development of new experimental systems that give insights into basic physiologic and pathologic processes including development, wound healing, and inflammation.

While there is a great need to understand how to reconstruct environments that can coax cells to function in a specified manner, such tools have only just begun to emerge, primarily from the transfer of semiconductor manufacturing technologies into the biomedical arena. As an illustration of how microfabrication tools can help engineer microenvironments, Fig. 18.1 shows two sets of cultured endothelial cells: one in a conventional culture environment where cell-cell adhesion or spreading are unconstrained, and the other in a microengineered system where the extent of cell-cell adhesion and spreading is controlled using microfabrication tools. There are clear advantages to biological experimentation using such well-defined systems in order to study cellular responses to shape and adhesion [56]. In addition to adhesion studies, many other problems in biomedical science and technology may prove to benefit by such control. In this chapter, we will describe studies that have utilized microfabrication techniques to engineer the cell culture environment. First we will discuss from a biological perspective, the key aspects of the microenvironment that are thought to regulate cell behavior. Then, the technological needs and demands for engineering cellular microenvironments will be described, and the tools that have been

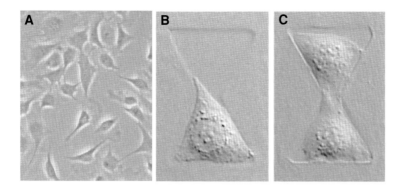

FIGURE 18.1. The figure illustrates three images of cells. A. a conventional cell culture where cell spreading and cell-cell contact are not regulated. B. a microfabricated cells culture system with constrained spreading and no cell-cell contacts and C. the same microfabricated cell culture system with constrained cell spreading and a single cell-cell contact. B and C are shown at a higher magnification than A (from [56]).

developed to meet those needs. Finally, we will describe areas of current and future work that are now emerging in the field of controlling cell adhesion, organization, and position.

18.2. REGULATING CELL FUNCTION VIA THE ADHESIVE MICROENVIRONMENT

Early in cell culture studies, it was found that cells in general require two classes of molecules—one is the class of basic metabolites such as ions and sugars, required for the building bocks of cells, and the other is that of signaling molecules, which can be both soluble and insoluble. Cells use signaling molecules, not directly as building blocks, but as elements of control systems that elicit appropriate changes in cell behavior in response to changes in the environment. For example, insulin is a soluble signaling protein secreted by the pancreas and forms a key element of the blood glucose control system in the body. Increases in blood glucose signal the release of insulin from the pancreas into the blood stream, and the subsequent regulation of blood glucose levels through a series of molecular events [23]. Cells sense their environment using different surface receptors, which bind specifically with different ligands.

In practice, biologists have experimentally manipulated soluble signals such as insulin simply by changing its concentration in the culture dish. However, cells also receive signals through adhesion receptors. Standard cell culture methods involve culturing cells, for example, in petri dishes. Changes in culture conditions, such as the ECM coating on the dish, are done at the level of the entire dish and not at the level of single cells. That is, all the cells are exposed to the same environment on the average. Experimentally manipulating cell-ECM adhesion, especially at single cell level, has been difficult with standard methods; methods were not available to localize the ECM at cell length scales. Because of the micrometer scale spatial control required, much of the progress made in the material science and semiconductor fabrication world has contributed to developing cell-patterning technologies. Before detailing these advances, we provide a brief background on current understanding of biological adhesion.

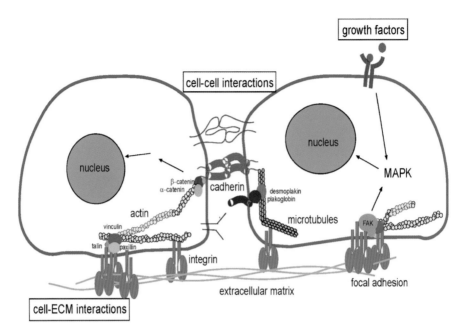

FIGURE 18.2. The figure illustrates some of the molecules both within and outside cells involved in cell adhesion.

Cell adhesion is a well-studied mechanism of cell communication. Adhesion is a form of mechanical linkage, of cells to cells and cells to the ECM, and is critically involved in cellular signaling events that control proliferation, survival, apoptosis, shape, polarity, motility, and differentiation. Adhesion is mediated by transmembrane proteins, which connect the interior of the cell to its extracellular environment (Fig. 18.2). One major feature biological adhesion that is different from non-biological adhesion such as with household and industrial adhesives is that the former is mediated by chemical signals that often positively or negatively feedback to adhesion and subsequent cell behavior. In other words, cell adhesion is more than a simple glue to hold cells and tissues together; it is also a critical signaling platform. Integrins and cadherins are two principal classes of molecules mediating primarily cell-ECM and cell-cell adhesion respectively.

Integrins are the best-understood class of adhesion receptors [31]. They mediate both cell-ECM and cell-cell adhesion, differing based on the cell type and the type of receptor used in the interaction. Integrins, composed of α-β subunit heterodimers, assemble into 24 distinct integrins and bind to proteins in the ECM such as fibronectin, collagen and laminin [31]. The binding of integrins initiates clustering of integrins to form structures and the recruitment of a host of signaling and adaptor proteins as well as the actin cytoskeleton [24]. These focal adhesions are involved in the signaling events that lead to proliferation, motility, cytoskeletal organization, and cell survival [24].

Cadherins mediate adhesive contact between cells in structures called adherens junctions, and play a vital role in morphogenic events during development [83]. The best studied of these are the classical cadherins, which mediate adhesion between adjacent cells by forming homotypic junctions at sites called adherens junctions, and are linked intracellularly to the actin cytoskeleton through beta-catenin and alpha-actinin (Fig. 18.2). Similar to integrins, cadherins too have a mechanical and signaling role. Along with the mechanical that adhesion plays in tissue cohesion, cadherins, like integrins, transmit specific signals to the cell interior through proteins at the adhesion site [69, 85].

Both integrin and cadherin adhesions are biochemically regulated and are both dynamic and reversible. Normal cell processes such as the rounding of spread cells during mitosis, cell sorting and migration during embryogenesis, or disease processes such as cancer cell metastasis involve active changes in adhesion strength between cell-ECM and cell-cell contacts. Also, adhesive signals regulate many of the same cell functions that soluble growth factors do. Adhesion and growth factor pathways are cooperative: for example, anchorage dependent cells do not grow in the absence of adhesion when they are placed in suspension even in the presence of saturating amounts of growth factors, neither do they grow in the absence of growth factors even when adherent. It is only when both adhesive and growth factor signals are present that growth pathways are optimally activated in cells [89].

The variables involved in controlling cell-environment interactions to achieve a desired phenotype for tissue engineering may be summarized as in Fig. 18.3. The bold arrows

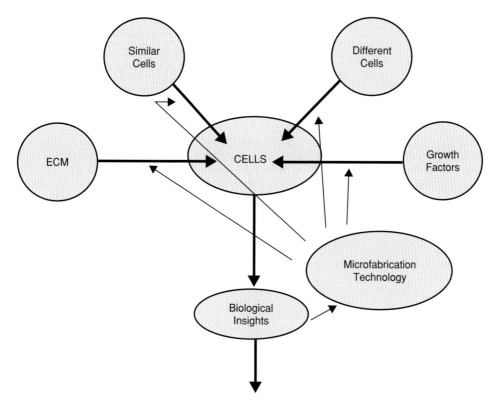

FIGURE 18.3. The figure indicates the role that microfabrication technologies can play in advancing cell engineering. The block arrows indicate the path to achieving a desired cell phenotype by changing the different adhesive and soluble inputs that a cell receives from its environment. The thin arrows represent the iterative feedback processes of understanding and refinement of experiments towards this goal using microfabrication technologies.

refer to possible inputs that each cell receives from the surroundings that are amenable to optimization by microfabrication technology. Such signals include adhesive interactions with the ECM, adhesion with similar neighboring cells (homotypic interactions), adhesion with dissimilar neighboring cells (heterotypic interactions), and signals from soluble growth factors in the environment. Researchers have demonstrated that each of these interactions is amenable to experimental control through microfabrication technology, as represented by the thinner arrows. Through iterative feedback of the resulting biological insights, it is thought that a desired phenotype can eventually be engineered explicitly with microsystems approaches.

18.3. CONTROLLING CELL INTERACTIONS WITH THE SURROUNDING ENVIRONMENT

Cell adhesion (to ECM and to other cells) is clearly important but is harder to control than soluble growth factor cues due to the difficulty in preparing defined surfaces. Most proteins are surface active due to the presence of hydrophobic and hydrophilic amino acids within their sequence and hence tend to accumulate at surfaces. This poses a challenge because cell culture media constituents such as serum are complex mixture of proteins, which differentially compete for adsorption to surfaces and leave a complex, ill-defined coating for experimental studies. The following section will describe two distinct technologies which have helped overcome the challenge of controlled cell adhesion: technologies that allowed preparing substrates with defined surface chemistry, and technologies that allowed the patterning of this surface chemistry on cell length scales.

18.3.1. Creating Defined Surface Chemistries

Traditional approaches to control surface chemistry have relied on pre-coating by adsorption of the culture surface with known concentrations of purified adhesion proteins such as fibronectin and subsequently blocking protein adsorptions to free surface areas with BSA (bovine serum albumin), which adsorbs well to hydrophobic surfaces. BSA repels further protein adsorption and is non-adhesive to cells. Using these methods to control cell adhesion and thereby spreading, researchers showed correlations between cell spreading and growth, differentiated function in microvascular endothelial cells [32] and hepatocytes [53]. However, this method to control cell adhesion suffered from the disadvantage that the physisorbed BSA is not stable, vulnerable to degradation by cell-derived proteases, and also necessitated the use of serum-free culture medium to minimize unwanted protein adsorption due to exchange with surface bound-BSA. Further, the methods did not allow varying local spatial organization of the ECM or the cells, as cells are randomly organized on the surface. Advances in surface engineering have allowed researchers to alleviate some of these limitations:

Several chemistries have been developed to immobilize proteins on activated surfaces using carboxy, amino, hydroxy, thiol, and other chemical moieties on amino acids [21]. For example, surface alcohols can be activated by substituting with the more reactive sulfonyl ester group, which can then be displaced by amino or thiol containing ligands [21]. Alternatively, light sensitive heterobifunctional bridges have been developed that link the substrate on one end and ligand on the other. For example, the photosensitive copolymer of styrene and vinylbenzyl N,N-diethyldithiocarbamate has been used to graft polymerize (N,N-dimethylacrylamide) onto poly(ethylene terephthalate) surfaces in the presence of UV light [55].

Besides protein immobilization, adhesion engineering also requires methods to block non-specific protein adsorption. This is achieved by localizing the blocking agent at the surface either by adsorption, or covalent coupling to alkanethiols or silanes. Hydrophilic polymers such as polyethylene oxide, polyethylene glycol, polyacrylamide, agarose, mannitol or the protein albumin have been used as blocking agents [36, 45]. Multiple factors appear to contribute to the non adhesivity of surfaces including the presence of surface charge, mobility of surface groups and ultimately the free energy cost of adsorption [3]. These chemistries have been critical to the development of surfaces with bio-specific adhesion, by eliminating non-specific, or undesired adhesion.

Self-assembled monolayers (SAMs) have been another important contribution to the development of defined surface chemistries. When certain long chained hydrocarbons are tethered to surfaces, they tend to present highly ordered surfaces due to thermodynamic and steric considerations [61]. The free end can be coupled to molecules of interest using suitable chemistries. Hence, surfaces can be functionalized by grafting a suitable end group on the free end of the molecules to engineer desired surface properties. For example, methyl terminated alkanethiols form extremely hydrophobic SAMs that permit protein adsorption whereas PEG-terminated (polyethylene glycol-terminated) alkanethiols form very hydrophilic surfaces that are resistant to protein adsorption [62]. Examples of physisorbed (i.e. not involving covalent bonding) SAMs include those prepared from Langmuir-Blodgett films (LB films) of synthetic double-tailed alkyl groups [59] and examples of chemisorbed (i.e. involving covalent bonding) SAMs include alkanethiolates on gold [62] and silanes on glass and silicon [47]. Alkanethiols and silanes have been extensively used for micropatterning and will be considered in further detail below.

Alkanethiols are hydrocarbons containing a sulfhydryl group on one end. When a solution of alkanethiols is exposed to gold, the sulfur atom forms covalent bonds with gold. Due to a combination of epitaxial registry with the underlying gold crystal structure and lateral stabilization of the hydrophobic alkyl chains, alkanethiols form highly oriented SAMs on gold [5], that can be exploited for engineering adhesion. The building blocks of silane SAMs are derivatives of the prototypical silane $SiH4$. One of the hydrogens attached to silicon is substituted with a long chain hydrocarbon terminating in a functional group such as an amine or a carboxyl group and the other three with alkoxy groups (-OR) or chlorine atoms. The alkoxy groups can be hydrolyzed in the presence of water or moisture to give silanols (-OH) which can further condense with similar groups on activated substrates as plasma treated glass or plastic, or amongst themselves, to create covalent crosslinks (O-Si-O). The amino or carboxy functional groups can be used for coupling of proteins [12, 47].

With these newfound controls over surfaces, researchers were able to study molecular aspects of cell adhesion previously not possible. Biologists had identified several recurring adhesion motifs in ECM proteins including the tripeptide RGD sequence found in a fibronectin, fibrinogen and collagens amongst others [67] and the YIGSR sequence in laminin [49]. The discovery of these specific adhesion sequences as parts of much larger proteins with multiple binding sites now allowed researchers to pose questions such as what the ligand density required for spreading is, or what the optimal molecular conformation is for receptor binding for specific ligands. Massia and Hubbell examined the question of the minimum density of ligand required for cell spreading and for focal adhesion formation [48]. Using covalent coupling of the integrin recognition sequence RGD to non adhesive glass surface, they reported that for human foreskin fibroblasts, the ligand spacing for maximal spreading was 440 nm and 140 nm for focal contact formation.

Adherent cells deposit their own matrix when plated on surfaces. Because of this, it was not known if short segments of ECM such as the RGD sequence were sufficient to maintain long-term cell health in the absence of augmentation by cell-deposited ECM. Roberts et al tested this question by making use of the exceptional protein repellent characteristics of oligoethyleneglycol-terminated thiols [66]. Surfaces created with RGD functionalized oligoethyleneglycol-terminated thiols (EG thiols) in a background of EG thiol were able to resist cell-secreted ECM deposition, and cells cultured on these surfaces survived for at least 24h [66]. They reported that microvascular endothelial cells were able to survive by attachment to RGD moieties in the absence of other proteins. A number of studies with immobilized RGD peptides had shown that RGD conformation determines its activity [52, 59]. Houseman et al examined the relation between RGD accessibility on a substrate and adhesion using the alkanethiol SAM system [28]. Using RGD terminated EG thiol in a bed of EG thiol of varying length (by changing the number of EG repeats, thereby changing the exposure of the RGD moiety), they showed by a variety of measures that cell responses such as spreading strongly depend on the background against which RGD is presented [28].

While these tools to present ECM ligands in better-defined systems have already provided important insights and advances to our understanding of a few key cell-surface adhesive interactions, the potential wealth of information to be gained from the much larger number of less well studied adhesive interactions remains untapped.

18.3.2. The Development of Surface Patterning

The technologies described so far allow uniform control of the adhesiveness of a surface. Generating surfaces with heterogeneous adhesiveness, where the pattern of adhesive and non-adhesive chemistries is prescribed, is called "patterning". Patterning allows adhesive control at cell length scales. Patterning allows the ability to direct cell adhesion to desired regions of the substrate, exclude cells from other parts, and comprises the most basic level of spatial engineering of adhesion. Early efforts for micropatterning of cells used the deposition of palladium through a nickel mask, to make micrometer-sized adhesive squares of palladium on a non-adhesive background [15–16]. While significant insights into regulation of cell migration [16], neuronal guidance [43], and proliferation [81] were obtained using this method, it had the disadvantages of that mechanism of adhesion was not well defined and it required metal evaporation and specialized masks; hence, the techniques were not easily adaptable for general use by the biological research community.

Surface chemical modification using silanes and alkanethiols remedy some of these shortcomings and have been widely adopted for patterning cell adhesion to surfaces. Patterning with SAMs has involved coupling of the silane or alkanethiol to the surface by masking the remaining areas and then removing the mask [47], or selective removal of parts of a uniformly SAM-coated surface using UV ablation through a photomask [53]. However, these patterning tools have some disadvantages. For example, the materials used in the patterning process damage proteins and are cytotoxic. These limitations greatly impeded the mass production of patterned substrates. The development of microcontact printing alleviated some of these shortcomings.

Microcontact printing in its essence involves patterning regions with differing surface properties on a surface using physical stamps rather than direct lithographic approaches [70]. This patterning is achieved in the following manner. Photolithography is used to make a mold of the pattern of interest in a photoresist coated on a silicon wafer. A pre-polymer of PDMS (polydimethylsiloxane) is then cured against this mold ('master') and peeled to reveal a PDMS 'stamp' containing a negative replica of the original master. The 'inks' used to stamp patterns on the gold surface are the silanes and alkanethiols [18, 82]. When the elastomeric stamp is placed in contact with a surface, the ink is transferred to it, forming a SAM wherever the stamp contacts the surface. Then a different silane or alkanethiol can be used to functionalize unstamped regions. In the end, the substrate now contains a pattern of two different SAMs. For patterning cells, chemistries are chosen such that one region is adhesive to ECM proteins and cells, while the other is resistant to protein adsorption, the latter typically being achieved by SAMs employing ethylene glycol functionalization. While microcontact printing with alkanethiols considerably lowered the barrier to micropatterning, it still requires equipment for gold coating of substrates and requires special chemicals such specially modified thiols and silanes. Direct printing of proteins might alleviate these dependencies by creating a general-purpose method to print proteins on suitable substrates, without the need for specialized equipment once a replica-molded stamp is available [73].

There have been several variations of surface patterning built around PDMS stamps molded from photolithographically generated 'masters', taking advantage the conformal contact with a flat surface. Using adjacent, parallel laminar flows in a PDMS channel sealed against a plastic dish, Takayama et al demonstrated the facility to pattern cells and their substrates using microfluidics [72]. If each channel contains a different protein for example, they will adsorb in parallel lines on the surface. In the absence of turbulent mixing between adjacent streams, the resolution of the technique essentially is determined by diffusion intermixing, which in turn is determined by the size of the particle being patterned. Using such devices, it has been shown possible to pattern cells over a uniformly coated surface [72], or partition the fluid phase over a single cell so that different parts of the cell are exposed to different aqueous environments [73]. Turbulence can be purposefully created to allow mixing when needed by patterning roughness on the channel walls [71]. Further refinements built around microfluidics patterning have allowed the aligned printing of multiple proteins on the same substrates [76] potentially enabling the study of cell interactions with more complex surface chemistries. These microfluidic approaches are described in greater detail in excellent reviews elsewhere [36–37].

18.3.3. Examples of Patterning-Based Studies on Cell-To-Cell Interactions

The tools for micropatterning described above have been used in several studies to examine the consequences of the organization of cells with respect to each other, and the resulting cell-to-cell interactions, on cell function. Some of these are described below.

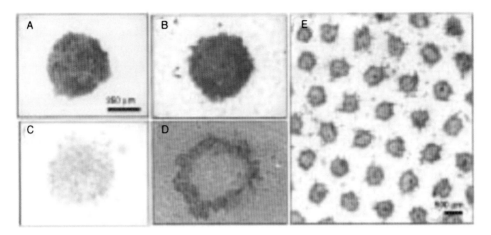

FIGURE 18.4. The figure demonstrates the effect of introducing heterotypic cell-cell contacts on hepatocyte differentiation in vitro. Red color indicates the synthesis of albumin, a marker for hepatocyte differentiation. A and B respectively are micropatterned hepatocytes cultured in isolation or in co-culture with fibroblasts. C and D show cultures under the same conditions after six days. Note the greater loss of albumin from isolated hepatocytes C compared to co-cultures D. E shows a low-magnification image of the co-culture (from [8]).

18.3.3.1. Hepatocyte Differentiation   In vivo, hepatocytes exist in heterotypic cell-cell contact with other cells such as sinusoidal endothelial cells and stellate cells. Hepatocytes isolated from the liver rapidly lose their differentiated phenotype, by measures such as albumin or urea secretion [53]. As part of the effort to maintain hepatocytes differentiated function in vitro, investigators have tried to recapitulate the in vivo context. In this effort, it was found that introducing heterotypic cell-cell contact by co-culture of hepatocytes with fibroblasts increased hepatocytes differentiated function [40]. However, because these conventional cell culture experiments did not allow the modulation of cell-cell contact independent of cell number, it was difficult to ascertain the mechanism of the upregulated function [8]. By using micropatterning methods, Bhatia et al. were able to culture hepatocytes in a defined pattern and fibroblasts in the intervening regions, thus restricting cell-cell contacts to the interface between regions [7]. This allowed the investigators to increase cell-cell contact at constant cell number and show that hepatocytes albumin synthesis is higher and is maintained for longer times with hepatocytes-fibroblast contact independent of cell density (Fig. 18.4). This study showed the potential for engineering enhanced differentiated function by using appropriate microfabricated substrates.

18.3.3.2. Angiogenesis   The growth of the small capillary blood vessels in the body, or angiogenesis, underlies many important normal physiological and pathological events such as wound healing, diabetes, atherosclerosis and the growth of cancerous tumors [14]. Angiogenesis is also a key element of tissue engineering, as organ replacements will require the development of ways to populate the devices with blood vessels to cater to the oxygen and nutrient needs of the cells within the device [35]. Hence, it is of great medical interest to understand and control the process of angiogenesis. Researchers have attempted to coax microvascular endothelial cells in culture dishes to form branched network of tubular

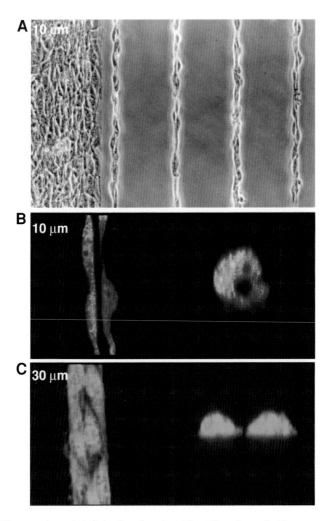

FIGURE 18.5. Microvascular endothelial cells cultured on 10 μm lines form tube-like structures. A. phase microscopy image of cells on 10 μm line. B. confocal sectioned image of the same cells showing a hollow lumen inside. C. cells on 30 μm lines which spread as ribbons and do not form a hollow lumen (from [20]).

vessels; enabling the study of the process in a convenient in vitro system. Conventional in vitro systems such as culturing endothelial cells on low densities of fibronectin [32] or within malleable gels of ECM [38], suffer from limitations such as a low yield of capillaries (low percentage of cells on the dish form capillaries) and random distribution disallowing convenient quantification of the process. Dike et al. have studied the role of the ECM and cell-cell contacts in angiogenesis by culturing endothelial cells on micropatterned lines of fibronectin [20]. While unspread endothelial cells apoptose in culture, cells cultured on wide (30 um) lines of ECM form extensive cell-cell contact but spread out as ribbons and do not form tubes (Fig. 18.5) [20]. Only cells cultured on an intermediate width (10 um) of fibronectin lines form both extensive cell-cell contacts and develop into tubular structures with a central lumen verified by microscopy [20]. Micropatterning angiogenesis models might in the future allow controlled studies of angiogenesis and in the design of substrates that promote angiogenesis in tissue engineered devices.

18.3.3.3. Cell Proliferation   The previous study examined the relation between cell-cell contact and cell shape on endothelial differentiation into capillaries. Nelson and Chen [56] examined the crosstalk between cell-cell contact and shape for endothelial proliferation using a novel patterning method that uses elements of microcontact printing (the use of a textured PDMS stamp) and microfluidic patterning (the use of flow though channels). The textured PDMS stamp is used to form channels against a glass surface while its elastomeric nature helps seal parts of the surface where the stamp is flush with it. This stamp was used as a mold to gel shallow wells of agarose on glass, within which pairs of cells were cultured (Fig. 18.1). The system allowed Nelson and Chen to independently control cell-cell contact and cell spreading. They used this system to demonstrate that cell-cell contact increases cell proliferation but that this effect is masked in conventional (unpatterned) cell culture conditions of unconstrained adhesion [56].

18.3.4. Examples of Patterning-Based Studies on Cell-Matrix Interactions

18.3.4.1. Cell Shape and Proliferation   Some of the key biological problems addressed by micropatterning derive from an earlier study on the growth of cells on surface of graded adhesivity [22]. Substrate adhesivity to cells was changed by coating tissue culture surfaces with graded concentrations of the non-adhesive hydrogel poly(2-hydroxyethyl methacrylate). Cell spreading and correspondingly DNA synthesis (also termed 'growth') were similarly found to increase with increased density of fibronectin and collagen coating on surfaces [32, 53]. Because changing ECM density not only changes cell shape but also integrin signaling events such as integrin clustering and focal adhesion formation, it was not clear if the observed proliferative effects were due to changes in the integrin signaling as was conventionally understood, or due to cell shape itself acting as an input for the cell regulation.

Singhvi et al examined this question using microcontact printing [70]. By printings adhesive islands of the ECM protein laminin, of small or large areas, they could change cell shape without local ligand density. They found that indeed rounded hepatocytes on small squares exhibit low growth and high differentiated function while spread cells on large islands displayed the opposite [70]. This study strengthened the hypothesis that cell shape acts as a signal distinct from integrin clustering. However, because the ligand density was the same between round and spread cells, the total amount of ligand seen by the spread versus cells round was higher, which could by itself explain the increase proliferation in spread cells. Chen et al addressed this question, again by using microcontact printing technology [17]. This time, they varied cell shape while keeping the ECM density *and* ECM area constant. Instead of using single large islands of ECM to spread cells, they used arrays of small ECM dots which were spaced so that cells could bridge the gaps between the dots and thereby spread to eh same extent as cells on single large islands, although the total ECM that the cells came in contact was smaller in the case of dot patterns. This study showed that cell shape independent of the extent of ECM contact, modifies cell behavior. Huang et al extended this study by examining the effect of cell shape on some key cell proliferation-related proteins, demonstrating that cell shape regulated cyclin D1 and p27 [29].

18.3.4.2. Cell Motility and the ECM   Cells sample their adhesive surroundings using different kinds of cell membrane projections such as lamellipodia and filopodia, which require specific signaling pathways for their generation [57, 65]. Such membrane projections are required for several basic cell processes such as spreading and migration. As part of sampling the surroundings, cells also exert forces through adhesions [30]. Parker et al examined the relationship between cell geometry, lamellipodial projections, and cell-generated forces used alkanethiol patterning together with a polyacrylamide patterning system [60]. They reported the interesting finding that square shaped cells exert the strongest forces along the diagonals and that lamellipodial projections are confined to the corners of such squares [60]. In a further study it was shown that increased lamellipodial activity at corners is true for a variety of polygonal shapes and interestingly, cells accumulate fibronectin at these corners [13]. Cellular remodeling of existing ECM and deposition of their own ECM is believed to aid migration. If the results from these patterned cell models are applicable to migrating cell, it may be possible as the study suggests, to design patterned ECMs that can direct cell migration, for example to the interior of a tissue-engineered matrix [13].

18.4. FUTURE WORK

While micro- and nanotechnologies have been invaluable in designing experiments that control the adhesive interactions of cells, the greatest potential of these methods is yet to be achieved, particularly for tissue engineering. Most current methods to control cell-environment interactions share a set of common shortcomings: patterning is largely limited to hard materials such as glass and silicon but in vivo applications will likely require the adoption of patterning to soft materials such as biodegradable gels; they are mostly applicable to two dimensional (2D) substrates, while tissues are intrinsically 3D; few methods have the capacity to actively move and position cells, limiting the complexity of structures that can be formed; and finally, in face of an abundance of evidence to suggest that substrate mechanics play a substantial role in cell function, few of the current methods allow the patterning of adhesion in conjunction with substrate mechanics. Some of the elements that may form the building blocks of such technologies exist in a rudimentary form now, but much work remains to be done before they can be employed for building cellular devices. New technologies that address these issues are predicted to encompass some of the most exciting advances over the next several years.

18.4.1. Developing New Materials

Novel substrates and surface chemistries can further advance the field of patterning. A qualitative advancement over the technique of alkanethiol microcontact printing has been the development of electroactive thiols. Microcontact printing of thiols on gold is essentially an irreversible process. So once a pattern is created, it is not possible to change it. This limitation has been addressed by designing thiol molecules with RGD moieties covalently incorporated in them through an electroactive quione ester [86, 87]. Applying a reducing potential to the gold substrate causes the reduction of the quione to hydroxyquinone and subsequent cyclization to a lactone with the release of the RGD moiety. Using this approach, the capability to detach of adherent cells upon RGD release using a reducing potential, or to pattern multiple cell types has been demonstrated [86, 87]. This capability should potentially allow the study of dynamic adhesion events after patterning.

Apart from the methods to modify surfaces, there is a need to adapt microfabrication techniques to biocompatible if not biodegradable substrates, for micropatterning to be widely employed in clinical applications. If the biodegradable materials were available in a photopolymerizable form, processing with standard microfabrication techniques such as photolithography and micromolding would be convenient. A promising biodegradable material that can be micromolded is PLGA, which can be solvated and then cast into desired shapes with micrometer scale resolution [79]. Finally, PEG-based polymers can be made photocurable, and thus amenable to standard photolithography [51]. Complete 3D structures can be perhaps formed by lamination [49] of 2D microfabricated layers.

18.4.2. Better Cell Positioning Technologies

The capacity to pick and place cells in configurations in combination with patterned surfaces would be valuable for basic science studies. The placement of cells using physical force is greatly complicated by the cells' small size scale, on the order of 10 μm. In theory, any of a variety of mechanical actuators could be employed for the purpose and have been demonstrated including micro actuators [34], electric forces [58], magnetic forces [63], and electromagnetic forces [4]. Electrical trapping in the form of dielectrophoresis (DEP) in particular has recently received much attention [77]. DEP is the movement of uncharged particles in inhomogeneous electric fields due to induced electric charges. This phenomenon has been used create electrical forces on cells and particles and thereby move and position them [2, 78]. In combination with microcontact printing, it has been possible to achieve the nontrivial task of positioning multiple cell types in complex geometries. While there are clear advantages to the use of force to trap and position cells, there is much further scope for combining the advantages of cell positioning with the patterning techniques described in this chapter.

18.4.3. Patterning in 3D Environments

In vivo tissues exist in three dimensions. Tissues culture systems recapitulating 3D environments have shown distinct differences in composition of cell adhesions and cell function [19] or tissue morphogenesis [68]. Because most cellular structures are intrinsically 3D, the need for 3D fabrication techniques to even just model such structures is great. With 2D fabrication, cells can simply be allowed to settle on the surface and attach to it. Although such cell compatible surfaces with topology have been engineered, they typically only support a single layer of cells. With 3D, cells must be somehow introduced into any pre-made structure. While cells are capable of invading porous scaffolds, this doesn't allow for patterning their locations and precise interactions. 3D microfabricated hydrogels containing cells have recently been demonstrated [44]. Future advances would likely involve higher resolution fabrication, and the combination of adhesive patterning with 3D microfabrication.

18.4.4. Patterning Substrate Mechanics

It has been known for some time now that substrate mechanics affect cell function. The development of load bearing tissue such as muscle and bone requires the presence of mechanical load on the cells, which atrophy in its absence [80, 84]. In cell culture models too, there are several examples of the influence of mechanical environment on cell function: breast cells cultured on rigid collagen gels attached to a surface, or malleable free-floating gels exhibit dramatic differences in the secretion breast specific proteins such as caseins [42]. Similarly hepatocytes cultured on rigid collagen films dedifferentiate whereas those cultured on malleable gels exhibit enhanced differentiated function [26]. Beside these metabolic responses, cells exhibit differences in mechanical functions such as contractile forces, motility and adhesion itself, in response to the stiffness of the substrate [25]. Thus, an abundance of material suggests that control of the mechanical properties of the environment is an important component of regulating cell function.

Cells exhibit other responses to substrate stiffness. Using adjacent islands of acrylamide on a glass coverslip, Lo et al reported that cells migrate faster on stiffer substrates, and tend to move toward the stiffer substrate [46]. Gray et al extended this by patterning stiffness on a micrometer scale: substrates of different stiffness were created using either polyacrylamide (kPa stiffness) or PDMS (kPa to MPa stiffness) coated with fibronectin [25]. Cells were found to accumulate on the stiffer surfaces due to migration and remodel the fibronectin [25]. In a different approach, Tan et al have created a bed of flexible PDMS posts on which cells can be cultured [75]. When cells are cultured on such surfaces, they adhere to and bend the posts, thereby giving readout of cell forces. The mechanical stiffness of the posts can be patterned by changing post height (shorter posts are stiffer), or by changing the cross-sectional aspect ratio [75]. Patterning mechanical properties of substrates is one of the least developed amongst the ways to pattern cell environment. Because of the abundance of information showing a connection between substrate mechanical properties and cell behavior, this avenue offers much scope for future advances in microfabricating materials with interesting mechanical properties on cellular length scales.

18.5. CONCLUSIONS

Cells are governed by the cellular and non-cellular components of their surrounding environment from first stages of development through adulthood. Cells are not only highly sensitive biosensors, but also capable of responding to inputs by changing their behaviors and phenotypes. Microfabrication technologies have the potential to precisely control the inputs that cells receive. Ultimately, the full potential of cells for therapeutics, drug discovery, biosensors, and basic science can be achieved by giving the correct inputs to correct cell type. It seems a truism that recreating the complexity of an in vivo cell- environment is an extremely challenging task. The likely realization of this dream will be through the creation of seed conditions by microfabrication, which will direct cells along pathways that lead them to create the microenvironment suitable to the specific tissue. To participate in these efforts in a meaningful way, researchers in the microfabrication field need to be aware of the important biological problems that microfabrication has and can be applied to, and biologists and clinicians without a microfabrication background must have widespread and convenient access to these exciting patterning tools.

REFERENCES

[1] M. Adham. Extracorporeal liver support: waiting for the deciding vote. *Asaio J.*, 49:621–632, 2003.

[2] D.R. Albrecht, R.L. Sah, and S.N. Bhatia. Dielectrophoretic Cell Patterning within Tissue Engineering Scaffolds. *Proceedings of the Second Joint EMBS/BMES Conference*, Houston, TX, USA, pp. 1708–1709.

[3] N.A. Alcantar E.S. Aydil, and J.N. Israelachvili. Polyethylene glycol-coated biocompatible surfaces. *J. Biomed. Mater. Res.*, 51:343–351, 2000.

[4] A. Ashkin, J.M. Dziedzic, and T. Yamane. Optical trapping and manipulation of single cells using infrared laser beams. *Nature*, 330:769–771, 1987.

[5] C.D. Bain, E.B. Troughton, Y. Tao, J. Evall, G.M. Whitesides, and R.G. Nuzzo. Formation of monolayer films by the spontaneous assembly of organic thiols from solution onto gold. *J. Am. Chem. Soc.*, 111:321–335, 1989.

[6] K.Bhadriraju and C.S. Chen. Engineering cellular microenvironments to improve cell-based drug testing. *Drug Discov. Today*, 7:612–620, 2002.

[7] S.N. Bhatia U.J. Balis, M.L. Yarmush, and M. Toner. Probing heterotypic cell interactions: hepatocyte function in microfabricated co-cultures. *J. Biomater. Sci. Polym. Ed.*, 9:1137–1160, 1998.

[8] S.N. Bhatia, U.J. Balis, M.L. Yarmush, and M. Toner. Effect of cell-cell interactions in preservation of cellular phenotype: cocultivation of hepatocytes and nonparenchymal cells. *Faseb. J.*, 13:1883–1900, 1999.

[9] S.N. Bhatia and C.S. Chen. Tissue engineering at the micro-scale. *Biomed. Microdev.*, 2:131–144.

[10] M.J. Bissell and M.H. Barcellos-Hoff. The influence of extracellular matrix on gene expression: is structure the message? *J. Cell Sci. Suppl.*, 8:327–343, 1987.

[11] M.J. Bissell, A. Rizki, and I.S. Mian. Tissue architecture: the ultimate regulator of breast epithelial function.*Curr. Opin. Cell Biol.*, 15:753–762, 2003.

[12] S. Britland, E. Perez-Arnaud, P. Clark, B. McGinn, P. Connolly, and G. Moores. Micropatterning proteins and synthetic peptides on solid supports: a novel application for microelectronics fabrication technology.*Biotechnol. Prog.*, 8:155–160, 1992.

[13] A. Brock, E. Chang, C.-C. Ho, P. LeDuc, X. Jiang, G.M. Whitesides, and D.E. Ingber. Geometric deteminants of cell motility revealed using microcontact printing. *Langmuir*, 19:1611–1617, 2003.

[14] P. Carmeliet and R.K. Jain. Angiogenesis in cancer and other diseases. *Nature*, 407:249–257, 2000.

[15] S.B. Carter. Haptotactic islands: a method of confining single cells to study individual cell reactions and clone formation. *Exp. Cell Res.*, 48:189–193, 1967a.

[16] S.B. Carter. Haptotaxis and the mechanism of cell motility. *Nature*, 213:256–260, 1967b.

[17] C.S. Chen, M. Mrksich, S. Huang, G.M. Whitesides, and D.E. Ingber. Geometric control of cell life and death. *Science*, 276:1425–1428, 1997.

[18] C.S. Chen, M. Mrksich, S. Huang, G.M. Whitesides, and D.E. Ingber. Micropatterned surfaces for control of cell shape, position, and function. *Biotechnol. Prog.*, 14:356–363, 1998.

[19] E. Cukierman, R. Pankov, D.R. Stevens, and K.M. Yamada. Taking cell-matrix adhesions to the third dimension. *Science*, 294:1708–1712, 2001.

[20] L.E. Dike, C.S. Chen, M. Mrksich, J.Tien, G.M. Whitesides, and D.E. Ingber. Geometric control of switching between growth, apoptosis, and differentiation during angiogenesis using micropatterned substrates. *In Vitro Cell Dev. Biol. Anim.*, 35:441–448, 1998.

[21] P. Drumheller and J.A. Hubbell. *The Biomedical Engineering Handbook*. CRC Press, Boca Raton, FL, pp. 110–111, 2000.

[22] J. Folkman and A. Moscona. Role of cell shape in growth control. *Nature*, 346:760–763, 1978.

[23] L.M. Furtado, R. Somwar, G. Sweeney, W. Niu, and A. Klip. Activation of the glucose transporter GLUT4 by insulin. *Biochem. Cell Biol.*, 80:569–578, 2002.

[24] B. Geiger and A. Bershadsky. Assembly and mechanosensory function of focal contacts. *Curr. Opin. Cell Biol.*, 13:584–592, 2001.

[25] D.S. Gray, J. Tien, and C.S. Chen. Repositioning of cells by mechanotaxis on surfaces with micropatterned Young's modulus. *J. Biomed. Mater. Res.*, 66A:605–614, 2003.

[26] L.K. Hansen and J.H. Albrecht. Regulation of the hepatocyte cell cycle by type I collagen matrix: role of cyclin D1. *J. Cell Sci.*, 112:2971–2981, 1999.

[27] R. Hoekstra. In N. Berry and Anthony M. Edwards (eds.), The Hepatocyte ReviewMichael. Kluwer Academic Publishers; EUR 224.50/USD 245.00; ISBN: 0-7923-6177-6. *J. Hepatol.*, 35:546, 2001.

[28] B.T. Houseman and M. Mrksich. The microenvironment of immobilized Arg-Gly-Asp peptides is an important determinant of cell adhesion. *Biomaterials*, 22:943–955, 2001.

[29] S. Huang, C.S. Chen, and D.E. Ingber. Control of cyclin D1, p27(Kip1), and cell cycle progression in human capillary endothelial cells by cell shape and cytoskeletal tension. *Mol. Biol. Cell*, 9:3179–3193, 1998.

[30] S. Huang and D.E. Ingber. The structural and mechanical complexity of cell-growth control. *Nat. Cell Biol.*, 1:E131–E138, 1999.

[31] R.O. Hynes. Integrins: bidirectional, allosteric signaling machines. *Cell*, 110:673–687, 2002.

[32] D.E. Ingber and J. Folkman. Mechanochemical switching between growth and differentiation during fibroblast growth factor-stimulated angiogenesis in vitro: role of extracellular matrix. *J. Cell Biol.*, 109:317–330, 1989.

[33] O. Ishii, M. Shin, T. Sueda, and J.P. Vacanti. In vitro tissue engineering of a cardiac graft using a degradable scaffold with an extracellular matrix-like topography. *J. Thorac. Cardiovasc. Surg.*, 130(5):1358–1363, Nov 2005.

[34] E.W. Jager, O. Inganas, and I. Lundstrom. Microrobots for micrometer-size objects in aqueous media: potential tools for single-cell manipulation. *Science*, 288:2335–2338, 2000.

[35] R.K. Jain Molecular regulation of vessel maturation. *Nat. Med.*, 9:685–693, 2003.

[36] R.S. Kane, S. Takayama, E. Ostuni, D.E. Ingber, and G.M. Whitesides. Patterning proteins and cells using soft lithography. *Biomaterials*, 20:2363–2376, 1999.

[37] P.J. Kenis, R.F. Ismagilov, S. Takayama, G.M. Whitesides, S. Li, and H.S. White. Fabrication inside microchannels using fluid flow. *Acc. Chem. Res.*, 33:841–847, 2000.

[38] Y. Kubota, H.K. Kleinman, G.R. Martin, and T.J. Lawley. Role of laminin and basement membrane in the morphological differentiation of human endothelial cells into capillary-like structures. *J. Cell Biol.*, 107:1589–1598, 1998.

[39] P. Landers. *Cost of Developing a New Drug Increases to About $1.7 Billion*. The Wall Street Journal, 2003.

[40] R. Langenbach, L. Malick, A. Tompa, C. Kuszynski, H. Freed, and E. Huberman. Maintenance of adult rat hepatocytes on C3H/10T1/2 cells. *Cancer Res.*, 39:3509–3514, 1979.

[41] R. Langer and J.P. Vacanti. Tissue engineering. *Science*, 260:920–926, 1993.

[42] E.Y.-H. Lee, G. Parry, and M.J. Bissell. Modulation of secreted proteins of mouse mammary epithelial cells by the collagenous substrata. *J. Cell Biol.*, 107:1589–1598, 1984.

[43] P.C. Letourneau. Cell-to-substratum adhesion and guidance of axonal elongation. *Dev. Biol.*, 44:92–101, 1975.

[44] V.A. Liu and S.N. Bhatia. Three dimensional photo-patterning of hydrogels containing living cells. *Biomed. Microdev.*, 4:256–266, 2002.

[45] V.A. Liu, W.E. Jastromb, and S.N. Bhatia. Engineering protein and cell adhesivity using PEO-terminated triblock polymers. *J. Biomed. Mater. Res.*, 60:126–134, 2002.

[46] C.M. Lo, H.B. Wang, M. Dembo, and Y.L. Wang. Cell movement is guided by the rigidity of the substrate. *Biophys. J.*, 79:144–152, 2000.

[47] B. Lom, K.E. Healy, and P.E. Hockberger. A versatile technique for patterning biomolecules onto glass coverslips. *J. Neurosci. Methods*, 50:385–397, 1993.

[48] S.P. Massia and J.A. Hubbell. An RGD spacing of 440 nm is sufficient for integrin alpha V beta 3-mediated fibroblast spreading and 140 nm for focal contact and stress fiber formation. *J. Cell Biol.*, 114:1089–1100, 1991.

[49] S.P. Massia, S.S. Rao, and J.A. Hubbell. Covalently immobilized laminin peptide Tyr-Ile-Gly-Ser-Arg (YIGSR) supports cell spreading and co-localization of the 67 kilodalton laminin receptor with alpha-actinin and vinculin. *J. Biol. Chem.*, 268:8053–8059, 1993.

[50] A.G. Mikos, G. Sarakinos, S.M. Leite, J.P. Vacanti, and R. Langer. Laminated three-dimensional biodegradable foams for use in tissue engineering. *Biomaterials*, 14:323–330, 1993.

[51] M. Mizutani and T. Matsuda. Liquid photocurable biodegradable copolymers: in vivo degradation of photocured poly(epsilon-caprolactone-co-trimethylene carbonate). *J. Biomed. Mater. Res.*, 61:53–60, 2002.

[52] H. Mohri, Y. Hashimoto, M. Ohba, H. Kumagai, and T. Ohkubo. Novel effect of cyclicization of the Arg-Gly-Asp-containing peptide on vitronectin binding to platelets. *Am. J. Hematol.*, 37:14–19, 1991.

[53] D.J. Mooney, L.K. Hansen, J.P. Vacanti, R. Langer, S.R. Farmer, and D.E. Ingber. Switching from differentiation to growth in hepatocytes: control by extracellular matrix. *J. Cell. Phys.*, 151:497–505, 1992.

[54] J.F. Mooney, A.J. Hunt, J.R. McIntosh, C.A. Liberko, D.M.Walba, and C.T. Rogers. Patterning of function-alantibodies and other proteins by photolithography of silane monolayers. *Proc. Natl. Acad. Sci. U.S.A.*, 93:12287–12291, 1996.

[55] Y. Nakayama, T. Matsuda, and M. Irie. A novel surface photo-graft polymerization method for fabricated devices. *ASAIO J.*, 39:M542, 1993.

[56] C.M. Nelson and C.S. Chen. VE-cadherin simultaneously stimulates and inhibits cell proliferation by altering cytoskeletal structure and tension. *J. Cell Sci.*, 116:3571–3581, 2003.

[57] C.D. Nobesand A. Hall. Rho, rac, and cdc42 GTPases regulate the assembly of multimolecular focalcom-plexes associated with actin stress fibers, lamellipodia, and filopodia. *Cell.*, 81:53–62, 1995.

[58] M. Ozkan, T. Pisanic, J. Scheel, S. Barlow, S. Esener, and S.N. Bhatia. Electro-optical platform for the manipulation of live cells. *Langmuir*, 19:1532–1538, 2003.

[59] T. Pakalns, K.L. Haverstick, G.B. Fields, J.B. McCarthy, D.L. Mooradian, and M. Tirrell. Cellular recognition of synthetic peptide amphiphiles in self-assembled monolayer films. *Biomaterials*, 20:2265–2279, 1999.

[60] K.K. Parker, A.L. Brock, C. Brangwynne, R.J. Mannix, N. Wang, E. Ostuni, N.A. Geisse, J.C. Adams, G.M. Whitesides, and D.E. Ingber. Directional control of lamellipodia extension by constraining cell shape and orienting cell tractional forces. *Faseb. J.*, 16:1195–1204, 2002.

[61] G.E. Poirier and E.D. Pylant. The self-assembly mechanism of alkanethiols on Au(111). *Science*, 272:1145–1148, 1996.

[62] K.L. Prime and G.M. Whitesides. Self-assembled organic monolayers: model systems for studying adsorption of proteins at surfaces. *Science*, 252:1164–1167, 1991.

[63] D.H. Reich M. Tanase, A. Hultgren, L.A. Bauer, C.S. Chen, and G.J. Meyer. Biological applications of multifunctional magnetic nanowires. *J. Appl. Phys.*, 93:7275, 2003.

[64] T.H. Rider, M.S. Petrovick, F.E. Nargi, J.D. Harper, E.D. Schwoebel, R.H. Mathews, D.J. Blanchard, L.T. Bortolin, A.M. Young, J. Chen, and M.A. Hollis. A B cell-based sensor for rapid identification of pathogens. *Science*, 301:213–215, 2003.

[65] Ridley, H.F. Paterson, C.L. Johnston, D. Diekmann, and A. Hall. The small GTP-binding protein rac regulates growth factor-induced membrane ruffling. *Cell*, 70:401–410, 1992.

[66] C. Roberts, C.S. Chen, M. Mrksich, M. V., D.E. Ingber, and G.M. Whitesides. Using mixed self-assembled monolayers presenting RGD and (EG)3OH groups to characterize long-term attachment of bovine capillary endothelial cells to surfaces. *J. Am. Chem. Soc.*, 120:6548–6555, 1998.

[67] E. Ruoslahti and and M.D. Pierschbacher. New perspectives in cell adhesion: RGD and integrins. *Science*, 238:491–497.

[68] K.L. Schmeichel and M.J. Bissell. Modeling tissue-specific signaling and organ function in three dimensions. *J. Cell Sci.*, 116:2377–2388, 2003.

[69] A. Shay-Salit, M. Shushy, E.Wolfovitz, H. Yahav, F. Breviario, E. Dejana, and N. Resnick. VEGF receptor 2 and the adherens junction as a mechanical transducer in vascular endothelial cells. *Proc. Natl. Acad. Sci. U.S.A.*, 99:9462–9467, 2002.

[70] R. Singhvi, A. Kumar, G.P. Lopez, G.N. Stephanopoulos, D.I.C. Wang, G.M. Whitesides, and D.E. Ingber. Engineering cell shape and function. *Science*, 264:696–698, 1994.

[71] A.D. Stroock, S.K. Dertinger, A. Ajdari, I. Mezic, H.A. Stone, and G.M. Whitesides. Chaotic mixer for microchannels. *Science*, 295:647–651, 2002.

[72] S. Takayama, J.C. McDonald, E. Ostuni, M.N. Liang, P.J. Kenis, R.F. Ismagilov, and G.M. Whitesides. Patterning cells and their environments using multiple laminar fluid flows in capillary networks. *Proc. Natl. Acad. Sci. U.S.A.*, 96:5545–5548, 1999.

[73] S. Takayama, E. Ostuni, P. LeDuc, K. Naruse, D.E. Ingber, and G.M. Whitesides. Selective chemical treatment of cellular microdomains using multiple laminar streams. *Chem. Biol.*, 10:123–130, 2003.

[74] J.L. Tan, J. Tien, and C.S. Chen. Microcontact printing of proteins on mixed self-assembled monolayers. *Langmuir*, 18:519–523, 2002.

[75] J.L. Tan, J. Tien, D.S. Pirone, D.S. Gray, K. Bhadriraju, and C.S. Chen. Cells lying on a bed of microneedles: an approach to isolate mechanical force. *Proc. Natl. Acad. Sci. U.S.A.*, 100:1484–1489, 2003.

[76] J. Tien, C.M. Nelson, and C.S. Chen. Fabrication of aligned microstructures with a single elastomeric stamp. *Proc. Natl. Acad. Sci. U.S.A.*, 99:1758–1762, 2002.

[77] J. Voldman. BioMEMS: building with cells. *Nat. Mater.*, 2:433–434, 2003.

[78] J. Voldman, M.L. Gray, M. Toner, and M.A. Schmidt. A microfabrication-based dynamic array cytometer. *Anal. Chem.*, 74:3984–3990, 2002.

[79] G. Vozzi, C. Flaim, A. Ahluwalia, and S. Bhatia. Fabrication of PLGA scaffolds using soft lithography and microsyringe deposition. *Biomaterials*, 24:2533–2540, 2003.

[80] J.S. Wayne. Load partitioning influences the mechanical response of articular cartilage. *Ann. Biomed. Eng.*, 23:40–47, 1995.

[81] B. Westermark. Growth control in miniclones of human glial cells. *Exp. Cell Res.*, 111:295–299, 1978.

[82] B.C. Wheeler, J.M. Corey, G.J. Brewer, and D.W. Branch. Microcontact printing for precise control of nerve cell growth in culture. *J. Biomech. Eng.*, 121:73–78, 1999.

[83] M.J. Wheelock and K.R. Johnson. Cadherins as modulators of cellular phenotype. *Annu. Rev. Cell Dev. Biol.*, 19:207–235, 2003.

[84] R.J., Wilkins, J.A. Browning, and J.P. Urban. Chondrocyte regulation by mechanical load. *Biorheology*, 37:67–74, 2000.

[85] A.S. Yap and E.M. Kovacs. Direct cadherin-activated cell signaling: a view from the plasma membrane. *J. Cell Biol.*, 160:11–16, 2003.

[86] W.S. Yeo, C.D. Hodneland, and M. Mrksich. Electroactive monolayer substrates that selectively release adherent cells. *Chembiochemistry*, 2:590–593, 2001.

[87] M.N. Yousaf, B.T. Houseman, and M. Mrksich. Using electroactive substrates to pattern the attachment of two different cell populations. *Proc. Natl. Acad. Sci. U.S.A.*, 98:5992–5996, 2001.

[88] F. Zaucke, R. Dinser, P. Maurer, and M. Paulsson. Cartilage oligomeric matrix protein (COMP) and collagen IX are sensitive markers for the differentiation state of articular primary chondrocytes. *Biochem. J.*, 358(Pt 1):17–24, 2001.

[89] X. Zhu, and R.K. Assoian. Integrin-dependent activation of MAP kinase: a link to shape-dependent cell proliferation. *Mol. Biol. Cell*, 6:273–282, 1995.

19

Cell Biology on a Chip: A Microfluidic Cell Culture Laboratory

Albert Folch and Anna Tourovskaia

Department of Bioengineering, University of Washington, Seattle, Washington 98195, USA

19.1. INTRODUCTION

Tissue function is modulated by the spatiotemporal organization of cells and biomolecules on a micrometer scale. In vivo, cells respond to a myriad signaling factors, either in the form of freely-diffusing molecules—e.g. enzymes, nutrients, small ions, growth factors secreted by cells that can be adjacent or as far as several meters for a large mammal—or in the form of relatively immobilized molecules that are anchored to the membrane of adjacent cells (e.g. membrane receptors, cadherins) or bound to extracellular scaffolds, such as the extracellular matrix (ECM) or bone; in addition, cells are also responsive to the physical topography and stiffness of the matrix. All these factors vary locally in smoothly graded or sharp concentration changes at the cellular and sub-cellular scale, as conceptually depicted in Fig. 19.1. This multi-signal changing environment is particularly dynamic during development, wound healing, and cancer.

The ability to culture cells outside of their natural organism—pioneered by Harrison[2] and Carrel [3]—has, despite constituting a simplistic simulation of the organism's inner workings, revolutionized hypothesis testing in basic cell and molecular biology research and become a standard methodology in drug testing and toxicology assays. Indeed, cell culture ("in vitro") systems inherently lack the three-dimensional, multicellular architecture found in an organism's tissue but offer precious advantages over whole-animal ("in vivo") experimentation: a) the parameters necessary for cell function can be isolated without interference from more complex, whole-organism or whole-organ responses; b) since many experimental conditions can be tested with the cells from only one sacrificed animal—or a small portion of it—, it reduces animal care expenses, human labor costs, and animal suffering; c) since the cells are distributed in a thin layer, optical observation under a microscope is unobstructed by other cell layers; and d) with cell lines, the researcher effectively circumvents the time necessary to raise the animal and its very sacrifice. A wide range of sophisticated medium formulations and cell lines from almost any type of tissue are now commercially available [5]; cell culture equipment is becoming increasingly ubiquitous: dissociating cells from their organ using enzymes, culturing them in humidified CO_2 incubators, and/or time-lapse imaging them in fluorescence microscopes, to name a few techniques, are now common procedures even in undergraduate laboratories. This accumulated wealth of knowledge has brought in-vitro experimentation closer to real animal research and has led to many important discoveries in fields ranging from basic biology to pharmaceutical screening.

However, the technology of cell culture is falling behind in the pace of progress. Genes can now be probed simultaneously by the thousands on a DNA chip as animal genomes are being fully sequenced [6]. Biochemists can synthesize a combinatorial variety of drug candidates as well as a myriad reporters that tag specific biomolecules and organic compounds which mimic the function of other biomolecules. Our molecular understanding of cell behavior is materializing as a picture of pathways of biochemical reactions intricately entangled with each other. Clearly, the biomedical field is entering an era where data retrieval and analysis has to deal with complex systems. Not surprisingly, social and political concerns are being raised which reflect the proportions that genetic engineering and animal experimentation are taking in modern science. In cell culture, due to recent advances in drug discovery and molecular cell biology, there is an increasing pressure for testing even more complex medium formulations that include putative drug candidates, growth factors, neuropeptides, genes, and retroviruses, to name a few. Since these various factors have nonlinear effects that may change when combined with other factors, the complexity of testing increases exponentially as new components are added to the cell culture medium.

Yet cell culture methodology has remained basically unchanged for almost a century: it consists essentially of the immersion of a large population of cells in a homogeneous fluid medium [5]. This requires at least one cell culture surface (such as a petri dish, a slide, or a well) for each cell culture condition to be investigated; in general, a few surfaces are typically used for each condition to account for sample variability and measurement errors. Hence, as cells need to be fed periodically with fresh medium, the combinatorial testing of a variety of medium conditions involves large numbers of cell culture surfaces, bulky incubators, large fluid volumes (~0.1–2 mL per sample), and expensive human labor and/or equipment (see Fig. 19.2).

19.2. THE LAB-ON-A-CHIP REVOLUTION

Microfluidic devices—and, more generally speaking, microsystems—promise to play a key role in circumventing the above limitations for several reasons: 1) they consume small quantities of precious/hazardous reagents (thus reducing cost of operation/disposal); 2) they can be straightforwardly integrated with other microfluidic devices; 3) they can be mass-produced in low-cost, portable units; 4) the dimensions of their microchannels can be comparable to or smaller than a single cell; and 5) due to the large surface-area-to-volume ratio of these microchannels, liquids flow in sheets or "laminarly", i.e. without turbulence [7, 8]. In the last five years, there has been an eruption of microfluidic implementations of a variety of traditional bioanalysis techniques encompassing capillary electrophoresis [9–12], DNA amplification [13–16] and hybridization [17], blood sampling [18], single-cell enzymatic analysis [19], mass-spectrometry [20–21], NMR spectroscopy [22], and immunoassays [23–28], among others. Considerable efforts are underway to integrate two or more of the above functions in "micro-total analysis systems" (Micro-TAS) [14, 29–32]. On the other hand, the integration of microfluidic systems with live cells, despite its great potential for cell culture and biosensor technologies, is still in its infancy. Recently, microfluidic devices for cell sorting [33, 34], cell separation [35], embryo labeling [36], blood capillary flow simulation [37], delivery of cells to selected areas of a substrate [38–40] or heterogeneous-flow stimulation of cultured cells [39, 41] have been reported. An important part of this revolution has been the advent of soft lithography, a family of sister techniques based on micromolding poly(dimethylsiloxane) (PDMS) [42] that was pioneered by Whitesides and co-workers. Soft lithography has become an enabling technique for lab-on-a-chip applications because PDMS is inexpensive (both the material and its replica-molding process, which requires no expertise except for the one-time fabrication of the master mold), transparent (i.e. compatible with optical microscopes), and biocompatible (cells may be cultured atop a PDMS surface). Importantly, PDMS microstructures can be sealed against almost any smooth dry surface, which allows for fabricating microfluidic devices [43] for a small fraction of the cost of traditional methods.

One may distinguish two classes of advances that microdevices may bring to cell culture technology. On one hand, microfluidic cell cultures promise to yield *higher throughputs* than their macrofluidic counterparts, since the latter are often limited by fluid handling issues. On the other hand, a new generation of cell biology experiments will benefit from microtechnology's ability to design *small areas and volumes* of a complex microenvironment (both substrate and fluid) around single cells, often exploiting the unique physicochemical behavior of fluids in microchannels.

19.3. INCREASING EXPERIMENTATION THROUGHPUT

Below we review how microfluidic devices may aid in increasing the throughput of cell culture technology and which challenges are ahead.

19.3.1. From Serial Pipetting to Highly Parallel Micromixers

A major throughput-limiting step in cell culture technology is fluid handling, in particular reagent mixing, medium delivery, and medium sampling. This is currently done by interrupting incubation and by subsequent, painstaking manual pipetting or very costly automated multi-pipetters equipped with complex tubing and valve systems—both processes being prone to contamination and spills. The advent of multi-well plates represents an attempt to mitigate these limitations, but it is not scalable because the required number of

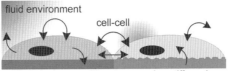

FIGURE 19.1. Cells respond to a variety of local, dynamic physichochemical signals, either from their fluidic environment, their cell neighbors, or their underlying substrate.

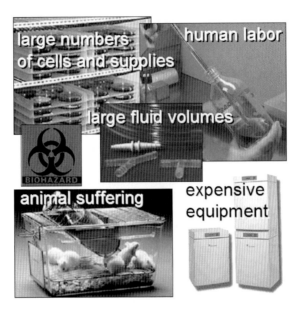

FIGURE 19.2. Central elements and limitations of cell culture technology. Cell culture is essentially based on the immersion of a monolayer of cells in a homogeneous bath, using costly human labor, bulky equipment, and large numbers of cells, supplies, and fluid volumes.

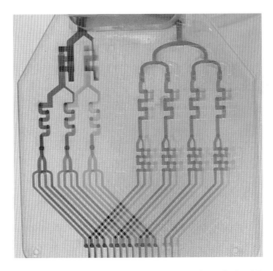

FIGURE 19.3. A combinatorial micromixer that allows for generating all the 16 mixture combinations of 4 dilutions of blue dye (left side of the device) and 4 dilutions of orange dye (right side of the device). Flow is from top to bottom. The outlet channels are 1 mm wide and 100 μm deep.

plates and fluid delivery steps grows geometrically as the number of cell culture parameters increases. For example, the very optimization of a defined cell culture medium (for
a particular cell type, a particular application, and even a particular phenotype), which
may involve studying the non-linear, non-additive effects of various hormones, growth factors, aminoacids, glucose, salts, etc. at different concentrations, as critical as it can be for
the success of an experiment, can become a monumental task in its own that only a few
cell culture laboratories can afford to undertake. As cell culture studies become increasingly sophisticated, the development of fast, inexpensive mixers that generate a combinatorial range of fluid mixtures becomes imperative. Previously, several groups have reported
micromixer designs that generate microfluidic dilutions. Dertinger et al. [44] have demonstrated a symmetric two-dimensional (2D) microfluidic network that continuously generates
certain combinations of dilutions of two compounds, although not all the combinations are
possible because mixing and dilution are not independent. Independent mixing and dilution
clearly require three-dimensional (3D) networks [45, 46] that allow channels to pass over
one another. Ismagilov et al. [47, 48] have demonstrated orthogonal microchannel arrays
that create a potential mixing point at every channel intersection. Another 3D device converted four inlets (A, B, C, D) into four undiluted binary mixtures (AC, AD, BC, BD) [49].
However, none of these devices allowed for combinatorial mixing of all the dilutions of
the input compounds. Another design dilutes and mixes stationary fluids in microchambers
connected by valves [50, 51]. We have devised a microfluidic mixer design that produces
all the mixture combinations of four dilutions of (two) input compounds and delivers the
sixteen mixture combinations in separate outlet microchannels (Fig. 19.3). The device features four different flow levels made by stacking nine laser-cut Mylar laminates. The fluidic
network has a symmetric design that guarantees that the flow rates are nearly identical at
all the outlets. Such systems should find uses in cell-based combinatorial screening.

19.3.2. From Incubators to "Chip-Cubators"

Cell culture equipment—comprising at least an incubator (required to preserve temperature, humidity, and gas concentration of the cell culture medium) and a tissue culture hood
(required to preserve sterile conditions)—is expensive and bulky; usually, a dedicated room
is recommended. Culturing cells in closed, microfluidic chambers circumvents the need for
constant-humidity systems, however it also calls for the development of gas exchange and
temperature control systems on chip if the miniaturization of the cell maintenance equipment is to be fully realized. Although seemingly trivial from an engineering point of view,
the full miniaturization of a cell maintenance system is yet to be achieved; this may reflect
not so much its technical difficulty but the fact that such a system has a small payoff for the
average researcher, who still needs the traditional cell culture room for intrinsically "bulky
activities" such as primary cell isolation and expansion of cell lines. Relatively small (yet
macrofluidic) perfusion chambers featuring incubation capabilities for long-term live cell
(e.g. time lapse) microscopy are commercially available.

19.3.3. From High Cell Numbers in Large Volumes (and Large Areas) to Low Cell Numbers in Small Volumes (and Small Areas)

A typical cell culture experiment uses large numbers of cells, which results in substantial
animal suffering—an increasing concern in our society—and supplies (the culture surfaces

as well as the fluids); this results in expensive experiments that take a lot of bench or incubator space. When the cultures are used for harvesting certain biochemicals, the need for large numbers of cells is most often not intrinsic to the experiment but mostly due to the need for collecting high concentrations of the biomolecule of interest (e.g. a cell-secreted product, DNA or mRNA content) and/or the need for collecting high volumes if the assay so requires (e.g. centrifugation, filling a well of a 96-well plate). In a microfluidic culture, compared to a traditional open-dish culture, both the area occupied by the cells and the volume bathing the cells are scaled down, but the volume can be (and typically is) scaled down by a larger factor than the area because of the reduced height of the fluid. Aside from considerations on the constancy of the cell culture environment (which will be addressed below), this means that, in microfluidic cultures compared to their non-microfluidic counterparts, the collected concentrations are higher (which contributes to increase the signal-to-noise ratio of the measurement), but the collected volumes are smaller (which often makes the fluid collection and the detection more challenging); thus, depending on the sensitivity of the assay, the microfluidic collection of fluids for use in a macrofluidic assay may not be beneficial nor practical, and a microengineered version of the assay may have to be developed. When possible, assays based on microscopic observation (e.g. calcium imaging using fluorescent markers, time-lapse imaging of changes in cell morphology), which can be done in situ, should prove more practical. However, microscopy assays have inherently low throughput (the microscope probes only a small field of view), so adopting a microscopy assay in a microfluidic cell culture hinders the high-throughput benefits of microfluidic perfusion; to achieve high throughputs, a programmable motorized stage can be used to automate image acquisition from multiple fields of view.

19.3.4. From Milliliters to Microliters or Nanoliters

A typical cell culture experiment also uses large volumes of reagents, some of which are extremely expensive (on the order of hundreds of dollars per microgram) and costly to dispose of. Microfluidic perfusion naturally requires smaller volumes and, as a result, is inexpensive compared to macrofluidic perfusion. In some cases, the benefits may not be huge; a macrofluidic culture can be stagnant (no medium change) for many hours, sometimes days, because the necessary nutrients, ions and gases required by the cell are amply drawn from the large volume of medium (typically on the order of a 1 mm-thick layer of fluid, or ~25 nL/cell for a confluent monolayer of cells, ~100 cells/mm^2), which contains enough nutrients and buffering capability to sustain a nearly-constant environment for periods up to 1–3 days. In contrast, in a microfluidic culture the fluid layer thickness and the volume per cell are typically one order of magnitude smaller. As a result, a microfluidic culture cannot be stagnant for long periods of time (depending on the cell density) because the cells rapidly acidify their environment and require a stable supply of nutrients and oxygen. Hence, long-term/dense microfluidic cultures may require frequent medium changes or continuous perfusion that can also consume large volumes if precautions are not taken to minimize the flow rate. The most convenient source of energy to drive flow is undoubtedly gravity (i.e. the inlet reservoir is placed higher than the outlet reservoir); however, given the high density of water, gravity-driven flow is not easily minimized, which results in accidental fluctuations and lack of reproducibility. We have achieved minimal flow rates of 1 mL/day for >7 days by placing the inlet(s) of fresh cell culture media lower than the outlet of the microchannel; a piece of paper tissue then pulls the fluid up the gravity slope from the outlet by capillarity (the fluid does not accumulate because it eventually evaporates from the wet paper surface). Long-term cultures are especially important in studies of cell differentiation, because many differentiation processes take several days to complete. As an example, our group is interested in studying how the microenvironment of muscle cells affects their differentiation. The central part of their differentiation program consists of fusing with each other to form multinucleated, tubular-shaped cells called myotubes; subsequently, the cells up-regulate the synthesis of acetylcholine receptors (AChRs), which end up forming clusters at the cell membrane through the action of agrin (in vivo, agrin is released by the nerve tip). The AChRs can be visualized by fluorescence microscopy after the cells are fluorescently labeled with α–bungarotoxin (BTX), a neurotoxin that blocks AChRs. In our efforts to control the microenvironment of myotubes, we seed C2C12 cells (a muscle cell line that is widely used as a model for myogenic differentiation) in microchannels and allow them to proliferate for a period of many days and subsequently fuse; importantly, the cells do not fuse unless their medium is switched to low-serum content after several days of serum-containing medium. Fig. 19.4 shows a microfluidic muscle cell culture where all solution changes (from the first, for seeding, till the last, for the fluorescence assay) were done through the microchannel inlet. Fig. 19.4a shows C2C12 cells in a PDMS microchannel (on fibronectin-coated glass) 4 hours after seeding. The images in Fig. 19.4b (phase-contrast micrograph) and 19.4c (corresponding fluorescence image of BTX-labeled AChRs) were taken at day 7, after the medium had been switched to low levels of serum (which arrested their proliferation and prompted them to fuse); the clusters of AChRs in Fig. 19.4c indicate an advanced level of muscle cell differentiation. We note that Fig. 19.4c represents a fully-microfluidic fluorescence assay, where the fixative and fluorescent labeling reagents were delivered via the microchannel. Minimization of the flow rate can be important not only to minimize reagent cost but also to minimize the area and volume taken up by the reservoir of fresh cell culture medium. Importantly, this long-term perfusion setup is simple and can be transferred to any biology laboratory equipped with a cell culture incubator.

19.3.5. From Manual/Robotic Pipetting to Microfluidic Pumps and Valves

Fluid metering is largely based on calibrated pipettes (volumes > 1 μL); this approach is clearly not adequate for nanoliter (or smaller) volumes and has an extremely low through-put unless expensive robotic dispensers are used. In a microfluidic device, although the volumes of microchambers can be known with high precision, the measurement of flow rates is challenging because it requires specialized flow visualization techniques; as a result, continuous flow systems such as micromixers are not easily metered,. However, stationary volumes (e.g. microchambers closed by microvalves) are straightforwardly known independently (roughly) of their size and number. The volume of the chamber may be difficult to predict if the microvalve is of the "pinch-type" [52–54] (where the chamber is created by flattening a given section of a microchannel); furthermore, in these pinch-type valves, the microchannels must be fabricated with a rounded profile (a challenging geometry in microfabrication, usually produced by the "photoresist reflow" method), otherwise the microchannel does not flatten completely [54].

An alternative design [55, 56] employs a PDMS membrane but only deforms the chamber when it opens, returning to the original geometry when it closes (see Fig. 19.5 for an

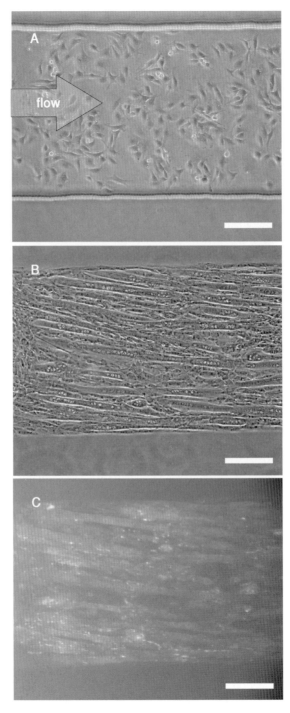

FIGURE 19.4. C2C12 cells cultured inside a PDMS microchannel. (A) phase-contrast image taken 4 hours after seeding. (B) Phase-contrast image taken after the cells have fused into myotubes and have been fixed. Note that the myotubes have aligned with the direction of the channel. (C) Fluorescence microscopy image of the same area as (B) demonstrating a myotube-specific staining assay performed in the microchannels; the small bright spots reveal the presence of clusters of acetylcholine receptors (bound to fluorescently-labeled a-bungarotoxin), a marker for muscle cell differentiation. Scale bar is ~50 μm.

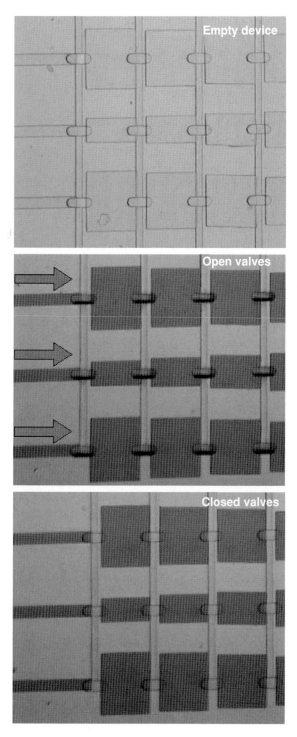

FIGURE 19.5. Elastomeric microvalves to control nanoliter-sized chambers. The largest chambers are 500 μm × 580 μm × 55 μm ≈ 16 nL, the smallest ∼ 6 nL in volume.

example). This geometry is more convenient for applications where the volume of the microchambers needs to be known with high accuracy: the shape of the microchambers is defined by photolithography and their height is given by the height of the photoresist layer. For example, the smallest chamber in Fig. 19.5 is 500 μm × 220 μm × 55 μm ≈ 6 nL; since it is possible to know the dimensions with an error of around 1% on each dimension, the volume can be known within approximately ±3% = ±20 pL.

19.3.6. Single-Cell Probing and Manipulation

In cell culture, both single-cell probing (e.g. "patch clamp" electrophysiological recordings [57]) and cell manipulation (e.g. oocyte handling, microinjection) are at present mostly based on glass micropipettes. While micropipettes are a powerful tool by virtue of the small size of their end apertures (∼1 μm or less in diameter for ion channel recordings), they require expensive vibration isolation equipment and micromanipulators (to gently approach the micropipette to the cell) and a considerable amount of manual skill and expert training, resulting in very low throughputs and high failure rates. Recently, a number of groups have reported microfluidic systems featuring arrays of micron- or submicron-sized holes onto which the cells are deposited, potentially allowing for the automation of large numbers of "patch-clamp on a chip" measurements simultaneously [58–60]; some systems even feature micronozzles [61], with potential for highly parallel microinjection applications. As a gentler alternative to micropipettes for positioning single cells, focused "laser traps" [62] are being implemented but have an exorbitant cost and cannot be scaled up to manipulate large numbers of cells. Dielectrophoretic traps, on the other hand, can be actuated in parallel for separating, immobilizing and/or releasing large arrays of single cells in a microfluidic environment [63–67].

In summary, for the traditional cell culture user, the issues related to cost and throughput are practical and can, in principle, be overcome with increased resources or funding. Nevertheless, most cell biologists will compromise on the number of conditions (different factors, substrates, cell types) that would ideally be probed, resulting in poor statistics (particularly when obtaining single-cell data) and non-quantitative results; in such cases, the microfluidic implementation of a cell culture experiment represents a *practical improvement* that results in lower overall cost, higher throughput, and/or more quantitative/single-cell results. Very large increases in throughput may *enable* certain classes of experiments—such as statistically demanding ones—that would otherwise be impractical or too expensive to be addressed with traditional cell culture methods.

19.4. INCREASING THE COMPLEXITY OF THE CELLULAR MICROENVIRONMENT

The interest in miniaturizing the cell culture laboratory goes beyond the need for higher-throughput technology: it represents an enabling tool that allows the researcher to quantitatively control the microenvironment of the cell at the single-cell level and hence address biological questions which cannot be addressed (in a practical, statistically-meaningful manner) with present tools. Cellular processes such as adhesion, migration, growth, secretion, and overall gene expression are triggered, controlled, or influenced by the three-dimensional biochemical and biophysical architecture of neighboring surfaces. This organization cannot be, to the present day, straightforwardly reproduced in the laboratory, but many advances have been made in that direction. Below we review how microtechnology may aid in increasing the complexity of the cellular microenvironment and discuss the issues that may arise:

19.4.1. From Random Cultures to Microengineered Substrates

In traditional cell culture, cells are randomly seeded over the whole substrate; thus, cellular interactions that are strongly dependent on the proximity, distribution, and/or relative position of cells or biomolecules can become confounded by the presence of other, similar interactions. The advent of cellular micropatterning methods has offered the possibility of tailoring the cellular and biochemical neighborhood of cells with resolution down to single cells [68]. In the past, cell biologists have resorted to clever approaches to recreate different degrees of tissue organization in the laboratory. Harrison, for example, used spider webs to study cell migration as early as 1912 [69]. Others studied cell behavior on surface features such as milled grooves on mica [70], polystyrene replicas of diffraction gratings [71], polyvinylchloride music records [72], dried protein spots [73, 74], crystals [75], and scratches in agar [76], in phospholipid films [77], or in extracellular matrix protein [78]. Albeit ingenious, the technology utilized in these studies could not address the structural dimensions, chemical heterogeneity and/or precise repeatability over large areas found in live tissue. With the adaptation of microfabrication technology to biological materials and biochemical processes, it has become possible to design surfaces that reproduce some of the aspects of that architecture. For example, Ingber, Whitesides and colleagues [79, 80] were able to constrain the shape of cells within cell-adhesive islands of microstamped self-assembled monolayers and they showed that cell function (albumin secretion in hepatocytes [79] and cell cycle progression in endothelial cells [80]) depends on the shape of the cell. Similarly, Healy and co-workers [81] directed the selective attachment and spreading of osteoblasts on micropatterns of cell-repellent polymeric thin films and observed that cytoskeletal organization was dictated by the shape of the adhesive regions. Toner's group [82] was able to create micropatterns of two cell types (hepatocyte and fibroblast co-cultures) on glass to show that albumin production and urea secretion increase with the amount of contact between the two cell populations, with the highest levels corresponding to single-hepatocyte islands [83]. Hundreds of references on cellular micropatterning can be found elsewhere [68]. Examples of cellular micropatterns from our lab are shown in Fig. 19.6; note that the muscle cell micropattern in the bottom left picture was created and maintained *within* a microfluidic channel. Unfortunately, the implementation of cellular micropatterning approaches by cell biologists has faced until recently an important hurdle: most cell biologists do not have the required microfabrication expertise; such technological gap has become narrower in recent years with the advent of straightforward, inexpensive micropatterning techniques (specially soft lithography [84]) and with the increasing population of scientists with cross-training in both biology and engineering. Cellular micropatterning techniques represent a powerful tool not only for studies of molecular and cell biology, but also for engineering cellular scaffolds [85–88] (potentially, for tissue replacement) and for developing cell-based gene expression screens [89] and/or cellular microsensors [58–60, 90–92].

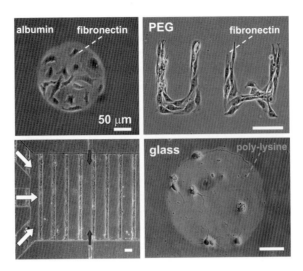

FIGURE 19.6. Cellular micropatterns. *Top left*: Fibroblasts attached to fibronectin islands surrounded by albumin; both fibronectin and albumin are adsorbed onto polystyrene and micropatterned using PDMS stencils [1]. *Top right*: Muscle cells attached to fibronectin micropatterns surrounded by a polymer thin film containing poly(ethylene glycol) (PEG); fibronectin is physisorbed onto glass and PEG is chemically grafted onto glass; the pattern was defined by photolithography and oxygen plasma etching. *Bottom right*: embryonic cortical neurons attached on islands of poly-lysine physisorbed on glass; the islands were defined with PDMS stencils. *Bottom left*: microfluidic culture of a cellular micropattern of muscle cells on lines of fibronectin (physisorbed on glass) surrounded by a PEG background; the lines were defined using oxygen plasma etching masked by PDMS microchannels [4]; the white arrows indicate the heterogeneous flow used to seed and to stimulate the cells and the dark gray arrows indicate the flow used to feed the cells for over a week. All scale bars are 50 μm.

19.4.2. From "Classical" to "Novel" Substrates

Micropatterning cells (and often also combining cells with microfluidic devices) naturally requires that the cell culture areas be contacted by chemical or physical obstacles that block the adhesion of cells or proteins on selected areas [68], so in the contact process undesired deposition of other materials may occur. For example, deposition of PDMS monomers may occur during microstamping of proteins or microfluidic patterning of cells, and methods based on photolithography processes may result in photoresist remains on the cell culture areas; these residues may be difficult to detect. This is a concern from a cell biologist's perspective because cells are exquisitely sensitive to submonolayer coverages of adsorbates (either directly or by affecting the adsorption of key proteins) and, as a result, contaminants may produce confounding results. Importantly, this cell biologist also faces a second, often-overlooked subtle dilemma: typically, the cell biologist considering the implementation of a cellular micropatterning approach chooses to do so based on a previous line of research that, until then, used a randomly-organized cell culture on homogeneously-adhesive substrate.

The micropatterned cell cultures thus represent the "next experiment" after the random cultures, and the random cultures serve as the "control experiments". Ideally, the control (random) cultures should use the same exact type of substrate and seeding/culture protocols as used previously for random cultures *in the past*—otherwise, only the cellular functions measured in that particular experiment are "controlled", and the available body of knowledge obtained on other cellular functions may not apply on the "novel" substrates. We stress that this preference is not merely for convenience; rather, given the high sensitivity of cells to minute changes in surface composition, changing the protocols and/or substrate may yield artifactual results in micropatterning experiments or affect future experiments probing different cellular functions. This need for the random-culture experiment to be an appropriate control for the micropatterned-culture experiment places stringent constraints on the choices of micropatterning techniques that are appealing to the cell biologist. The substrates universally used for cell culture in molecular and cell biology research are either glass or polystyrene—either coated with proteins or bare. Cell micropatterning approaches largely utilize one of two strategies to deposit cells on designated areas of the cell culture substrates (for a review, see [68]): 1) selective cell attachment is guided by differential adhesiveness of the substrate (a very simple and widely used method to deter cell attachment consists of adsorbing albumin, a protein that lacks cell adhesion motifs, on the cell culture substrate); or 2) cell attachment to a homogenously adhesive substrate is blocked in selected areas with a removable physical barrier. Recently, soft lithographic methods have been developed to selectively deposit cell/protein-adhesive [93] or repellent [94] coatings from solution [93], by microcontact printing [79, 95], and to deposit cells directly using physical masks (microchannels [38, 40] or stencils [1, 96]). While techniques relying on albumin backgrounds (e.g. Fig. 19.6 top left image of cells on a fibronectin circle) are suitable for short-term studies, the albumin-coated areas turn into cell-adhesive over a period of few hours to days when cells are cultured in serum-containing medium, probably due to elution of albumin or its displacement by adhesive serum proteins. For long-term cultures of patterned cells, surface chemistry methods are necessary to prevent protein adsorption and cell adhesion onto undesired areas of the substrate. Surface modification has been employed to produce protein-repellent coatings and to dictate protein adsorption and cell attachment onto artificial materials [97] (for a review, see [98]). The most successful chemistries for engineering long-term cell/protein repellent surfaces have consisted of ethyleneglycol-terminated self-assembled monolayers (SAMs) [79, 99, 100], polymeric thin films containing poly(ethyleneglycol) (PEG) [81, 101, 102], PEG aldehydes covalently bound to amine-functionalized glass [103], or a commercial copolymer of poly(ethyleneoxide) and poly(propyleneoxide) (Pluronic^{TM}) [94]. Among those techniques, the interpenetrated network of poly(acrylamide-co-ethyleneglycol) or "PEG IPN" developed by Healy and co-workers [81, 101, 102] is especially attractive because: 1) it resists non-specific protein adsorption and has maintained cell patterns for the longest time periods demonstrated to date (>60 days) [81], 2) it is compatible with glass substrates and it can be deposited on polymers such as polystyrene, and 3) it utilizes only off-the-shelf reagents. Previously, cellular micropatterns on PEG IPN background were created by defining the formation of PEG IPN on glass using a photolithographically-defined photoresist mask [81]. However, photolithographic patterning of PEG IPN can be expensive and laborious. For all these reasons combined (need for a glass surface, choice of PEG IPN for robustness, and inconvenience of photolithographic patterning), we have developed a soft-lithographic method based on the selective complete etching of PEG IPN using an oxygen plasma and a removable PDMS mask [4], as shown in Fig. 19.7. The mask can be a stencil (not shown) or a set of microchannels (Fig. 19.7), both of which are cast once from a photolithographic master and are re-usable (so access to a photolithographic facility is only needed to fabricate the master). After the stencil or mold is peeled off the surface, the remaining PEG IPN separates areas of

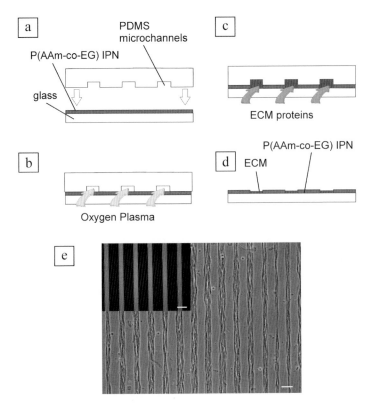

FIGURE 19.7. Schematic illustration (a–d) of the procedure for cellular micropatterning based on the use of PDMS microchannels as masks for selective surface etching [4]: (a) The PDMS mold is placed onto a glass substrate homogeneously grafted with the P(AAm-co-EG) interpenetrating polymer network (IPN); (b) Selective etching: the IPN is selectively removed after exposure to oxygen plasma, which leaves alternating lines of IPN-free glass and IPN; (c) Selective deposition: the microchannels are filled with extracellular matrix (ECM) protein to promote cell attachment. After flushing with PBS, the microchannels are removed, leaving lines of ECM protein surrounded by IPN. The ECM protein/IPN patterned glass is incubated with a cell suspension and unattached cells are removed by exchanging the medium; (e) Phase-contrast image of NIH 3T3 fibroblasts attached to the lines of fibronectin conjugated to Alexa Fluor 488 dye. Scale bars are 100 μm.

bare glass; the bare glass areas are amenable to protein adsorption and cell attachment using traditional cell seeding protocols. This method allows for producing cellular micropatterns with isolated or connected features, with sub-cellular resolution, and containing more than one cell type. The cellular micropatterns exhibit excellent temporal stability over periods of time longer than two weeks even when serum is present in the culture medium and at high cell densities. Other examples of PEG IPN patterns can be seen in Fig. 19.6 (top right pattern reading "UW" and bottom left image of myotubes in a microchannel). The PEG IPN surface chemistry is specially important in our long-term muscle cell differentiation studies; muscle cells are seeded at low densities on lanes of fibronectin (physisorbed on glass) surrounded by PEG IPN and, over the following days, the cells proliferate until they reach confluence. Importantly, after cell division the daughter cells also stay confined to the fibronectin lines and finally achieve confluence and fuse, forming a "myotube". The process of division until contact and fusion occur is paralleled in non-microfluidic, non-micropatterned cultures; however, in non-micropatterned cultures the myotubes form at seemingly random orientations rather than at the orientation dictated by the surface chemistry.

19.4.3. From Cells in Large Static Volumes to Cells in Small Flowing Volumes

As pointed out above, in high-density microfluidic cell cultures the microenvironment's nutrients, pH and gas concentrations cannot be kept constant for long periods of time, and replenishing the volume of the microfluidic chamber becomes necessary. An extreme deviation of a "proper" cell culture environment results in obvious cell death. However, the effects can be much subtler, due to the fact that a) cells are exposed to shear stress from the flow, and/or b) at the cell membrane surface, the actual concentration of growth factors (that either bind to cells or are secreted by them) "seen" by the cell depends on the speed of the flow, which effectively "washes away" the growth factor molecules that happen to diffuse into the stream. In our studies of muscle cell fusion in microfluidic environments (Fig. 19.4), we have indeed observed that muscle cell differentiation (as ascertained by the ability of cells to fuse) proceeds at different paces depending on the flow rates; at high flow rates, cell fusion takes longer to occur, likely due to the dilution of cell-secreted growth factors into the flow.

19.4.4. From a Homogeneous Bath to Microfluidic Delivery of Biochemical Factors

In traditional cell culture, cells are bathed in a homogeneous medium; thus, any cellular response that relies on graded or focal exposure of the cell to a given factor will not be observable, including many cell growth and motility phenomena. Traditionally, focal stimulation of cells with fluids has been possible only using micropipettes, which eject fluid upon application of a pressure pulse ("puffing") [104] or voltage pulse ("iontophoresis") [104–106], or using caged compounds that are uncaged by a laser pulse [107]. Clearly, these techniques result in very low throughputs and/or poorly-characterized volumes or gradients, require bulky equipment and substantial manual skill, and are unscalable (stimulation at more than 3 or 4 sites is not practical). Microfluidic systems, on the other hand, constitute a technology for directing many different fluids to many different (small) cell populations [39, 41, 108] that is scalable, amenable to fluid dynamics modeling [109], and where the delivery system is pre-aligned with the cells. Takayama et al. first demonstrated heterogeneous labeling of internal cellular structures and subcellular enzymatic treatment using laminar flows [39, 41, 108] in microchannels. In our lab, we have exposed microengineered myotubes to a heterogeneous flow consisting of a fluorescently-labeled bungarotoxin (BTX) stream sided by two BTX-free streams; PEG-IPN patterns were defined at the floor of the microchannel to produce (after seeding and culturing muscle cells for several days in the microchannel) myotubes perpendicular to the direction of the flow. As shown in Fig. 19.8, only the central portions of the myotubes were actually labeled with BTX. Thus, with microfluidic systems it is possible to confine a membrane receptor-labeling assay to a region smaller than a cell; importantly, the fact that the assay is subcellular allows for using lethal, high-affinity neurotoxin labels like BTX. This type of subcellular labeling should allow for the live tracking of a variety of membrane proteins.

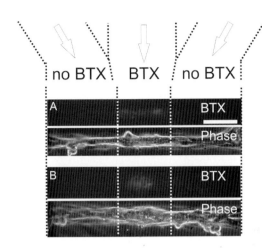

FIGURE 19.8. Acetylcholine receptor (AChR) clusters visualized by binding of a-bungarotoxin conjugated with Alexa Fluor 488 dye (BTX). The labeling was confined to a central portion of the myotubes (between the white dotted lines) and the myotubes span the entire microchannel. Each pair of BTX/Phase images (labeled A and B) correspond to the same myotube, with A being upstream of B; the images labeled "Phase" are phase-contrast images. The scale bar is 100 μm.

19.5. CONCLUSION

In sum, following a trend similar to the technology forces that pressed for the miniaturization of electronic devices, there is a growing interest in "going micro" in cell culture technology. First and foremost, microfluidic cell culture systems enable the modulation of signaling factors on a scale ranging millimiters to sub-cellular dimensions, thus providing for a more reproducible in-vivo environment and allowing for a new class of experiments that require heterogeneous and/or local stimulation of cells. For the same price, microfluidic cultures are also attractive in that 1) by virtue of being batch-fabricated, they allow for addressing large numbers of "units" (either cells, groups of cells, microchambers, etc.) at high-throughput and at low cost, and 2) by virtue of their size, they consume small reagent volumes and require small numbers of cells compared to traditional macrofluidic cultures.

REFERENCES

[1] A. Folch, B.-H. Jo, D. Beebe, and M. Toner. *J. Biomed. Mater. Res.*, 52:346, 2000.
[2] R.G. Harrison. *Proc. Soc. Exp. Biol. Med.*, 4:140, 1907.
[3] A. Carrel. *J. Exp. Med.*, 15:516, 1912.
[4] A. Tourovskaia, T. Barber, B.T. Wickes, D. Hirdes, B. Grin, D.G. Castner, K.E. Healy, and A. Folch. *Langmuir*, 19:4754, 2003.
[5] R.I. Freshney. *Culture of Animal Cells: A Manual of Basic Technique*, (4th ed.). Wiley-Liss, New York, 2000.
[6] R.F. Service. *Science*, 282:396, 1998,
[7] P. Gravesen, J. Branebjerg, and O.S. Jensen. *J. Micromech. Microeng.*, 3:168, 1993.
[8] J.P. Brody, P. Yager, R.E. Goldstein, and R.H. Austin. *Biophys. J.*, 71:3430, 1996.
[9] S.C. Jacobson, R. Hergenroder, L.B. Koutny, and J.M. Ramsey. *Anal. Chem.*, 66:1114, 1994.

[10] C.S. Effenhauser, G.J. M. Bruin, A. Paulus, and M. Ehrat. *Anal. Chem.*, 69:3451, 1997.

[11] R.M. McCormick, R.J. Nelson, M.G. Alonso-Amigo, D.J. Benvegnu, and H.H. Hooper. *Anal. Chem.*, 69:2626, 1997.

[12] G.H.W. Sanders and A. Manz. *Trac-Trends Anal. Chem.*, 19:364, 2000.

[13] M.A. Shoffner, J. Cheng, G.E. Hvichia, L.J. Kricka, and P. Wilding. *Nucleic Acids Res.*, 24:375, 1996.

[14] A.T. Woolley, D. Hadley, P. Landre, A.J. deMello, R.A. Mathies, and M.A. Northrup. *Anal. Chem.* 68:4081, 1996.

[15] D.T. Burke, M.A. Burns, and C. Mastrangelo. *Genome Res.*, 7:189, 1997.

[16] M.U. Kopp, A.J. Mello, and A. Manz. *Science*, 280:1046, 1998.

[17] Z.H.Fan, S. Mangru, R. Granzow, P. Heaney, W. Ho, Q.P. Dong, and R. Kumar. *Anal. Chem.*, 71:4851, 1999.

[18] E. Altendorf, D. Zebert, M. Holl, A. Vannelli, C. Wu, and T. Schulte. In D. Harrison, and A. van den Berg (eds.). *Micro Total Analysis Syst. '98* , Kluwer Academic, Banff, Canada, p. 73, 1998.

[19] G. Ocvirk, H. Salimi-Moosavi, R.J. Szarka, E. Arriaga, P.E. Andersson, R. Smith, N.J. Dovichi, and D.J Harrison. In D. Harrison, and A. van den Berg (eds.). *Micro Total Analysis Syst. '98*, Kluwer Academic, Banff,Canada, p. 203, 1998.

[20] Q. Xue, F. Foret, Y.M. Dunayevskiy, P.M. Zavracky, N.E. McGruer, and B.L. Karger. *Anal. Chem.*, 69:426, 1997.

[21] R.D. Oleschuk and D.J. Harrison. *Trac-Trends in Anal. Chem.*, 19:379, 2000.

[22] J.D. Trumbull, I.K. Glasgow, D.J. Beebe, and R.L. Magin. *IEEE Trans. Biomed. Engin.*, 47:3, 2000.

[23] L.B. Koutny, D. Schmalzing, T.A. Taylor, and M. Fuchs. *Anal. Chem.*, 68:18, 1996.

[24] N. Chiem and D.J. Harrison. *Anal. Chem.*, 69:373, 1997.

[25] N.H. Chiem and D.J. Harrison. *Clin. Chem.*, 44:591, 1998.

[26] A.E. Kamholz, B.H. Weigl, B.A. Finlayson, and P. Yager. *Anal. Chem.*, 71:5340, 1999.

[27] B.H. Weigl and P. Yager. *Sens. Actu. B-Chem.*, 39:452, 1997.

[28] B.H. Weigl and P. Yager. *Science*, 283:346, 1999.

[29] E.T. Lagally, P.C. Simpson, and R.A. Mathies. *Sens. Actua. B-Chem.*, 63:138, 2000.

[30] L.C. Waters, S.C. Jacobson, N. Kroutchinina, J. Khandurina, R.S. Foote, and J.M. Ramsey. *Anal. Chem.*, 70:158, 1998.

[31] A.T. Woolley, K. Lao, A.N. Glazer, and R.A. Mathies. *Anal. Chem.*, 70:684, 1998.

[32] J. Cheng, E.L. Sheldon, L. Wu, A. Uribe, L.O. Gerrue, J. Carrino, M.J. Heller, and J.P. O'Connell. *Nat. Biotech.*, 16:541, 1998.

[33] P.C. Li and D.J. Harrison. *Anal. Chem.*, 69:1564, 1997.

[34] E. Altendorf, D. Zebert, M. Holl, and P. Yager. *Proc. Int. Solid State Sensors and Actuators Conf. (Transducers '97)*, IEEE, Chicago, IL, U.S.A., p. 531, 1997.

[35] P.R.C. Gascoyne, W. Xiao-Bo, H. Ying, and F.F. Becker. *IEEE Trans. Ind. Appl.*, 33:670, 1997.

[36] S.J. Choi, I. Glasgow, H. Zeringue, D.J. Beebe, and M.B. Wheeler. *Biol. Reprod.*, 58:96, 1998.

[37] N. Sutton, M.C. Tracey, I.D. Johnston, R.S. Greenaway, and M.W. Rampling. *Microvascular Res.*, 53:272, 1997.

[38] A. Folch, A. Ayon, O. Hurtado, M.A. Schmidt, and M. Toner. *J. Biomech. Eng.*, 121:28, 1999.

[39] S. Takayama, J.C. McDonald, E. Ostuni, M.N. Liang, P.J.A. Kenis, R.F. Ismagilov, and G.M. Whitesides. *Proc. Natl. Acad. Sci. U.S.A.*, 96:5545, 1999.

[40] D.T. Chiu, N.L. Jeon, S. Huang, R.S. Kane, C.J.Wargo, I.S. Choi, D.E. Ingber, and G.M. Whitesides. *Proc. Natl. Acad. Sci. U.S.A.*, 97:2408, 2000.

[41] S. Takayama, E. Ostuni, P. LeDuc, K. Naruse, D.E. Ingber, and G.M. Whitesides. *Nature*, 411:1016, 2001.

[42] Y. Xia, E. Kim, and G.M. Whitesides. *J. Electrochem. Soc.*, 143:1070, 1996.

[43] J.C. McDonald, D.C. Duffy, J.R. Anderson, D.T. Chiu, H.K. Wu, O.J.A. Schueller, and G.M. Whitesides. *Electrophoresis*, 21:27, 2000.

[44] S.K.W. Dertinger, D.T. Chiu, N.L. Jeon, and G.M. Whitesides. *Anal. Chem.*, 73:1240, 2001.

[45] B.H. Jo, L.M. Van Lerberghe, K.M. Motsegood, and D.J. Beebe. *J. Microelectromech. Syst.*, 9:76, 2000.

[46] J.R. Anderson, D.T. Chiu, R.J. Jackman, O. Cherniavskaya, J.C. McDonald, H.K.Wu, S.H. Whitesides, and G.M. Whitesides. *Anal. Chem.*, 72:3158, 2000.

[47] R.F. Ismagilov, J.M.K. Ng, P.J.A. Kenis, and G.M. Whitesides. *Anal. Chem.*, 73:5207, 2001.

[48] R.F. Ismagilov, D. Rosmarin, P.J. A.Kenis, D.T. Chiu, W. Zhang, H.A. Stone, and G.M. Whitesides. *Anal.Chem.*, 73:4682, 2001.

[49] Y. Kikutani, H. Hisamoto, M. Tokeshi, and T. Kitamori. In J. M. Ramsey and A. v. d. Berg (eds.). *Micro Total Analysis Systems*. Kluwer Academic, Monterrey, CA, p. 161, 2001.

[50] H.-P. Chou, M.A. Unger, and S.R. Quake. *Biomed. Microdev.*, 3:323, 2001.

[51] J. Liu, M. Enzelberger, and S. Quake. *Electrophoresis*, 23:1531, 2002.

[52] M.A. Unger, H.P. Chou, T. Thorsen, A. Scherer, and S.R. Quake. *Science*, 288:113, 2000.

[53] T. Thorsen, S.J. Maerkl, and S.R. Quake. *Science*, 298:580, 2002.

[54] V. Studer, G. Hang, A. Pandolfi, M. Ortiz, W.F. Anderson, and S.R. Quake. *J Appl. Phys.*, 95:393, 2004.

[55] B.H. Jo, J. Moorthy, and D.J. Beebe. In v. d. B. et. al. (ed.), *Micro Total Analysis Systems 2000* Kluwer Academic Publishers, p. 335, 2000.

[56] K. Hosokawa and R. Maeda. *J. Micromechan. Microeng.*, 10:415, 2000.

[57] E. Neher. *Science*, 256:498, 1992.

[58] N. Fertig, R.H. Blick, and J.C. Behrends. *Biophys. J.*, 82:3056, 2002.

[59] N. Fertig, M. Klau, M. George, R.H. Blick, and J.C. Behrends. *Appl. Phys. Lett.*, 81:4865, 2002.

[60] K.G. Klemic, J.F. Klemic, M.A. Reed, and F.J. Sigworth. *Biosen. Bioelectron.*, 17:597, 2002.

[61] T. Lehnert, M.A.M. Gijs, R. Netzer, and U. Bischoff. *Appl. Phys. Lett.*, 81:5063, 2002.

[62] C.G. Galbraith and M.P. Sheetz. *J. Cell Biol.*, 147:1313, 1999.

[63] R. Pethig. *Crit. Rev. Biotechnol.*, 16:331, 1996.

[64] E. Richter, G. Fuhr, T. Muller, S. Shirley, S. Rogaschewski, K. Reimer, and C. Dell. *J. Mater. Science-Mater. Med.*, 7:85, 1996.

[65] T. Matsue, N. Matsumoto, and I. Uchida. *Electrochim. Acta*, 42:3251, 1997.

[66] J. Voldman, R.A. Braff, M. Toner, M.L. Gray, and M.A. Schmidt. *Biophys. J.*, 80:531, 2001.

[67] J. Voldman, M.L. Gray, M. Toner, and M.A. Schmidt. *Anal. Chem.*, 74:3984, 2002.

[68] A. Folch and M. Toner. *Ann. Rev. Biomed. Eng.*, 2:227, 2000.

[69] R.G. Harrison. *Anat. Rec.*, 6:181, 1912.

[70] P. Weiss. *J. Exp. Zool.*, 100:353, 1945.

[71] A.S.G. Curtis and M. Valverde. *J. Natl. Cancer Inst.*, 33:15, 1964.

[72] Y.A. Rovensky, I.L. Slavnaja, and J.M. Vasiliev. *Exp. Cell Res.*, 65:193, 1971.

[73] E.J. Furshpan, P.R. MacLeish, P.H.O'Lague, and D.D. Potter. *Proc. Natl. Acad. Sci. U.S.A.*, 73:4225, 1976.

[74] S. Grumbacher-Reinert. *Proc. Natl. Acad. Sci. U.S.A.*, 86:7270, 1989.

[75] D.C. Turner, J. Lawton, P. Dollenmeier, R. Ehrismann, and M. Chiquet. *Dev. Biol.*, 95:497, 1983.

[76] M. Lieberman, A.E. Roggeveen, J.E. Purdy, and E.A. Johnson. *Science*, 175:909, 1972.

[77] O.Y. Ivanova and L.B. Margolis. *Nature*, 242:200, 1973.

[78] R.A. Rovasio, A. Delouvee, K.M. Yamada, R. Timpl, and J.P. Thiery. *J. Cell. Biol.*, 96:462, 1983.

[79] R. Singhvi, A. Kumar, G.P. Lopez, G.N. Stephanopoulos, D.I. Wang, G.M.Whitesides, and D.E. Ingber. *Science*, 264:696, 1994.

[80] C.S. Chen, M. Mrksich, S. Huang, G.M. Whitesides, and D.E. Ingber. *Science*, 276:1425, 1997.

[81] C.H. Thomas, J.B. Lhoest, D.G. Castner, C.D. McFarland, and K.E. Healy. *J. Biomech. Eng.*, 121:40, 1999.

[82] S.N. Bhatia, M.L. Yarmush, and M. Toner. *J. Biomed. Mater. Res.*, 34:189, 1997.

[83] S.N. Bhatia, U.J. Balis, M.L. Yarmush, and M. Toner. *Biotechnol. Prog.*, 14:378, 1998.

[84] Y.N. Xia and G.M. Whitesides. *Angew. Chem.-Int. Edit. Engl.*, 37:551, 1998.

[85] A. Folch, S. Mezzour, M. During, O. Hurtado, M. Toner, and R. Mueller. *Biomed. Microdev.*, 2:207, 2000.

[86] V.A. Liu and S.N. Bhatia. *Biomed. Microdev.*, 4:257, 2002.

[87] S. Petronis, K.L. Eckert, J. Gold, and E. Wintermantel. *J. Mater. Sci.-Mater. Med.*, 12:523, 2001.

[88] G.D. Pins, M. Toner, and J.R. Morgan. *FASEB J.*, 14:593, 2000.

[89] J. Ziauddin and D.M. Sabatini. *Nature*, 411:107, 2001.

[90] H.M. McConnell, J.C. Owicki, J.W. Parce, D.L. Miller, G.T. Baxter, H.G.Wada, and S. Pitchford. *Science*, 257:1906, 1992.

[91] L. Bousse and W. Parce. *IEEE Eng. Med. Biol. Mag.*, 13:396, 1994.

[92] P. Fromherz and A. Stett. *Phys. Rev. Lett.*, 75:1670, 1995.

[93] A. Folch and M. Toner. *Biotechnol. Prog.*, 14:388, 1998.

[94] V.A. Liu, W.E. Jastromb, and S.N. Bhatia. *J. Biomed. Mater. Res.*, 60:126, 2002.

[95] M. Mrksich, L.E. Dike, J. Tien, D.E. Ingber, and G.M. Whitesides. *Exp. Cell Res.*, 235:305, 1997.

[96] E. Ostuni, R. Kane, C.S. Chen, D.E. Ingber, and G.M. Whitesides. *Langmuir*, 16:7811, 2000.

[97] P.D. Drumheller and J.A. Hubbell. In J. D. Bronzino (ed.) *The Biomedical Engineering Handbook*. CRC-Press, Boca Raton, FL, p. 1583, 1995.

[98] M. Mrksich. *Cell. Mol. Life Sci.*, 54:653, 1998.

[99] N. Patel, R. Padera, G.H.W. Sanders, S.M. Cannizzaro, M.C. Davies, R. Langer, C.J. Roberts, S.J.B. Tendler, P.M. Williams, and K.M. Shakesheff. *FASEB J.* 12:1447, 1998.

[100] C.B. Herbert, T.L. McLernon, C.L. Hypolite, D.N. Adams, L. Pikus, C.C. Huang, G.B. Fields, P.C. Letourneau, M.D. Distefano, and W.S. Hu. *Chem. Biol.*, 4:731, 1997.

[101] J.P. Bearinger, D.G. Castner, S.L. Golledge, A. Rezania, S. Hubchak, and K.E.Healy. *Langmuir*, 13:5175, 1997.

[102] J.P. Bearinger and K.E. Healy. *J. Dent. Res.*, 76:2948, 1997.

[103] S. Saneinejad and M.S. Shoichet. *J. Biomed. Mater. Res.* 42:13, 1998.

[104] T.W. Stone. *Microiontophoresis and Pressure Ejection, Vol. 8*, (1st Ed.). Wiley, John & Sons, Inc., 1985.

[105] W.L. Nastuk. *J. Cell. Comp. Physiol.*, 42:249, 1953.

[106] J. del Castillo and B. Katz. *J. Physiol.*, 128:157, 1955.

[107] K. Kandler, L.C. Katz, and J.A. Kauer. *Nat. Neurosci.*, 1:119, 1998.

[108] S. Takayama, E. Ostuni, P. LeDuc, K. Naruse, D.E. Ingber, and G.M. Whitesides. *Chem. Biol.*, 10:123, 2003.

[109] G.A. Ledezma, A. Folch, S.N. Bhatia, M.L. Yarmush, and M. Toner, *J. Biomech. Eng.*, 121:58, 1999.

About the Editors

Professor Mauro Ferrari is a pioneer in the fields of bioMEMS and biomedical nanotechnology. As a leading academic, a dedicated entrepreneur, and a vision setter for the Nation's premier Federal programs in nanomedicine, he brings a three-fold vantage perspective to his roles as Editor-in-Chief for this work. Dr. Ferrari has authored or co-authored over 150 scientific publications, 6 books, and over 20 US and International patents. Dr. Ferrari is also Editor-in-Chief of Biomedical Microdevices and series editor of the new Springer series on Emerging Biomedical Technologies.

Several private sector companies originated from his laboratories at the Ohio State University and the University of California at Berkeley over the years. On a Federal assignment as Special Expert in Nanotechnology and Eminent Scholar, he has provided the scientific leadership for the development of the Alliance for Cancer Nanotechnology of the National Cancer Institute, the world-largest medical nanotechnology operation to date. Dr. Ferrari trained in mathematical physics in Italy, obtained his Master's and Ph.D. in Mechanical Engineering at Berkeley, attended medical school at The Ohio State University, and served in faculty positions in Materials Science and Engineering, and Civil and Environmental Engineering in Berkeley, where he was first tenured. At Ohio State he currently serves as Professor of Internal Medicine, Division of Hematology and Oncology, as Edgar Hendrickson Professor of Biomedical Engineering, and as Professor of Mechanical Engineering. He is Associate Director of the Dorothy M. Davis Heart and Lung Research Institute, and the University's Associate Vice President for Health Science, Technology and Commercialization.

Dr. Tejal Desai is currently an Associate Professor of Physiology and Bioengineering at the University of California, San Francisco. She is also a member of the California Institute for Quantitative Biomedical Research and the UCSF/UCB Bioengineering Graduate Group. Prior to joining UCSF, she was a professor of Biomedical Engineering at Boston University and Associate Director of the Center for Nanoscience and Nanobiotechnology at BU. She received the Sc.B. degree in Biomedical Engineering from Brown University (Providence, RI) in 1994 and the Ph.D. degree in bioengineering from the joint graduate program at University of California, Berkeley and the University of California, San Francisco, in 1998. Dr. Tejal Desai directs the Laboratory of Therapeutic Micro and Nanotechnology. In addition to authoring over 60 technical papers, she is presently an associate editor of Langmuir, Biomedical Microdevices, and Sensors Letters and is editing a book on Therapeutic Microtechnology. She has chaired and organized several conferences and symposia in the area of bioMEMS, micro and nanofabricated biomaterials, and micro/nanoscale tissue engineering.

Desai's research efforts have earned her numerous awards. In 1999, she was recognized by Crain's Chicago Business magazine with their annual "40 Under 40" award for leadership. She was also named that year by Technology Review Magazine as one of the nation's "Top 100 Young Innovators" and Popular Science's Brilliant 10. Desai's teaching efforts were recognized when she won the College of Engineering Best Advisor/Teacher Award. She also won the National Science Foundation's "New Century Scholar" award and the NSF Faculty Early Career Development Program "CAREER" award, which recognizes teacher-scholars most likely to become the academic leaders of the 21st century. Her research in therapeutic microtechnology has also earned her the Visionary Science Award from the International Society of BioMEMS and Nanotechnology in 2001, a World Technology Award Finalist in 2004, and the EURAND award for innovative advances in drug delivery.

Sangeeta N. Bhatia, M.D., Ph.D. is an Associate Professor at the Massachusetts Institute of Technology. Her work focuses on using micro- and nanotechnology tools to repair damaged tissues. Dr. Bhatia trained at Brown, MIT, and Harvard. After postdoctoral training at the Massachusetts General Hospital, she was a member of the Bioengineering Department at University of California at San Diego for 6 years. In 2005, she returned to Boston to join the MIT faculty. She has been awarded the David and Lucile Packard Fellowship given to 'the nation's most promising young professors in science and engineering,' the MIT TR100 Young Innovators Award, and been named one of San Diego's '50 People to Watch in 2004'. Her research portfolio includes funding from NIH, NSF, DARPA, NASA, the Whitaker Foundation, the Packard Foundation, and private industry. She co-authored the first undergraduate textbook on tissue engineering and is a frequent advisor to governmental organizations on cell-based sensing, nanobiotechnology, and tissue engineering. She holds 12 issued or pending patents and has worked in industry at Pfizer, Genetics Institute, ICI Pharmaceuticals, and Organogenesis.

Index

Abbreviated Table of Contents

Printed in Singapore